Medical Mycology

Medical Mycology

Cellular and Molecular Techniques

Edited by

Kevin Kavanagh

Department of Biology, National University of Ireland, Maynooth, Co. Kildare, Ireland

BICENTENNIAL 1807 WILEY 2007 BICENTENNIAL

John Wiley & Sons, Ltd

Telephone (+44) 1243 779777

Email (for orders and customer service enquiries): cs-books@wiley.co.uk
Visit our Home Page on www.wileyeurope.com or www.wiley.com

Other Wiley Editorial Offices

John Wiley & Sons Inc., 111 River Street, Hoboken, NJ 07030, USA
Jossey-Bass, 989 Market Street, San Francisco, CA 94103-1741, USA
Wiley-VCH Verlag GmbH, Boschstr. 12, D-69469 Weinheim, Germany
John Wiley & Sons Australia Ltd, 42 McDougall Street, Milton, Queensland 4064, Australia
John Wiley & Sons (Asia) Pte Ltd, 2 Clementi Loop #02-01, Jin Xing Distripark, Singapore 129809
John Wiley & Sons Canada Ltd, 6045 Freemont Blvd, Mississauga, ONT, L5R 4J3

Wiley also publishes its books in a variety of electronic formats. Some content that appears in print may not be available in electronic books.

Library of Congress Cataloging-in-Publication Data

Medical mycology: cellular and molecular techniques / edited by Kevin Kavanagh.
p. ; cm.
Includes bibliographical references and index.
ISBN-13: 978-0-470-01923-8 (alk. paper)
ISBN-10: 0-470-01923-9 (alk. paper)
1. Medical mycology–Laboratory manuals. 2. Mycoses–Molecular aspects–Laboratory manuals. I. Kavanagh, Kevin.
[DNLM: 1. Candidiasis–diagnosis–Laboratory Manuals. 2. Candidiasis–immunology–Laboratory Manuals. 3. Antibodies, Monoclonal–diagnostic use–Laboratory Manuals. 4. Clinical Protocols–Laboratory Manuals. 5. Immunohistochemistry–methods–Laboratory Manuals. 6. Immunologic Factors–diagnostic use–Laboratory Manuals. 7. Mycology–methods–Laboratory Manuals. WC 470 M489 2006]
QR245.M434 2006
616.9'6901–dc22 2006028045

British Library Cataloguing in Publication Data

A catalogue record for this book is available from the British Library

ISBN-13 978-0-470-01923-8 ISBN-10 0-470-01923-9

Typeset in 10.5/12.5pt Times by Thomson Digital
Printed and bound in Great Britain by Antony Rowe Ltd., Chippenham, Wiltshire
This book is printed on acid-free paper responsibly manufactured from sustainable forestry in which at least two trees are planted for each one used for paper production.

Contents

Preface xiii

List of Contributors xv

1 Diagnosis of *Candida* Infection in Tissue by Immunohistochemistry 1

Malcolm D. Richardson, Riina Rautemaa and Jarkko Hietanen

1.1 Introduction 1

1.2 Specificity of monoclonal antibody 3H8 for *C. albicans* 3

Protocol 1.1 Testing of specificity of monoclonal antibody 3H8 4

1.3 Evaluation of monoclonal antibody 3H8 for the detection of *C. albicans* morphological forms 6

Protocol 1.2 Evaluation of monoclonal antibody 3H8 for the detection of *C. albicans* morphological forms 6

1.4 Application of immunohistochemistry in the diagnosis of *Candida* periodontal disease 7

Protocol 1.3 Use of monoclonal antibody 3H8 in the detection of *C. albicans* in tissue 8

1.5 References 11

2 Transmission Electron Microscopy of Pathogenic Fungi 13

Guy Tronchin and Jean-Philippe Bouchara

2.1 Introduction 13

2.2 Glutaraldehyde-potassium-permanganate or glutaraldehyde-osmium-tetroxide fixation for ultrastructural analysis 16

Protocol 2.1 Glutaraldehyde-osmium tetroxide (#) or glutaraldehyde-potassium permanganate (*) fixation for ultrastructural analysis 17

2.3 Identification of the different compartments of the secretory pathway in yeasts 18

Protocol 2.2 Identification of the different compartments of the secretory pathway in yeasts 19

2.4 Cytochemical localization of acid phosphatase in yeasts 20

Protocol 2.3 Cytochemical localization of acid phosphatase in yeasts 21

2.5 Detection of anionic sites 23

Protocol 2.4 Detection of anionic sites 23

2.6 Detection of glycoconjugates by the periodic acid-thiocarbohydrazide-silver proteinate technique (PATAg) 25

Protocol 2.5 Detection of glycoconjugates by the periodic acid-thiocarbohydrazide-silver proteinate technique (PATAg) 26

2.7 Enzyme-gold approach for the detection of polysaccharides in the cell wall 28

Protocol 2.6 Enzyme-gold approach for the detection of polysaccharides in the cell wall 29

2.8 Detection of glycoconjugates by the lectin-gold technique 30

Protocol 2.7 Detection of glycoconjugates by the lectin-gold technique 31

2.9 Immunogold detection of antigens on ultrathin sections of acrylic resin 33

Protocol 2.8 Immunogold detection of antigens on ultrathin sections of acrylic resin 34

2.10 Cryofixation and freeze substitution for ultrastructural analysis and immunodetection 36

Protocol 2.9 Cryofixation and freeze substitution for ultrastructural analysis and immunodetection 37

2.11 Overview 38

2.12 References 39

3 Evaluation of Molecular Responses and Antifungal Activity of Phagocytes to Opportunistic Fungi 43

Maria Simitsopoulou and Emmanuel Roilides

3.1 Introduction 43

3.2 Preparation of conidia and hyphae of opportunistic fungi 45

Protocol 3.1 Preparation of conidia and hyphae of opportunistic fungi 45

Protocol 3.2 Preparation of hyphal fragments 47

3.3 Isolation of human monocytes from whole blood 48

Protocol 3.3 Isolation of human MNCs from whole blood 48

3.4 Analysis of human MNC function in response to fungal infection 51

Protocol 3.4 XTT microassay 52

Protocol 3.5 Superoxide anion assay in 96-well plate 53

Protocol 3.6 Hydrogen peroxide-rhodamine assay 55

Protocol 3.7 Phagocytosis assay 55

3.5 Evaluation of immunomodulators in response to fungal infection 57

Protocol 3.8 Analysis of gene expression by RT-PCR 58

Protocol 3.9 Analysis of pathway-specific gene expression by microarray technology 63

Protocol 3.10 Assessment of cytokines and chemokines by ELISA 66

3.6 Overview 67

3.7 References 68

4 Determination of the Virulence Factors of *Candida Albicans* and Related Yeast Species 69

Khaled H. Abu-Elteen and Mawieh Hamad

4.1 Introduction 69

4.2 Measurement of *Candida* species adhesion *in vitro* 70

Protocol 4.1 The visual assessment of candidal adhesion to BECs 70
Protocol 4.2 The radiometric measurement of candidal adhesion 73

4.3 Adhesion to inanimate surfaces 75

Protocol 4.3 Assessment of candidal adhesion to denture acrylic material 75
Protocol 4.4 Adherence of *Candida* to plastic catheter surfaces 76

4.4 *C. albicans* strain differentiation 77

Protocol 4.5 Resistogram typing 77
Protocol 4.6 The slide agglutination technique 79
Protocol 4.7 Serotyping of *C. albicans* by flow cytomerty 80

4.5 Phenotypic switching in *C. albicans* 81

Protocol 4.8 Evaluation of phenotype switching in *C. albicans* 82

4.6 Extracellular enzymes secreted by *C. albicans* 83

Protocol 4.9 Measurement of extracellular proteinase production by *C. albicans* 85
Protocol 4.10 Measurement of extracellular proteinase produced by *C. albicans* (staib method) 86
Protocol 4.11 Measurement of extracellular phospholipases of *C. albicans* 87

4.7 Germ-tube formation in *C. albicans* 88

Protocol 4.12 Germ-tube formation assay 88

4.8 References 89

5 Analysis of Drug Resistance in Pathogenic Fungi 93

Gary P. Moran, Emmanuelle Pinjon, David C. Coleman and Derek J. Sullivan

5.1 Introduction 93
5.2 Method for the determination of minimum inhibitory concentrations (MICs) of antifungal agents for yeasts 96

Protocol 5.1 Method for the determination of minimum inhibitory concentrations (MICs) of antifungal agents for yeasts 97

5.3 Measurement of Rhodamine 6G uptake and glucose-induced efflux by ABC transporters 102

Protocol 5.2 Measurement of rhodamine 6G uptake and glucose-induced efflux 102

5.4 Analysis of expression of multidrug transporters in pathogenic fungi 105
5.5 Analysis of point mutations in genes encoding cytochrome P-450 lanosterol demethylase 106
5.6 Qualitative detection of alterations in membrane sterol contents 108

Protocol 5.3 Qualitative detection of alterations in membrane sterol contents 109

5.7 Overview 110
5.8 References 110

6 Animal Models for Evaluation of Antifungal Efficacy Against Filamentous Fungi 115

Eric Dannaoui

6.1 Introduction 115
6.2 Disseminated zygomycosis in non-immunosuppressed mice 118

Protocol 6.1 Disseminated zygomycosis in non-immunosuppressed mice 118

6.3 Animal model of disseminated aspergillosis 125
Protocol 6.2 Disseminated aspergillosis in neutropoenic mice 126
6.4 Study design for evaluation of antifungal combinations therapy in animal models 130
Protocol 6.3 Study design for the evaluation of combination therapy in animal models 131
6.5 References 133

7 Proteomic Analysis of Pathogenic Fungi 137
Alan Murphy
7.1 Introduction 137
Protocol 7.1 2D SDS-PAGE of protein samples 139
7.2 Protein digestion in preparation for mass spectrometry by MALDI-TOF 140
Protocol 7.2 Peptide mass fingerprinting (PMF) by MALDI-TOF mass spectrometry 141
Protocol 7.2a In-gel digestion 142
Protocol 7.2b In-solution digestion 143
7.3 MALDI-TOF mass spectrometry 145
Protocol 7.3 Preparation of matrix for MALDI-TOF 147
7.4 Peptide mass fingerprinting (PMF) 149
Protocol 7.4 Post-source decay (PSD) and chemically assisted fragmentation (CAF) 150
7.5 Interpreting MALDI-TOF result spectra 152
7.6 Overview 156
7.7 References 157

8 Extraction and Detection of DNA and RNA from Yeast 159
Patrick Geraghty and Kevin Kavanagh
8.1 Introduction 159
8.2 The extraction of yeast DNA with the aid of phenol: chloroform 161
Protocol 8.1 Whole-cell DNA extraction from *C. albicans* using phenol: chloroform 161
Protocol 8.2 Rapid extraction of DNA from *C. albicans* colonies for PCR 164
8.3 Detection of yeast DNA using radio-labelled probes 165
Protocol 8.3 DNA detection by Southern blotting 165
8.4 Extraction of whole-cell RNA using two different protocols 169
Protocol 8.4 The extraction of whole-cell RNA from yeast using phenol: chloroform 170
Protocol 8.5 Rapid extraction of whole-yeast-cell RNA 172
8.5 Detection and expression levels of specific genes by the examination of mRNA in yeast 174
Protocol 8.6 Examining mRNA content as a means of investigating gene-expression profile by northern blot analysis 175
Protocol 8.7 Examining mRNA content as a means of investigating gene-expression profile by RT-PCR analysis 176
8.6 References 179

9 **Microarrays for Studying Pathogenicity in *Candida Albicans*** **181**
André Nantel, Tracey Rigby, Hervé Hogues and Malcolm Whiteway
9.1 Introduction 181
9.2 DNA microarrays 182
9.3 Building a second-generation 2-colour long oligonucleotide microarray for *C. albicans* 183
Protocol 9.1 Isolation of *C. albicans* RNA 185
Protocol 9.2 Isolation of total RNA using the hot phenol method 186
Protocol 9.3 Isolation of Poly-A+ mRNA 187
Protocol 9.4 Determination of the efficiency of incorporation of the probe 192
Protocol 9.5 Hybridization to DNA microarrays 194
9.4 Experiment design 196
9.5 Microarray-based studies in *C. albicans* 200
9.6 Conclusion 205
9.7 References 205

10 **Molecular Techniques for Application with *Aspergillus Fumigatus*** **211**
Nir Osherov and Jacob Romano
10.1 Introduction 211
10.2 Preparation of knockout vectors for gene disruption and deletion in *A. fumigatus* 212
Protocol 10.1 Preparation of knockout vectors for gene disruption and deletion in *A. fumigatus* 213
10.3 Transformation of *A. fumigatus* 217
Protocol 10.2 Chemical transformation of *A. fumigatus* 218
10.4 Molecular verification of correct single integration (PCR-based) 220
Protocol 10.3 Molecular verification of correct integration by PCR 221
10.5 General strategies for the phenotypic characterization of *A. fumigatus* mutant strains 223
Protocol 10.4 General strategies for the phenotypic characterization of mutants 223
10.6 References 229

11 **Promoter Analysis and Generation of Knock-out Mutants in *Aspergillus Fumigatus*** **231**
Matthias Brock, Alexander Gehrke, Venelina Sugareva and Axel A. Brakhage
11.1 Introduction 231
11.2 Site-directed mutagenesis of promoter elements 233
Protocol 11.1 Site-directed mutation of promoter elements 233
11.3 *lacZ* as a reporter gene 236
Protocol 11.2 *lacZ* as a reporter gene: discontinuous determination of β galactosidase activity 237
Protocol 11.3 *lacZ* as a reporter gene: continuous determination of β-galactosidase activity 239

11.4 Transformation of *A. fumigatus* 241
Protocol 11.4 Transformation of *A. fumigatus* 242
11.5 Hygromycin B as a selection marker for transformation 246
Protocol 11.5 Hygromycin B as a selection marker for transformation 247
11.6 *pyrG* as a selection marker for transformation 249
Protocol 11.6 *pyrG* as a selection marker for transformation 249
11.7 URA-blaster (*A. niger pyrG*) as a reusable selection marker system for gene deletions/disruptions 251
Protocol 11.7 URA-blaster (*A. niger pyrG*) as a reusable selection marker system for gene deletions/disruptions 253
11.8 References 255

12 Microarray Technology for Studying the Virulence of *Aspergillus Fumigatus* **257**
Darius Armstrong-James and Thomas Rogers
12.1 Introduction 257
12.2 Isolation of RNA from *A. fumigatus* 259
Protocol 12.1 Isolation of total RNA from *A. fumigatus* 260
12.3 Reverse transcription of RNA and fluorescent labelling of cDNA 263
Protocol 12.2 Indirect labelling of cDNA with fluorescent dyes 264
12.4 Hybridization of fluorescent probes to DNA microarrays and post-hybridization washing 266
Protocol 12.3 Hybridization of fluorescent probes to DNA microarrays and post-hybridization washing 267
12.5 Image acquisition from hybridized microarrays 269
Protocol 12.4 Image acquisition from hybridized microarrays 269
12.6 Microarray image analysis 270
Protocol 12.5 Microarray image analysis 271
12.7 References 272

13 Techniques and Strategies for Studying Virulence Factors in *Cryptococcus Neoformans* **275**
Nancy Lee and Guilhem Janbon
13.1 Introduction 275
13.2 Construction of *C. neoformans* gene-disruption cassettes 278
Protocol 13.1 Construction of *C. neoformans* gene-disruption cassettes 278
13.3 Genetic transformation of *C. neoformans* 283
Protocol 13.2 Biolistic transformation of *C. neoformans* 283
Protocol 13.3 Transformation via electroporation 286
13.4 Extraction of genomic DNA from *C. neoformans* 287
Protocol 13.4 DNA for use in library construction and hybridization analysis 288
Protocol 13.5 DNA for use in PCR 291
13.5 Screening and identification of deletion strains 292
Protocol 13.6 Screening and identification of deletion strains 292

13.6 Measuring capsule size in *C. neoformans* 297
Protocol 13.7 Measuring capsule size in *C. neoformans* 298
13.7 Purification of glucuronoxylomannan (GXM) 298
Protocol 13.8 Purification of glucuronoxylomannan (GXM) 299
13.8 References 301

14 Genetic Manipulation of Zygomycetes **305**
Ashraf S. Ibrahim and Christopher D. Skory
14.1 Introduction 305
14.2 Genetic tools to manipulate mucorales 306
14.3 Selectable markers used with mucorales fungi 307
14.4 Introduction of DNA used for transformation 308
Protocol 14.1 Protoplasting of *R. oryzae* 309
Protocol 14.2 Biolistic delivery system transformation of *R. oryzae* 313
Protocol 14.3 *A. tumefaciens*-mediated transformation 316
14.5 Molecular analysis of transformants 319
14.6 Overview 322
14.7 References 323

Index **327**

Preface

Pathogenic fungi represent a serious threat to the life and health of immunocompromised patients and are responsible for the deaths of up to 5% of all those who die in hospitals in the developed world. In addition to such life-threatening infections, fungi are also responsible for a wide range of superficial infections (e.g. 'ringworm', 'thrush') which can affect apparently healthy individuals. As the numbers of patients immunocompromised due to underlying diseases or medical therapies continue to rise, the incidence of fungal infections will increase. Despite their impact on human health, our understanding of the factors that allow fungi to colonize the body and surmount the defences of the immune system is poorly developed. In addition, our armamentarium for treating fungal infections, until recently, has been stocked with drugs (e.g. amphotericin B) which induce severe side effects in the host and which target, in the case of the azoles and polyenes, the same fungal biosynthetic pathway. Another problem hampering the treatment of fungal infections is the difficulty in diagnosing systemic infections promptly so that chemotherapy can be initiated with maximum benefit to the patient.

The aim of this book was to assemble a range of cellular and molecular techniques in order to facilitate an enhanced understanding of fungal virulence to aid in the development of improved diagnostic and chemotherapeutic regimens. Each chapter is written by internationally recognized experts who have direct experience of the relevant techniques and in a format that should allow those new to the field of research to master procedures with the minimum of delay. Each protocol is a self-contained unit so the reader is not required to search through an entire chapter to understand the requirements of a specific protocol.

The book starts with a description of techniques for diagnosing *Candida* infections in tissue using a variety of immunochemical methods. Chapter 2 details procedures for examining pathogenic fungi by transmission electron microscopy and shows how analysis of this type gives an understanding of the interaction of fungal cells with host tissue. Chapter 3 describes protocols for studying the interaction of phagocytes with fungi and offers an insight into how the immune system deals with pathogenic fungi. A number of methods for studying the virulence of *Candida albicans* and related species are described in Chapter 4, while protocols for assessing the response of yeast to antifungal drugs are highlighted in Chapter 5. Chapter 6 continues this theme and describes the use of animal models for evaluating the efficacy of antifungal drugs. In recent years, proteomic analysis has revolutionized our study of the structure and function of cellular proteins, and the application of this technique to medical mycology is described in Chapter 7. Chapter 8 is dedicated to detailing protocols for extracting and detecting nucleic acids from fungi. Chapter 9 describes the application of microarrays to the study of the pathogenicity of *C. albicans*. Chapter 10 details transformation protocols for use with the pulmonary pathogen *Aspergillus fumigatus*. This chapter is followed by another dedicated to *A. fumigatus*, where protocols for the generation of knock-out mutants are described, while Chapter 12 describes the use of microarray technology for studying the virulence of *A. fumigatus*. The penultimate chapter details techniques for studying the virulence of *Cryptococcus neoformans*, while Chapter 14 is a detailed description of the procedures for genetically manipulating the zygomycetes.

In a book of this size and scope it is inevitable that some techniques will be described in more than one place. A decision was taken not to remove multiple descriptions of techniques, so that the reader would be able to follow the specific guidelines of each chapter without having to refer to a standardized protocol that would undoubtedly miss some of the nuances necessary for a specific strain, species or experimental condition.

It is the hope of all the contributors to this book that the use of the protocols described here will increase our understanding of fungal virulence and ultimately enable the development of new and improved methods to diagnose and treat fungal infections.

Kevin Kavanagh

List of Contributors

Khaled H. Abu-Elteen
Department of Biological Science
Hashemite University
Zarqa 13133, Jordan

Darius Armstrong-James
Department of Infectious Diseases
Imperial College, London W12 0NN
England

Jean-Philippe Bouchara
Groupe d'Etude des Interactions
Hôte-Parasite, UPRES-EA 3142
Faculté de Pharmacie
16, boulevard Daviers 49045
Angers cedex, France

Axel A. Brakhage
Leibniz Institut für Naturstoff-Forschung
und Infektionsbilogie-Hans-Knöll-
Institut-Molekulare und angewandte
Mikrobiologie Beutenbergstr.11a
07745 Jena, Germany

Mathias Brock
Leibniz Institut für Naturstoff-Forschung
und Infektionsbilogie -Hans-Knöll-Institut-
Molekulare und angewandte
Mikrobiologie Beutenbergstr.11a
07745 Jena, Germany

David C. Coleman
Microbiology Division
Dublin Dental School and Hospital
Trinity College Dublin
Dublin 2, Ireland

Eric Dannaoui
Institut Pasteur, 25 rue du Dr Roux
75015 Paris, France

Alexander Gehrke
Leibniz Institut für Naturstoff-Forschung und
Infektionsbilogie-Hans-Knöll-Institut-
Molekulare und angewandte Mikrobiologie
Beutenbergstr.11a
07745 Jena, Germany

Pat Geraghty
Research Laboratory, RCSI Building
Beaumont Hospital, Beaumont, Dublin 9
Ireland

Mawieh Hamad
Department of Biological Science
Hashemite University, Zarqa 13133
Jordan

Jaako Hietanen
Department of Bacteriology & Immunology
University of Helsinki, Haartmaninkatu 3
00014 Helsinki, Finland

Hervé Hogues
Biotechnology Research Institute
National Research Council of Canada
6100 Royalmount Avenue, Montreal, QC
Canada H4P 2R2

Ashraf S. Ibrahim
David Geffen School of Medicine at UCLA
Division of Infectious Diseases (RB-2)
Los Angeles Biomedical Research Institute at
Harbor-UCLA Medical Center, 1124 W
Carson Street, Torrance, CA 90502, USA

Guilhem Janbon
Unité de Mycologie Moléculaire
CNRS FRE2849, Institut Pasteur
25 rue du Dr Roux
75015 Paris, France

Kevin Kavanagh
Department of Biology
National University of Ireland Maynooth
Co. Kildare, Ireland

Nancy Lee
Unité de Mycologie Moléculaire
CNRS FRE2849, Institut Pasteur
25 rue du Dr Roux
75015 Paris, France

Gary Moran
Microbiology Division
Dublin Dental School and Hospital
Trinity College Dublin
Dublin 2, Ireland

Alan Murphy
Department of Biology
National University of Ireland Maynooth
Co. Kildare, Ireland

André Nantel
Biotechnology Research Institute
National Research Council of Canada
6100 Royalmount Avenue, Montreal, QC
Canada H4P 2R2

Nir Osherov
Dept of Human Microbiology
Sackler School of Medicine
Tel Aviv University Ramat Aviv
Tel-Aviv 69978, Israel

Emmanuelle Pinjon
Microbiology Division
Dublin Dental School and Hospital
Trinity College Dublin
Dublin 2, Ireland

Riina Rautemaa
Department of Bacteriology & Immunology
University of Helsinki, Haartmaninkatu 3
00014 Helsinki, Finland

Malcolm Richardson
Department of Bacteriology & Immunology
University of Helsinki, Haartmaninkatu 3
00014 Helsinki, Finland

Tracey Rigby
Biotechnology Research Institute
National Research Council of Canada
6100 Royalmount Avenue, Montreal, QC
Canada H4P 2R2

Tom Rogers
Department of Clinical Microbiology
St James Hospital, Trinity College
Dublin, Ireland

Emmanuel Roilides
3rd Dept Pediatrics, Aristotle University
Hippokration Hospital
Konstantinoupoleos 49
GR-54642 Thessaloniki, Greece

Jacob Romano
Dept of Human Microbiology
Sackler School of Medicine
Tel Aviv University, Ramat Aviv
Tel-Aviv 69978, Israel

Maria Simitsopoulou
3rd Dept Pediatrics, Aristotle University
Hippokration Hospital
Konstantinoupoleos 49
GR-54642 Thessaloniki, Greece

Christopher D. Skory
National Center for Agricultural Utilization
Research, USDA
Agricultural Research Service
1815 North University Street
Peoria IL 61604, USA

Venelina Sugareva
Leibniz Institut für Naturstoff-Forschung und Infektionsbilogie-Hans-Knöll-Institut-Molekulare und angewandte Mikrobiologie
Beutenbergstr.11a
07745 Jena, Germany

Derek J. Sullivan
Microbiology Division
Dublin Dental School and Hospital
Trinity College Dublin
Dublin 2, Ireland

Guy Tronchin
Groupe d'Etude des Interactions
Hôte-Parasite, UPRES-EA 3142
Faculté de Pharmacie, 16
boulevard Daviers
49045 Angers cedex, France

Malcolm Whiteway
Biotechnology Research Institute
National Research Council of Canada
6100 Royalmount Avenue, Montreal, QC
Canada H4P 2R2

1

Diagnosis of *Candida* infection in tissue by immunohistochemistry

Malcolm D. Richardson, Riina Rautemaa and **Jarkko Hietanen**

1.1 Introduction

As a result of their large size, polysaccharide content and morphological diversity, fungi can be detected readily and may also to a certain level be identified in histological sections by conventional light microscopy. Application of immunohistochemical techniques may, in several cases, be the only means of establishing an accurate aetiological diagnosis in fixed-tissue sections because of morphological similarities among the tissue forms of several fungal genera (e.g. aspergillosis, fusariosis and scedosporiosis), when atypical forms of the fungus are present or when fungal elements are sparse (Jensen *et al.*, 1996; Jensen and Chandler, 2005).

The prerequisite for all immunohistochemical staining systems is a primary antibody properly characterized, especially in terms of specificity under optimal conditions. The technique used for obtaining an immunohistochemical diagnosis of mycoses may be either direct (conjugated primary antibodies) or indirect (conjugated secondary antibodies or tertiary enzyme complexes). In accordance with other immunohistochemical staining systems, the reaction complexes may be visualized with the help of fluorchromes (e.g. fluorescein isothiocyanate (FITC) and tetramethylrhodamine isothiocyanate R (TRITC), gold-silver complexes or complexes of enzymes (e.g. peroxidase-antiperoxidase (PAP) techniques, and alkaline phosphatase-antialkaline phosphatase (APAAP) techniques. In addition, avidin-biotin enzyme complex (ABC) methods may be used with horseradish peroxidase (HRP) and galactosidase. The major advantage of using enzyme immunohistochemistry compared with immunofluorescence techniques is that permanent sections are

Medical Mycology: Cellular and Molecular Techniques. Edited by Kevin Kavanagh

provided, specialized microscopes are not needed and pathological reactions may be assessed simultaneously during the evaluation of the immunoreactivity. A range of different fungal forms or fungal elements are frequently observed within lesions, especially when more than one organ is studied, and it must be ascertained whether these elements belong to a single or more taxa. In such cases, dual immunostaining techniques are useful tools for obtaining a reliable diagnosis.

Apart from tissue sections, immunohistochemical techniques can also be used to identify fungi in smears of lesional exudates, bronchial washings, bronchoalveolar lavage fluids, blood, bone marrow, cerebral-spinal fluid and in sputum specimens that have been enzymatically or chemically digested. An important limitation to the widespread application of immunochemical techniques and their use in the routine diagnosis of mycoses lies in the fact that sensitive and specific reagents are usually derived from multiple heterologously adsorbed polyclonal antisera which are not commercially available. However, in recent years more specific monoclonal antibodies have been developed, some of which are available commercially.

Before immunostaining, localization and presumptive identification of fungal elements in Gomori-methenamine-silver (GMS)-stained tissue sections enable the pathologist to narrow the aetiologic possibilities so that the most appropriate panel of immunoreagents can be employed. A number of appropriate control procedures are essential in the evaluation of immunohistochemical diagnostic assays (e.g. replacement of antifungal antibodies by antibodies raised against irrelevant antigens and including both positive and negative control sections with each batch of slides). As the somatic and cell-wall antigens of most pathogenic fungi survive formalin fixation and processing into paraffin quite well, immunohistochemical techniques can be performed retrospectively on archival material. Newly acquired antibodies can be evaluated using known fungal species cultured in agar and fixed and sectioned, as for tissue material.

Yeasts are opportunistic pathogens and common members of the normal oral flora in humans. *Candida albicans* is the most common yeast species in the human oral cavity. The transition from a commensal to a harmful pathogenic state eventually depends on a decrease in host resistance, changes in the local ecology or changes in intrinsic fungal virulence. The histopathologic changes related to the infection include inflammatory cell infiltration of the epithelium and connective tissue, intraepithelial microabscesses, epithelial intercellular oedema and epithelial atrophy, hyperplasia and dysplasia. In the oral cavity, yeasts can be found on mucosal surfaces and in saliva but also in the inflamed periodontal pockets (Järvensivu *et al.*, 2004). Even though the role of yeast in periodontitis is largely unknown, there is some evidence to suggest that yeasts may be implicated in the disease process.

The monoclonal antibody 3H8 has been shown to recognize mannoproteins of high molecular mass present in the cell walls of *C. albicans*. By ELISA (enzyme-linked immunosorbent assay), it has been shown that the presence of the epitope recognized by monoclonal 3H8 was similar in both the yeast and mycelial cell walls of *C. albicans* (Marcilla *et al*., 1999). Immunohistochemical studies using this antibody have demonstrated its usefulness in specifically recognizing *C. albicans* in kidney, lung, thyroid, oesophagus, small bowel and gingival tissues (Marcilla *et al*., 1999; Järvensivu *et al*., 2004). To explore further the usefulness of this monoclonal antibody in detecting *Candida* in periodontal diseases, we wished to test the antibody against a wide range of yeasts strains and species and various morphological forms. Furthermore, considering the location of the 3H8 epitope on the external cell wall of some *C. albicans* strains (Marcilla *et al*., 1999), it seemed reasonable to determine whether the epitope could be expressed into the surrounding environment, further aiding the recognition of the organism in the tissues.

C. albicans has been shown to be involved in the pathogenesis of adult periodontitis (AP). The potential candidal diagnosis of AP depends largely on the identification of yeast and pseudomycelial forms in gingival tissue samples by using periodic acid-Schiff and GMS stains. However, these stains are non-specific and also reveal confusing artefacts seemingly rather difficult to distinguish from yeasts. With the recent development and availability of monoclonal antibodies (mabs) to various epitopes of *C. albicans*, for example mab 3H8 which recognizes a mannoprotein, it is now possible to identify *Candida* in human-tissue biopsies.

The following protocols describe the evaluation of a relatively new monoclonal antibody for immunohistochemistry and its use in the detection of *C. albicans* morphological forms in tissue, using adult periodontitis as a model.

1.2 Specificity of monoclonal antibody 3H8 for *C. albicans*

Because of the limited number of tissue biopsies from patients with *Candida*-associated periodontal diseases (Järvensivu *et al*., 2004), it is very convenient to use an agar block technique where *Candida* species are grown under varying environmental conditions in agar, which is then processed in a similar manner to that used for pathology tissue specimens. The 3H8 epitope appears to be located at the external surface of some *C. albicans* strains but is partially cryptic in the cell wall of other strains. It is highly probable that the epitope is expressed extracellularly. The use of agar blocks makes it easier to detect these epitopes because there are not any interfering tissue structures as background.

To evaluate the usefulness of mabs in detecting *Candida* in periodontal disease, the antibody should be tested against a wide range of yeast species and strains and various morphological forms, grown in agar blocks at various temperatures and for various time periods. Furthermore, considering the location of epitopes on the external cell wall of certain *C. albicans* strains, it is useful to determine whether the epitope can be expressed into the surrounding environment, further aiding the recognition of the organism in tissue.

Protocol 1.1
Testing of specificity of monoclonal antibody 3H8

Equipment, materials and reagents

Candida isolates maintained on glucose-peptone agar

RPMI-1640 medium (Sigma-Aldrich, Poole, Dorset, UK) supplemented with 2% glucose

Agarose (Sigma-Aldrich)

Repli-dishes

Pepsin

Phosphate-buffered saline

Antibodies. A mouse monoclonal antibody, for example 3H8, is used as the primary antibody (IgG1; Société de Recherche et de Réalisations Biotechnologiques, Paris, France). It is raised against a zymolyase-solubilized preparation from the blastoconidial cell walls of *C. albicans* ATCC26555 and recognizes high-molecular-weight mannoproteins present in the cell wall. Biotinylated anti-mouse IgG as a component of the Vectastain® kit (Vector Laboratories, Burlingame, CA, USA) is used as a secondary antibody. Phosphate-buffered saline (PBS) is used as a buffer. In some instances, bovine serum albumin (BSA; Behringverke Gmbh, Germany) can be added to PBS (PBS-BSA) to reduce non-specific reactions.

Method

1. **Agar block culture and histochemical processing.** To test the specificity of the monoclonal antibody 3H8 to detect high-molecular-weight mannoproteins expressed by different *Candida* species, use, for example, strains of *C. albicans*, *C. lusitaniae*, *C. glabrata*, *C. parapsilosis*, *C. krusei* and *C. tropicalis*. Culture the isolates in RPMI-1640 medium (Sigma-Aldrich) supplemented with 2% glucose. Add 1 ml of the suspensions to 1 ml of 2% molten agarose (held at 50 °C) and pour into individual compartments of plastic replica dishes. After solidifying, the embedded agarose material is fixed in 10% formal saline for 24 h before processing through graded concentrations of alcohol and xylene and embedded in paraffin wax. Cut 4-μm-thick sections.

2. **Immunohistochemical staining.** De-paraffinize sections in xylene and rehydrate in graded alcohol series and in water. Incubate sections in pepsin (5 mg pepsin

$+ 5$ ml $H_2O + 50$ µl 1 N HCl) for 45 min in a humid chamber and wash 3 times for 5 min in PBS. Inhibit endogenous peroxidase activity with 0.3% H_2O_2 in methanol for 30 min, and then wash the sections with PBS 3 times for 5 min. First incubate the sections in normal horse blocking serum from the kit diluted 1:50 in 2% PBS-BSA. Then incubate the sections with the monoclonal primary antibody (3H8, 1:500, diluted in 1% PBS-BSA) for 30 min at 37 °C and keep overnight at 4 °C in a humid chamber. Controls are performed by omitting the primary antibody. The following day, after washing 3 times, incubate the sections for 30 min at 37 °C with biotinylated anti-mouse IgG secondary antibody solution from the kit diluted 1:200 into 0.1% PBS-BSA. After three washes, the sections are then incubated with the kit reagent for 30 min at 37 °C and then washed 3 times with buffer. Peroxidase binding sites are revealed with 3-amino-9-ethylcarbazole (AEC) with 0.03% hydrogen peroxidase. Finally, the slides are washed with tap water and then counter-stained with Mayer's haematoxylin for 4 min and again rinsed with tap water before mounting with glysergel (DAKO Corporation, CA, USA).

C. albicans is positively stained by the monoclonal antibody, whereas the yeast cell morphology of *C. glabrata*, and the dimorphic growth forms of *C. krusei*, *C. parapsilosis*, *C. lusitaniae* and *C. tropicalis* are negatively stained indicating further that the mab is specific for the mannoprotein of *C. albicans* (Figure 1.1).

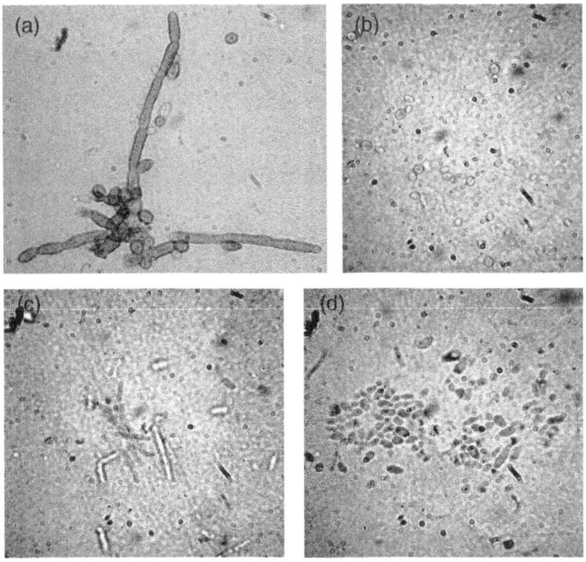

Figure 1.1 Expression of 3H8 epitope by *Candida* species (immunohistochemical staining with mab 3H8): (a) For *C. albicans*, very intensive staining of blastospore cell wall is seen. Small buds and cell-daughter junctions are negatively stained. Other *Candida* species are negatively stained: (b) *C. glabrata*, (c) *C. krusei*, (d) *C. parapsilosis*. However, it is still possible to see their typical morphology

1.3 Evaluation of monoclonal antibody 3H8 for the detection of *C. albicans* morphological forms

A range of different fungal morphological forms are frequently observed within lesions and it must be ascertained whether these elements belong to a single or more taxa. All species of *Candida* produce oval yeast-like cells, 3–6 μm, and mycelial elements composed of pseudohyphae and true hyphae. Pseudohyphae are composed of elongated yeast-like cells that remain attached end-to-end in chains. They are distinguished from true hyphae by the presence of prominent constrictions at points of attachment between adjacent cells, whereas the septate hyphae, 3 to 5 μm in width, are tubular and have parallel contours. Since the distinction between pseudohyphae and true hyphae is rarely of practical significance in histopathological diagnosis, they can be referred to collectively as mycelial elements.

The combination of yeast-like cells, pseudohyphae and true hyphae distinguishes *Candida* species from most other yeast-like pathogens in histopathological sections. *Trichosporon* species, which produce both yeast-like cells and mycelial elements, may provide a vexing problem in differential diagnosis. The yeast-like cells of *Trichosporon* are somewhat larger and more pleomorphic than those of *Candida*, and their hyphae produce rectangular arthroconidia. Weakly pigmented agents of phaeohyphomycosis can also be mistaken for *Candida* in histopathological sections. In difficult cases, a provisional morphological diagnosis of candidosis can be confirmed by culture or immunohistochemistry.

Protocol 1.2
Evaluation of monoclonal antibody 3H8 for the detection of *C. albicans* morphological forms

Equipment, materials and reagents

Candida albicans ATCC 28366

Normal human serum

Highly purified agarose (electrophoresis grade)

Method

Expression of antigens by morphological forms of *C. albicans*. To investigate the expression of extracellular antigenic mannoproteins by blastospores, pseudohyphae or hyphae of *C. albicans*, normal human serum is inoculated with blastospores of reference strain ATCC 28366 and incubated at either 30 °C or 37 °C for up to 30 h. After 3, 6, 24 and 30 h remove an aliquot of the inoculated serum and add this to molten 2% agarose and allow to set in compartmentalized replica dishes, or similar. Fix the samples, embed and stain as described in Protocol 1.1.

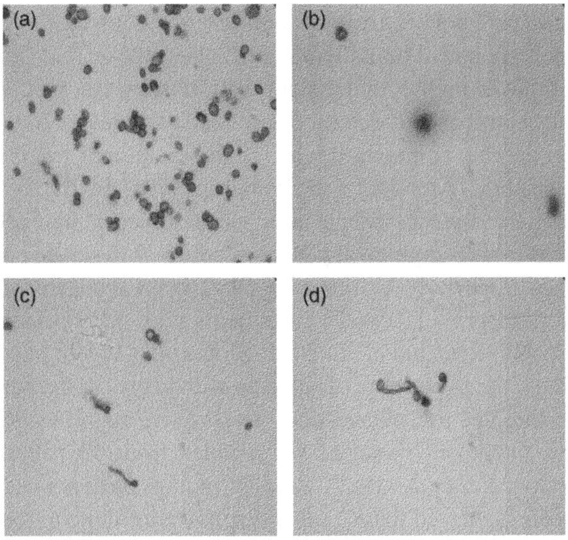

Figure 1.2 Expression of antigens by morphological forms of *C. albicans*. (a) When *C. albicans* cells are incubated for 3 h at 30 °C, cell-wall antigens are preferentially stained. (b) After incubation for 30 h at 30 °C, yeast cells can be seen to express antigen extracellularly. The cells incubated at 37 °C for 24 h (c) and 30 h (d) form hyphae which are positively stained

After incubation for 3 h at 30 °C, yeast cells are intensely stained indicating that the mannoprotein epitope has accumulated homogenously over the cell wall surface (Figure 1.2a). A positive signal in the parent yeast cells and particularly at the bud scar region will be seen. After 30 h incubation at 30 °C antigen expression will be observed extracellularly in the region immediately around the cell (Figure 1.2b).

Yeast cells incubated at 37 °C for 24 h and 30 h will form hyphae (Figure 1.2c, Figure 1.2d), and the staining reaction will indicate that the mannoprotein epitope is expressed uniformly on the cell wall of both yeast and hyphal growth forms.

1.4 Application of immunohistochemistry in the diagnosis of *Candida* periodontal disease

Periodontitis is an infection-induced inflammatory disease characterized by irreversible destruction of the tooth-supporting tissues (epithelium, gingival connective tissue and alveolar bone). Untreated periodontitis eventually may lead to tooth loss. Human periodontitis is associated with a widely diverse and complex subgingival microbiota encompassing both gram-positive and gram-negative bacteria, facultative and anaerobic organisms and possibly yeasts. At least 500 bacterial strains have been recovered from the subgingival crevice, a particularly well-studied microbial niche (Kroes *et al.*, 1999). Most of these strains are thought to be commensals, and a smaller number, potential opportunistic pathogens. The ability of one microbe to cause disease is greatly affected by the composition of the microbiota of the site.

Adult periodontitis (AP) results from a complex interplay of the mixed microbial infection and host response. The adherent microbes evoke release of a number of inflammatory mediators in the underlying soft tissues. In fact, these activation products ultimately result in the destruction of host tissue.

C. albicans is an aerobic commensal which can be cultured from the oral cavity of nearly every other adult (Arendorf and Walker, 1979). In the oral cavity, yeasts commonly colonize the tongue, palate and buccal mucosa (Arendorf and Walker, 1980) and may occur in the subgingival plaque of adults with severe periodontitis (Slots *et al.*, 1988). Yeasts, especially *C. albicans*, have been recovered from periodontal pockets in a large number (7.1–19.6%) of patients with AP (Dahlén and Wikström, 1995; Rams *et al.*, 1997; Reynaud *et al.*, 2001; Slots *et al.*, 1988). In a recent survey by Reynaud *et al.* (2001), the prevalence of subjects with yeasts in the periodontal pockets was 15.6%. Using the electron microscope, yeasts were found to be invading in the gingival connective tissue of 26 out of 60 samples from 12 patients with juvenile periodontitis (JP) (González *et al.*, 1987). Similar findings have not been reported of AP until now. It has been suggested that *C. albicans* may contribute to the development of necrotizing periodontal diseases in HIV-infected patients (Odden *et al.*, 1994).

The diagnosis of cutaneous and mucosal candidosis depends largely on the identification of yeast pseudomycelial forms in tissue samples by using periodic acid-Schiff (PAS) and Gomori methenamine silver (GMS) stains. However, these stains are non-specific and also reveal confusing artefacts seemingly rather difficult to distinguish from yeasts. The structural similarities between different fungi are a further source of diagnostic difficulties. Histopathological techniques evidently do not identify pathogenic fungi to the species level. To enhance the *in situ* identification of fungi in clinical specimens, a number of both direct and indirect immunohistochemical techniques have been developed, notably immunoperoxidase-based methods (Jensen *et al.*, 1996; Jensen and Chandler, 2005). With the recent development and availability of monoclonal antibodies to various epitopes of *C. albicans* and *C. dubliniensis* (Järvensivu *et al.*, 2004; Marcilla *et al.*, 1999), it is now possible to identify *Candida* in human tissue biopsies to the species level. Even though the role of yeasts in AP is largely unclear, there is evidence to suggest that yeasts can be implicated in the pathogenesis of the tissue-destructive periodontal disease process.

Protocol 1.3
Use of monoclonal antibody 3H8 in the detection of *C. albicans* in tissue

Equipment, materials and reagents

Patient tissue samples

Monoclonal antibody 3H8: IgG1; 2.5 mg/ml; Société de Recherche et de Réalisations Biotechnologiques, Paris, France

Rabbit polyclonal antibody 158: 4.5 mg/ml; Biodesign International, Saco, Maine, USA

Biotinylated anti-mouse or anti-rabbit IgG: Vectastain® kit (Vector Laboratories, Burlingame, CA, USA)

BSA

10% formalin

Method

1. Fix patient tissue samples in formalin and paraffin-embed.

2. For periodic acid-Schiff (PAS) staining de-paraffinize 4 μm-thick, formalin-fixed, paraffin-embedded sections in xylene and rehydrate in graded alcohol series and in water. Before staining with Schiff's leucofuchsin reagents, expose the sections to periodic acid.

3. For immunohistochemical staining, de-paraffinize 4 μm-thick, formalin-fixed, paraffin-embedded sections as described above. Incubate the sections with pepsin for 45 min in a humid chamber and wash 3 times for 5 min with PBS. To inhibit endogenous peroxidase activity, the sections are incubated with 0.3% H_2O_2 in methanol for 30 min followed by three washes with PBS. A modification of the Vectastain kit protocol can be used. To inhibit non-specific staining, incubate the sections with normal horse serum from the kit (1:50 in 2% PBS-BSA). The sections are incubated with the primary antibody (3H8 or 158) against *C. albicans* (1:500 and 1:5000 accordingly in 1% PBS-BSA) for 30 min at 37 °C and kept overnight at 4 °C in a humid chamber. Control stainings are performed by omitting the primary antibody. The next day, after three washes, incubate sections for 30 min at 37 °C with the corresponding biotinylated anti-mouse or anti-rabbit secondary antibody from the kit (1:200 in 0.1% PBS-BSA). The sections are washed and incubated with the kit reagent for 30 min at 37 °C and then washed again. Peroxidase binding sites are revealed with 3-amino-9-ethylcarbazole (AEC) with 0.03% hydrogen peroxidase. Finally, wash the slides with tap water and counterstain with Mayer's haematoxylin for 4 min and then rinse with tap water before mounting with glycerol (DAKO Corporation, CA, USA).

Typically, positive staining for *C. albicans* either with the polyclonal or monoclonal antibody is seen (Figure 1.3). Predominantly, pseudomycelial forms are immunoreactive. In general, the polyclonal antibody gives a stronger signal than the monoclonal antibody. The polyclonal antibody used here will also detect *Candida*-derived antigens in the periodontal tissues in addition to the specific staining of hyphal elements and candidal cells.

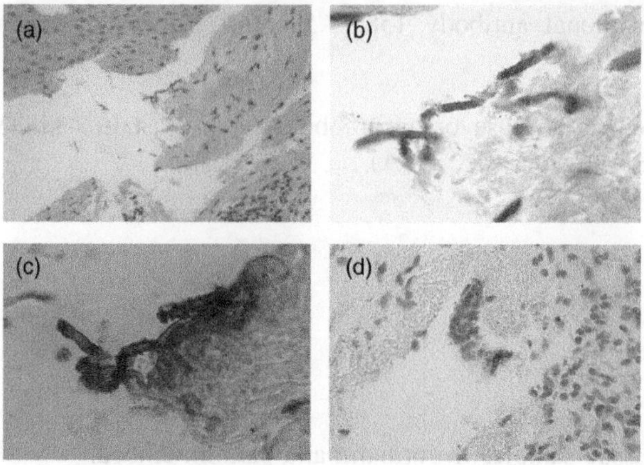

Figure 1.3 Adherence of *C. albicans* hyphae to periodontal connective tissue. With the mab 3H8, three to four *Candida* hyphae can be seen on the surface of connective tissue which has become detached from the overlying epithelium. This group of *Candida* hyphae is surrounded by a mild inflammatory reaction (Figure 3A, ×200 and Figure 3B, ×1000). Staining with the polyclonal antibody demonstrates six to eight *Candida* hyphae in almost the same localization also without a remarkable inflammatory reaction (Figure 3C, ×1000). In a deeper part of the specimen one hypha surrounded by a rather strong compact inflammatory cell infiltrate can be seen (Figure 3D, ×400). This infiltrate seems to consist mainly of mononuclear inflammatory cells

C. albicans is regarded as potentially the most pathogenic fungus normally found in the oral cavity. In tissues it appears mostly in pseudohyphal form and can tentatively be distinguished from other fungi by its morphology. *C. albicans* can also appear as a yeast cell form, especially when colonizing epithelial surfaces. Therefore, more-specific methods like immunohistochemistry are necessary for precise identification. Changes in the environmental conditions trigger germination. The gingival pocket and gingival crevicular fluid are favourable environments for the germination and hyphal growth of *Candida*. *Candida* hyphae have the ability to penetrate host tissue, and hyphae adhere to host surfaces to a greater extent compared to yeast cells. Thus, hyphae are important in the attachment and disease process.

In one evaluation of the monoclonal antibody 3H8, *C. albicans* was detected by immunohistochemistry in the tissues of four (16%) adult periodontitis patients. Three of these tissue samples also contained subgingival plaque in which candidal structures could be detected in every case. Staining of hyphal elements with the specific monoclonal antibody verified that the organism seen was *C. albicans*. *C. albicans* was often seen in the outer layers of the plaque and some hyphae reached towards the more central parts. It is clear that hyphal germination had already started in the gingival pocket. These findings suggest that *C. albicans* could have a role in the infrastructure of periodontal microbial plaque and in its adherence to the periodontal tissues.

These protocols show that the use of specific reagents avoids the complication of identifying other pathogenic yeasts to distinguish them from *C. albicans*, such as *C. lusitaniae*, *C. parapsilosis*, *C. glabrata*, *C. krusei* and *C. tropicalis*. The absence of any immunoreactivity of normal human tissues (Järvensivu *et al.*, 2004; Marcilla *et al.*, 1999) together with the fact that no immunostaining of other yeasts had been found supports the value of this antibody in the specific diagnosis of candidosis caused by *C. albicans*. The ability to identify *Candida* organisms upon their penetration in tissue lesions in cases of oral candidosis (e.g. in *Candida*-associated periodontitis, mucosal candidosis or chronic hyperplastic candidosis) will help to clarify the role of individual *Candida* spp. in the pathogenesis of oral candidosis. *Candida* spp. are frequently associated with the superficial infection of mucosal membranes, although fatal systemic infection in debilitated individuals may also occur. Chronic hyperplastic candidosis is a particularly important form of oral candidosis as it is associated with the development of epithelial dysplasia and intraoral squamous cell carcinoma.

1.5 References

Arendorf, T.M. and Walker, D.M. (1979) Oral candidal populations in health and disease. *Br Dent J.* **147**(10): 267–272.

Arendorf, T.M. and Walker, D.M. (1980) The prevalence and intra-oral distribution of *Candida albicans* in man. *Arch Oral Biol.* **25**(1): 1–10.

Dahlén. G, and Wikström, M. (1995) Occurrence of enteric rods, staphylococci and *Candida* in subgingival samples. *Oral Microbiol Immunol.* **10**(1): 42–46.

González, S., Lobos, I., Guajardo, A., Celis, A. *et al.* (1987) Yeasts in juvenile periodontitis: Preliminary observations by scanning electron microscopy. *J Periodontol.* **58**(2): 119–124.

Järvensivu, A., Hietanen, J., Rautemaa, R., Sorsa, T. and Richardson, M. (2004) Candida yeasts in chronic periodontitis tissues and subgingival microbial biofilms in vivo. *Oral Dis.* **10**(2): 106–112.

Järvensivu. A., Hietanen, J., Rautemaa, R., Sorsa, T. and Richardson, M. (2006) Differential cell surface expression of mannoproteins in yeast and mycelical forms of *Candida albicans* by immunohistochemical staining. *Oral Dis.* (in press).

Jensen, H.E., Schonheyder, H.C., Hotchi, M. and Kaufman, L. (1996) Diagnosis of systemic mycoses by specific immunohistochemical tests. *APMIS.* **104**(4): 241–258.

Jensen, H.E. and Chandler, F.W. (2005) Histopathological diagnosis of mycotic diseases. In: Medical Mycology (eds W.G. Merz and R.J. Hay), London, Hodder Arnold, pp. 121–143.

Kroes, I., Lepp, P.W. and Relman, D.A. (1999) Bacterial diversity within the human subgingival crevice. *Proc Natl Acad Sci USA.* **96**(25): 14547–14552.

Marcilla, A., Monteagudo, C., Mormeneo, S. and Sentandreu, R. (1999) Monoclonal antibody 3H8: a useful tool in the diagnosis of candidiasis. *Microbiology.* **145**(3): 695–701.

Odden, K., Schenick, K., Koppang, H.S. and Hurlen, B. (1994) Candidal infection of the gingiva in HIV-infected persons. *J Oral Pathol Med.* **23**(4): 178–183.

Rams, T.E., Flynn, M.J. and Slots, J. (1997) Subgingival microbial associations in severe human periodontitis. *Clin Infect Dis.* **25**(suppl 2): S224–226.

Reynaud, A.H., Nygaard-Østby, B., Bøygard, G.-K., Eribe, E.R. *et al.* (2001) Yeasts in periodontal pockets. *J Clin Periodontol.* **28**(9): 860–864.

Slots, J., Rams, T.E. and Listgarten, M.A. (1988) Yeasts, enteric rods and pseudomonads in the subgingival flora of severe adult periodontitis. *Oral Microbiol Immunol.* **3**(2): 47–52.

2

Transmission electron microscopy of pathogenic fungi

Guy Tronchin and **Jean-Philippe Bouchara**

2.1 Introduction

Since the development of electron microscopy in the 1950s, ultrastructural studies have considerably extended our knowledge of pathogenic fungi. Earliest research in transmission electron microscopy was first limited to fine structural observations of subcellular organelles. Good morphological details were often lacking in such studies due to the presence of a particular thick structure surrounding the fungal cell, the cell wall, which was the major challenge in specimen preparation. The cell wall plays important roles in pathogenesis, particularly in phagocytosis and in adherence to the host tissues, two important steps in virulence of pathogenic fungi. Expression at the fungal surface of specific receptors for some host ligands is a prerequisite for interaction with tissue and colonization. Apart from pathogenesis, the cell wall is a target for many antifungal agents. So, the localization of cell-wall components and the analysis of the mechanisms of their assembly, as well as the influence of environmental conditions or of swelling and germination on the cell surface's ultrastructural characteristics, and particularly on the expression of surface adhesins, are therefore important issues in both basic and applied biology.

In the 1960s, improved fixation using aldehydes as primary fixatives (particularly glutaraldehyde) considerably increased the cell preservation and the quality of fungal fine structure. It allowed detailed description of fungal subcellular organelles and cell-wall organization, spore development and germination, interactions with the host cells and effects of antifungal agents (Klomparens, 1990). Considering the structure

Medical Mycology: Cellular and Molecular Techniques. Edited by Kevin Kavanagh

and the composition of the fungal cell wall (mainly glycoproteinic in nature), a variety of treatments have been developed to facilitate the penetration of fixatives and to improve ultrastructural morphology. They include the use of different fixative solutions (potassium or lithium permanganate, acrolein, dimethylsulfoxide, tannic acid, glutaraldehyde, paraformaldehyde, osmium tetroxide) used separately or sometimes mixed. In some circumstances, the preparation of cryostat sections of fungal pellets, chemical (sodium periodate) or enzymatic (zymolyase, lyticase, glucanase) cell-wall digestion and production of sphero-protoplasts, give appreciable results. For ultrastructural analysis, the most frequently used chemical fixation method consists of treatment with glutaraldehyde followed by either potassium permanganate or osmium tetroxide (Osumi, 1998). Potassium permanganate is adequate to preserve membrane systems, but it is not efficient for (immuno)cytochemistry, whereas osmium tetroxide is valuable for cell-wall ultrastructure (particularly fibrillar structures) and makes possible the development of cytochemical techniques. In some cases, however, a good compromise between the ultrastructural preservation of cytoplasmic organelles and that of the cell wall has been obtained whatever the protocol used. This may depend on the nature of the fungus and on the culture conditions. For example, it has been demonstrated that the permeability of the cell wall of *Candida albicans* is maximal when the cells are fixed during early log phase and are grown on a rich medium (Wright, 2000).

Ultrastructural characterization of fungal organelles, particularly those involved in the secretory pathway, is often fastidious by standard fixation techniques. Several methodologies have been developed to avoid these difficulties. For example, osmium tetroxide reduced by potassium ferrocyanide intensely stains the endoplasmic reticulum cisternae, Golgi elements and secretion granules of yeasts (Rambourg *et al.*, 1995). In addition, this treatment clearly delineates the membranes of organelles, thus facilitating their identification. On the other hand, some enzymes like acid phosphatase are generally considered as typical lysosomal markers. Acid phosphatase can easily be demonstrated at the ultrastructural level, and it has been located in Golgi saccules and in the digestive vacuoles of several fungal cell types (Mahvi *et al.*, 1974). Likewise, carbohydrates are present in cell organelles involved in secretion and they could be responsible for their staining. So, these structures were usually well revealed in osmium-reduced sections with thiocarbohydrazide (Rambourg *et al.*, 2001) or by the periodic acid-thiocarbohydrazide-silver proteinate technique (Tronchin *et al.*, 1982).

The 1970s represented an important step in research about fungal ultrastructure with the development of cytochemical techniques, particularly those related to the carbohydrate cytochemistry of the cell wall. Carbohydrates present in polysaccharides or glycoproteins are the main components of the fungal cell wall, and they play important roles in numerous cellular or biological processes including cell agglutination, host-cell recognition and adherence, and elicitation of the host immune response. This explains the need to localize carbohydrate moieties at the cellular and/or molecular level. Several cytochemical methods have been developed for electron microscopy in relation to the biochemistry of the cell wall. They include the periodate oxidation of carbohydrate moieties or the use of gold-conjugated probes like lectins. The detection of alcohol groups can be achieved by using periodate oxidation of the carbohydrate moieties, a method that derives from the classic PAS technique (Thiéry, 1967). Likewise, lectins, which are a group of carbohydrate-binding proteins, have become essential tools for carbohydrate cytochemistry because of their specific binding properties. They are readily available from several commercial sources. In addition, anionic sites (especially carboxyl and sulphate groups) and enzymes like acid phosphatase (as a representative of extracellular mannoproteins) can be easily detected in fungi (Tronchin *et al.*, 1989). Together, these methods have provided some answers to the dynamic changes of the cell wall associated with swelling, germination or adherence.

Finally, another important milestone in ultrastructural studies of fungal cells was reached during the 1980s with the development of (immuno)cytochemical approaches based on the affinity of enzymes for their specific substrates (Bendayan and Benhamou, 1987) or of antibodies for their respective antigens (Benhamou, 1995). These methods were very efficent to determine the precise location of fungal molecules. The use of enzyme-gold complexes is a sensitive method allowing highly specific identification of various fungal carbohydrates such as chitin and glucans. Protein A-gold (which interacts with the Fc of immunoglobulins, especially IgG) or gold-conjugate secondary antibodies (specific for the primary antibody), were used to visualized the antibody–antigen interaction. Labelled protein A and labelled secondary antibodies are commercially available, with various gold particle sizes. Tissue processing is essential for the preservation of ultrastructure and of protein antigenicity so that the antibody can access intracellular molecules in tissue sections. Among the different techniques that have been proposed, those using acrylic resin as an embedding medium offer a good compromise between the preservation of ultrastructure and antigenicity. They allowed for the precise localization of fungal molecules in respective cell compartments and also contributed to a better understanding of their function during adherence or colonization (Esnault *et al.*, 1999).

Chemical fixation can affect the ultrastructure of the cell membrane and results in alterations of antigens. The methods of choice for both ultrastructural preservation of fungi and immunolabelling are cryofixation techniques based on physical fixation principles. Among them, rapid freezing combined with free-substitution allows excellent preservation of ultrastructure and antigenicity (McKerracher and Heath, 1985). The adequate preservation of biological materials requires the immobilization of cell structures within milliseconds. This can be obtained using specific equipment such as a high-pressure freezing machine or by rapidly plunge-freezing in liquid propane. Then the ice in the frozen specimen is substituted at low temperatures (−80 °C to −90 °C) with an organic solvent with or without chemical fixatives.

In this Chapter, we describe detailed protocols currently applied in transmission electron microscopy for ultrastructural studies, cytochemical analysis and immunolabelling of pathogenic fungi.

2.2 Glutaraldehyde-potassium-permanganate or glutaraldehyde-osmium-tetroxide fixation for ultrastructural analysis

The most commonly used chemical fixation for ultrastructural analysis of fungal cells involves treatment with glutaraldehyde (pre-fixation) followed either by potassium permanganate or osmium tetroxide (post-fixation). For standard ultrastructural observation, incubation of aldehyde-fixed sample with potassium permanganate ($KMnO_4$, 0.5 to 6% for 1 to 24 h) allows the extraction of cytoplasmic components and therefore facilitates the visualization of cell membranes. So permanganate fixation is extremely useful for an analysis of the overall organization of fungal cells. Membrane profiles of organelles like mitochondria, vacuoles, the nucleus, endoplasmic reticulum and microbodies are well preserved (Osumi, 1974). Osmium tetroxide (OsO_4, 1 to 2% for 30 min to 2 h), which strongly reacts with unsaturated fats, is another common post-fixative that preserves membranes (Figure 2.1). Because it extracts much less material than potassium permanganate, microtubules, microfilaments, ribosomes and chromatin remain visible following osmium fixation.

However, unlike potassium permanganate, good penetration of osmium inside fungal organisms sometimes requires removal or permeabilization of the cell wall. So, specimen preparation by cryostat sectioning and/or enzyme treatment resulting in sphero-protoplasts has been shown to increase markedly the quality of ultrastructural observations (Tronchin *et al.*, 1981). In our experience, osmium post-fixation gives better results for common ultrastructural analysis of germ tubes, hyphae and

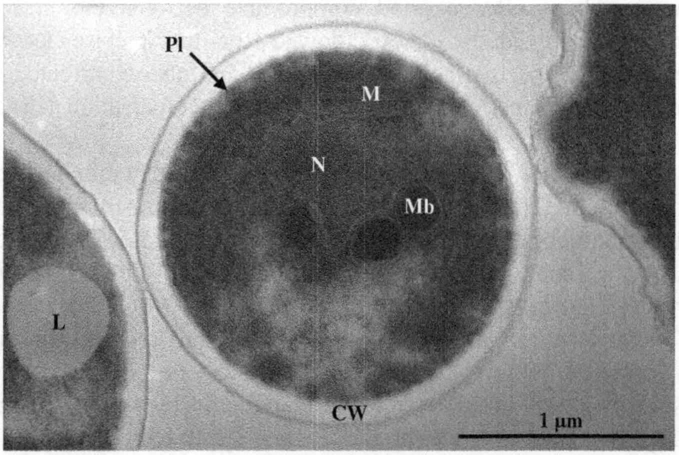

Figure 2.1 Ultrastructure of *C. glabrata* fixed with glutaraldehyde and osmium tetroxide and embedded in EPON 812® (Shell Chemicals Ltd., London, UK). Intracellular organelles like mitochondria (M), microbodies (Mb), nuclei (N) and lipid droplet (L) are clearly visible. The plasmalemma (Pl) and its invaginations are well delineated. Note the classical aspect of the yeast cell wall (CW) with an electron-transparent inner layer covered by an electron-dense outer layer

pseudohyphae which have a thin cell wall, compared to blastospores and conidia, whose cell walls are usually denser and thicker. In addition, because of its low viscosity, Spurr's resin is the resin of # choice for infiltration and embedding in ultrastructural studies of pathogenic fungi.

Protocol 2.1 Glutaraldehyde-osmium tetroxide (#) or glutaraldehyde-potassium permanganate (*) fixation for ultrastructural analysis

Equipment, materials and reagents

1.5 ml Eppendorf-type tubes

Cells to be analysed

Glutaraldehyde

Potassium permanganate

Osmium tetroxide

Cacodylate buffer 0.2 M, pH 7.4

PIPES buffer 0.2 M, pH 6.8

70%, 95% and absolute ethanol

Spurr's resin

Uranyl acetate

Lead citrate

Copper grids

Method

1. Wash cells (1 ml – 1×10^8 cells/ml) in distilled water.

2. Resuspend the cells in 2% glutaraldehyde solution buffered at pH 7.4 with 0.1 M cacodylate or 4% glutaraldehyde solution in 0.2 M PIPES buffer pH 6.8 (*) and incubate for 2 h at room temperature (#) or overnight at 4 °C (*).

3. Wash overnight at 4 °C in cacodylate (#) buffer (#) or water (*).
4. Post-fix the cells for 1 h with 1% osmium tetroxide solution buffered at pH 7.4 with 0.1 M cacodylate (#) or with 2% aqueous solution of potassium permanganate (*).
5. Dehydrate the cells in a graded series of ethanol dilutions (70% for 10 min, 95% for 10 min and finally 100% for 2 × 10 min).
6. Infiltrate in Spurr's resin and embed.
7. Incubate at 60 °C for at least 24 h.
8. Prepare 80–100 nm-thin sections.
9. Contrast ultrathin sections with uranyl acetate for 10 min and lead citrate for 5 min.
10. View under electron microscope.

Note: Steps 1–7 could be performed in the same Eppendorf-type tube. The working solution of potassium permanganate must be used immediately after preparation. After post-fixation with potassium permanganate, wash several times carefully with distilled water until no purple colour is observed. To increase membrane contrast, treatment with 1% aqueous uranyl acetate for 1 h may be performed before dehydration ('*en bloc*' staining). In this case, uranyl acetate staining of ultrathin sections is not necessary. Moreover, addition of 0.5 to 1% ferrocyanide to osmium also increases membrane contrast. Prepare resin according to the recommendations of the manufacturer.

2.3 Identification of the different compartments of the secretory pathway in yeasts

Ultrastructural characterization of yeast organelles, particularly those involved in the secretory pathway, is often difficult due to the lack of appropriate staining techniques. By immunolabelling of glutaraldehyde-fixed cells infiltrated with LR resin, Preuss *et al.* (1991, 1992) localized endoplasmic reticulum and Golgi markers in small single isolated cisternae. However, owing to the small size of the labelled structures and to the lack of a good contrast required for immunological analysis, they could not visualize accurately their ultrastructural characteristics. Rambourg *et al.* (1995) have developed a method in which cells are fixed by glutaraldehyde, then treated with sodium metaperiodate, post-fixed with a mix of osmium tetroxide and potassium ferrocyanide and embedded in agarose before dehydration. Treatment of intact glutaraldehyde-fixed yeast cells with sodium metaperiodate and pre-embedding in agarose improves the preservation of cell organelles after embedding in EPON 812® (Shell Chemicals Ltd., London, UK) and prevent retraction of the cytoplasm from the cell wall. In yeasts fixed with glutaraldehyde, reduced osmium delineates selectively intracytoplasmic membranes, and also stains the content of ER cisternae, Golgi elements and secretion granules. Tubular networks appear as separate elements or units dispersed throughout the cytoplasm. In addition, a three-dimensional configuration of these structures can be obtained from adjusted stereopairs of micrographs of the same field. Such a technique facilitates the identification of the various compartments of the secretory pathway and is very efficient at revealing the dynamic of the secretory machinery of yeasts.

Protocol 2.2
Identification of the different compartments of the secretory pathway in yeasts

Equipment, materials and reagents

1.5 ml Eppendorf-type tubes

Cells to be analysed

Agarose

Glutaraldehyde

Osmium tetroxide

Cacodylate buffer 0.2 M, pH 7.4

Sorbitol

Sodium metaperiodate

Potassium ferrocyanide

70%, 95% and absolute ethanol

EPON 812® (Shell Chemicals Ltd., London, UK)

Lead citrate

Copper grids

Method

1. Initiate cell fixation by adding 2% glutaraldehyde directly to the culture, the growth conditions being maintained for a further 5 min.
2. Collect the cells and incubate them overnight at 4 °C in a fixative containing 2% glutaraldehyde and 0.8 M sorbitol in 0.1 M cacodylate buffer pH 7.4.
3. Wash the cells twice in cacodylate buffer and then twice in distilled water.
4. Treat the cells for 15 min at room temperature in 1% sodium metaperiodate.
5. Wash the cells twice in distilled water.
6. Post-fix the cells for 1 h at room temperature in a 1:1 mixture of 2% aqueous osmium tetroxide and 3% aqueous potassium ferrocyanide.
7. Rinse twice in distilled water.
8. Embed the cells in 3% agarose.
9. Cut small blocks of approximately 1 mm^3.
10. Dehydrate in 70% ethanol for 20 min, then in 80%, 90%, 95% ethanol for 5 min each, and finally in two baths of 100% ethanol for 10 min each.
11. Embed the agar blocks in EPON at 60 °C for 17–20 h.
12. Counterstain 0.08–0.2 μm sections for 2 min with lead citrate.
13. View under electron microscope.

Note: Staining of endoplasmic reticulum and Golgi elements can be significantly increased by counterstaining osmium-reduced sections with thiocarbohydrazide instead of lead citrate (Rambourg *et al.*, 2001). For stereoscopy, grids are placed on the goniometric stage of the electron microscope and stereopairs are obtained by taking pictures of the same field after tilting the specimen at −15° and +15° from the 0° position.

2.4 Cytochemical localization of acid phosphatase in yeasts

Acid phosphatase can easily be demonstrated at the ultrastructural level using $Pb(NO_3)_2$ and Na-β-glycerophosphate at pH 5 as substrate media. This enzyme is generally considered as a typical lysosomal enzyme and is a reliable lysosomal

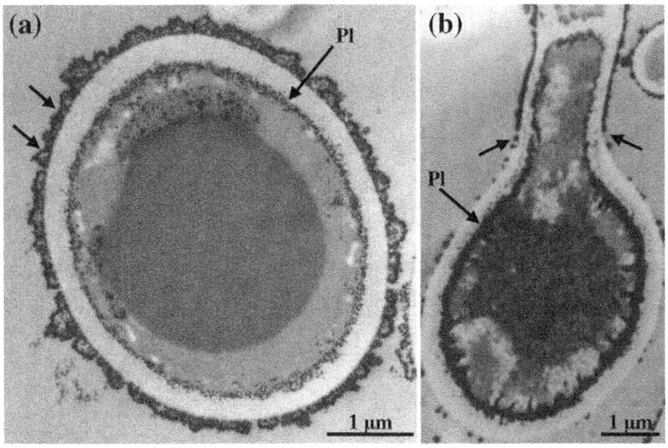

Figure 2.2 Detection of acid phosphatase in *C. albicans*. Note the contrast of the reaction between nongerminating (a) and germinating (b) blastoconidia. Heavy precipitates of lead phosphate decorate the cell surface (arrows) and the plasmalemma (Pl) of nongerminating blastoconidia and of mycelium, but they do not coat the surface of germinating ones

marker. Acid phosphatase activity has been located in Golgi saccules and in the digestive vacuoles of several cell types. In yeasts, it has been demonstrated that acid phosphatase is also a representative of extracellular mannoproteins, since it is associated with the mannan layers deposited during cell-wall formation (Linnemans *et al.*, 1977). Cryostat sectioning of intact glutaraldehyde-fixed yeast cells, which greatly improves the preservation of cell organelles after embedding in EPON® (Shell Chemicals Ltd., London, UK), has been used successfully for the demonstration of acid phophatase (Tronchin *et al.*, 1980). Intacytoplasmic membranes like nuclear envelope and plasma membrane are clearly delineated. The content of vacuoles, ER cisternae, Golgi structures and secretion granules is heavily stained. In addition, the dynamic of the cell-wall formation during budding and germination can be followed (Figure 2.2).

Protocol 2.3
Cytochemical localization of acid phosphatase in yeasts

Equipment, materials and reagents

1.5 ml Eppendorf-type tubes

Cells to be analysed

Glutaraldehyde

Osmium tetroxide

Cacodylate buffer 0.2 M, pH 7.4

Lead nitrate, $Pb(NO_3)_2$

Sodium-β-glycerophosphate

Sodium fluoride (NaF)

Acetate buffer 0.05 M, pH 7.2 and pH 5

Formaldehyde

Sucrose

70%, 95% and absolute ethanol

Spurr's resin

Copper grids

Method

1. Wash cells (1 ml – 1×10^8 cells/ml) in distilled water.

2. Fix the cells for 30 min at 4 °C in 5% glutaraldehyde solution in 0.1 M cacodylate buffer pH 7.4 (1.5 ml per tube).

3. Perform 5–7 μm cryostat sections of the pellet.

4. Fix again cryostat sections for 30 min in glutaraldehyde.

5. Wash overnight at 4 °C in cacodylate buffer supplemented with 7.5% sucrose.

6. Incubate for 30 min at 37 °C in Gomori medium (Gomori, 1952).

7. Wash with 0.05 M acetate buffer pH 7.2 containing formaldehyde 4% and 7% sucrose overnight.

8. Post-fix the cells for 30 min at 4 °C with 1% osmium tetroxide solution in 0.1 M cacodylate buffer pH 7.4.

9. Dehydrate the cells in a graded series of ethanol dilutions (70% for 10 min, 95% for 10 min and finally 100% for 2×10 min).

10. Embed material in Spurr's resin.

11. Do not counterstain sections.

12. Two types of controls must be performed: (1) Incubation of cells in Gomori medium without substrate. (2) Incubation of cells in Gomori medium with an enzyme inhibitor, NaF 1% (w/v).

13. View under electron microscope.

Note: Gomori medium must be freshly prepared before each experiment by dissolving 0.12 mg $Pb(NO_3)_2$ in 100 ml sodium acetate buffer 50 mM, pH 5.0, containing 7.5% sucrose (0.22 M) followed by slow addition of 10 ml of a 3% Na-β-glycerophosphate solution. Before use, the mixture must be warmed at 60 °C for 1 h, cooled at room temperature and filtered to eliminate the small precipitate that usually forms.

2.5 Detection of anionic sites

The electrostatic charge of the cell surface is thought to play an important role in host-tissue recognition and adherence to target cells or inert surfaces. The cell surface of fungal cells is negatively charged. Cationized ferritin is a positively charged ligand allowing the visualization of anionic sites at the surface of living cells under physiological pH and ionic strength. Owing to the simplicity of the reaction, the cationized ferritin method appears to be a useful tool for investigating the cell wall of fungi and its modifications during different biological processes like growth or adherence. For example, the localization of anionic sites, together with the detection of concanavalin A binding sites and cell-wall acid phosphatase, allowed us to analyse the dynamic changes of the cell-wall surface of *C. albicans* associated with germination and adherence to inert surfaces (Tronchin *et al.*, 1989). Likewise, using cationized ferritin labelling (Figure 2.3) associated with the determination of electronegative charge and cell surface hydrophobicity, we have analysed the cell surface characteristics of *Aspergillus fumigatus* conidia (Bouchara *et al.*, 1999).

Protocol 2.4
Detection of anionic sites

Equipment, materials and reagents

1.5 ml Eppendorf-type tubes

Cells to be analysed

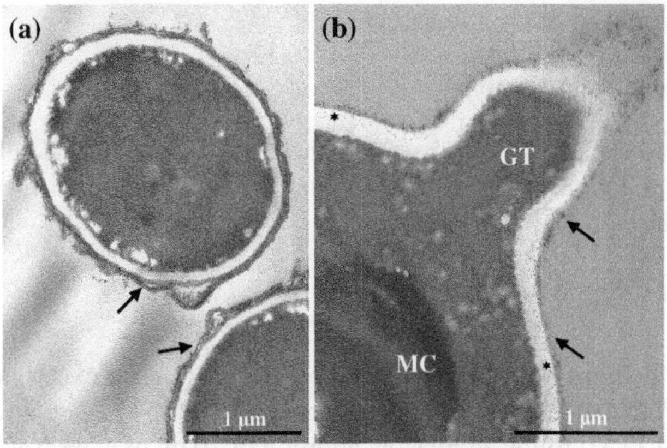

Figure 2.3 Localization of anionic sites in *A. fumigatus* conidia (a) and *C. albicans* germinating blastospore (b). A uniform layer of ferritin particles recovers the whole surface of *A. fumigatus* conidia (arrows). The surface of mother cell (MC) and germ tube (GT) of *C. albicans* is densely labelled (arrows). Ferritin particles are also detected inside the cell wall of mother cells (asterisks).

Glutaraldehyde

Osmium tetroxide

Cacodylate buffer 0.2 M, pH 7.4

70%, 95% and absolute ethanol

Cationized ferritin

Native ferritin

Neuraminidase (Sigma-Aldrich type X)

Acetate buffer 0.1 M, pH 5

Phosphate-buffered saline 0.1 M, pH 7.4 (PBS)

Calcium chloride ($CaCl_2$)

EPON 812® (Shell Chemicals Ltd., London, UK)

Uranyl acetate

Copper grids

Method

1. Wash cells (1 ml – 1×10^8 cells/ml) in distilled water.
2. Fix the cells for 1 h at 4 °C with 1% glutaraldehyde solution buffered at pH 7.4 with 0.1 M cacodylate.
3. Incubate the cells at room temperature for 1 h with shaking in the presence of cationized ferritin at a concentration of 1 mg/ml in PBS.
4. Wash overnight at 4 °C in cacodylate buffer.
5. Post-fix the cells for 30 min with 1% osmium tetroxide solution in 0.1 M cacodylate buffer pH 7.4.
6. Dehydrate the cells in a graded series of ethanol dilutions (70% for 10 min, 95% for 10 min and finally 100% for 2 × 10 min).
7. Embed in EPON 812.
8. Perform two types of controls: (1) Incubate the cells with native ferritin (1 mg/ml in PBS) instead of cationized ferritin. (2) Treat the cells with neuraminidase (Sigma-Aldrich type X, 1 unit/ml in 0.1 M acetate buffer pH 5, supplemented with 0.04 M $CaCl_2$) and then incubate with cationized ferritin.
9. Contrast ultrathin sections with uranyl acetate 5 min.
10. View under electron microscope.

Note: Poly-L-lysine-gold complex (Vorbrodt, 1987) and polycationic colloidal gold-chitosan complexes (Horisberger and Clerc, 1988) may be valuable alternatives to cationized ferritin.

2.6 Detection of glycoconjugates by the periodic acid-thiocarbohydrazide-silver proteinate technique (PATAg)

This technique derives from the periodic acid-Schiff (PAS) method for light microscopy. Aldehyde groups generated by oxidation of the vic-glycol groups present in the carbohydrate molecules are revealed by the thiocarbohydrazide (TCH)-silver proteinate technique. Silver proteinate reacts with thiocarbohydrazones, which are insoluble products formed by TCH with aldehydes, to give an opaque complex (Thiéry, 1967). The reaction results in selective and very good staining of all the PAS-positive structures: glycogen particles, vacuoles, endoplasmic reticulum, plasma membrane, Golgi apparatus and derived vesicles. Glycoconjugates are the main

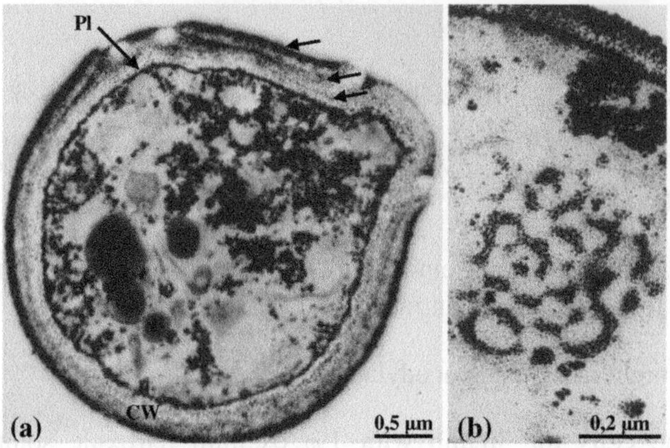

Figure 2.4 Detection of polysaccharides in *C. albicans* blastoconidia by the periodic acid-thiocarbohydrazide-silver proteinate technique (PATAg). (a) Silver proteinate deposits delineate concentric layers in the cell wall according to their specific staining reaction (arrows). The plasmalemma (Pl) is also strongly reactive. (b) Golgi elements appear as a tubular network whose cisternae are intensively stained

components of the fungal cell wall. Given the high selectivity of the staining reaction, the PATAg method appears to be a useful tool to use to investigate the structural organization of PAS-positive glycoconjugates of the cell wall (Figure 2.4) and to detect the most delicate changes in their distribution during biological (growth, germination) or pathological (colonization) events.

Applied to *C. albicans*, the PATAg technique provides evidence for a multilayered organization of the cell wall (Poulain *et al.*, 1978). This arrangement consists in an outer layer composed of amorphous material whose intense staining shows the glycoproteinic nature, and a weakly contrasted inner layer. In addition, subcellular modifications associated with sphero-protoplast regeneration could be followed (Tronchin *et al.*, 1982). The combination of the PATAg reaction with other techniques such as the use of specific lectins allowed us to demonstrate the presence of a cell-wall coat surrounding *C. albicans* cells (Tronchin *et al.*, 1981).

Protocol 2.5
Detection of glycoconjugates by the periodic acid-thiocarbohydrazide-silver proteinate technique (PATAg)

Equipment, materials and reagents

1.5 ml Eppendorf-type tubes

Cells to be analysed

Glutaraldehyde

Osmium tetroxide

Cacodylate buffer 0.2 M, pH 7.4

70%, 95% and absolute ethanol

Spurr's resin

Periodic acid

Thiocarbohydrazide

Acetic acid

Silver proteinate

Hydrogen peroxide

Gold grids

Method

1. Wash cells (1 ml – 1×10^8cells/ml) 3 times in distilled water.
2. Fix the cells for 30 min at 4 °C in 5% glutaraldehyde solution in 0.1 M cacodylate buffer pH 7.4.
3. Wash the cells overnight at 4 °C.
4. Post-fix the cells for 30 min at 4 °C with 1% osmium tetroxide solution in 0.1 M cacodylate buffer pH 7.4 (1.5 ml per tube).
5. Dehydrate the cells in a graded series of ethanol dilutions (70% for 10 min, 95% for 10 min and finally 100% 2 × 10 min).
6. Infiltrate and embed material in Spurr's resin.
7. Perform ultrathin sections and collect them on gold grids.
8. Incubate the sections on drops of periodic acid 1% in distilled water for 30 min.
9. Wash carefully in distilled water.

10. Incubate the sections on drops of thiocarbohydrazide 0.2% in 20% acetic acid solution for 1 to 24 h in a moist chamber.

11. Wash the sections twice for 10 min in 10% acetic acid, and several times in distilled water.

12. Incubate the sections on drops of 1% silver proteinate during 30 min in the dark.

13. Thoroughly rinse in distilled water and dry the grids.

14. View under electron microscope.

Three types of controls could be performed:

(1) omission of periodic acid treatment

(2) omission of TCH treatment

(3) incubation with hydrogen peroxide 10% instead of periodic acid

Note: A 24 h exposure to thiocarbohydrazide is required for *C. albicans* (Poulain *et al.*, 1978).

2.7 Enzyme-gold approach for the detection of polysaccharides in the cell wall

The enzyme-gold technique was introduced in the 1980s as a new approach in cytochemistry (Bendayan and Benhamou, 1987). This method is based on the affinity of enzymes for their corresponding substrates. Among the variety of molecules that can be specifically detected by enzyme-gold complexes, polysaccharides have received particular attention. Carbohydrate macromolecules exposed at the cell surface are known to fulfil specific biological functions, such as host-cell recognition and cell attachment. This may explain the growing interest in their cytochemical localization. In pathogenic fungi, enzyme-gold complexes allowed the localization of various polysaccharides such as chitin, polygalacturonic acid, β-(1-4)-D-glucans or carbohydrate like fucose, mannose and galactose (Benhamou and Ouellette, 1987; Benhamou *et al.*, 1987). Compared to the PATAg technique, which lacks specificity, the use of enzyme-gold complexes allows a sensitive and highly specific identification of cell-wall carbohydrates. Together with the use of lectin-gold complexes for the ultrastructural detection of oligosaccharides in the cell wall of fungi, the use of the enzyme-gold approach undoubtedly contributes to a better understanding of the organization of the cell wall of pathogenic fungi.

Protocol 2.6
Enzyme-gold approach for the detection of polysaccharides in the cell wall

Equipment, materials and reagents

1.5 ml Eppendorf-type tubes

Cells to be analysed

Glutaraldehyde

Osmium tetroxide

Cacodylate buffer 0.2 M, pH 7.4

70%, 95% and absolute ethanol

EPON 812® (Shell Chemicals Ltd., London, UK)

Purified exoglucanase

Colloidal gold (15 nm)

Phosphate-buffered saline 0.1 M, pH 7.4 (PBS)

Polyethylene glycol (PEG) 20 000

Uranyl acetate

Lead citrate

Nickel grids

Method

Note: The following protocol was used for ultrastructural localization of β-(1-4)-D-glucans by means of an exoglucanase-gold complex (Benhamou *et al.*, 1987).

1. Fix fungal suspension in 3% glutaraldehyde in 0.1 M sodium cacodylate buffer, pH 7.4 for 2 h.

2. Post-fix the cells for 30 min at 4 °C with 1% osmium tetroxide solution in 0.1 M cacodylate buffer for 1 h.

3. Dehydrate the cells in a graded series of ethanol dilutions (70% for 10 min, 95% for 10 min and finally 100% for 2 × 10 min).

4. Infiltrate and embed material in EPON 812.

5. Cut ultrathin sections and collect them on nickel grids.

6. Incubate the sections on drops of PBS-PEG pH 6 for 5 min.

7. Incubate the sections on drops of the exoglucanase-gold complex for 30 min at room temperature in a moist chamber.

8. Wash the grids 3 times in PBS.

9. Rinse with distilled water.

10. Contrast with uranyl acetate and lead citrate.

11. View under electron microscope

Note: Prepare the exoglucanase-gold complex by dissolving 100 μg of the purified enzyme in 100 μl of distilled water (Benhamou *et al.*, 1987). Then add 10 ml of the gold suspension at pH 9.0 to the enzyme. Centrifuge at 20 000 x *g* for 45 min at 4 °C, and resuspend the red pellet in 0.5 ml phosphate buffered saline (PBS) pH 6.0, containing 0.02% polyethylene glycol 20 000 (PEG 20 000). The exoglucanase-gold complex is stable for several days at 4 °C.

Use antimagnetic forceps to handle nickel grids, which becomes easily magnetized.

Cytochemical controls

1. Add 5 mg/ml of *β*-(1->4)-D-glucans from barley to the enzyme-gold complex prior to incubation of the sections.

2. Incubate the sections with a non-enzymatic protein-gold complex (BSA-gold complex).

3. Incubate the sections with stabilized gold suspension alone.

2.8 Detection of glycoconjugates by the lectin-gold technique

Lectins are a group of specific carbohydrate-binding proteins. Because of their high specificity combined with the large number of lectins now available from several

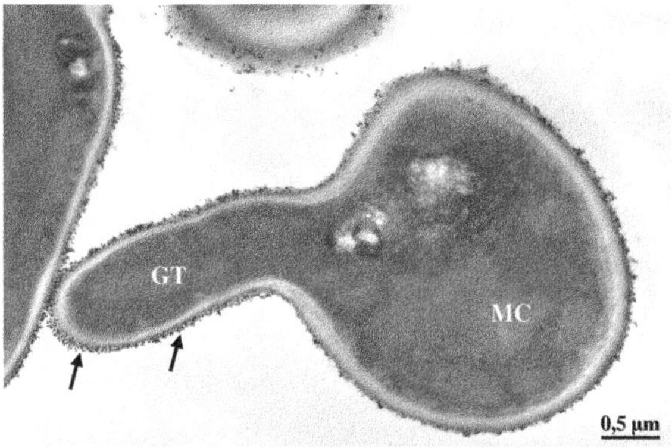

Figure 2.5 *C. albicans* cells treated with concanavalin A-gold showing an intense labelling of the fibrillar outer layer of the cell wall of germ tubes (GT, arrows), as compared with mother cells (MC)

commercial sources, they have become essential tools for the detection of carbohydrates in the cell wall of fungi. Among lectins that have been applied to fungal cells, wheat germ agglutinin (WGA), Concanavalin A (Con A) and peanut agglutinin (PNA), which are specific for N-acetylglucosamine residues, mannose residues and galactose residues respectively, are the most widely used. Since lectins are not electron-opaque, they cannot be visualized at the electron microscope level without electron-dense marker. The introduction of horseradish peroxidase, ferritin and more recently colloidal gold as markers of high electron density has contributed to the extensive use of lectin cytochemistry in a variety of biological systems including fungi (Tronchin *et al.*, 1979, 1981; Benhamou and Asselin, 1989). Lectins can be directly complexed to the marker and applied to fungal cells in a single-step procedure. A two-step procedure in which the marker is complexed to specific sugar of the lectin can also be used. Moreover, lectins can be applied to ultrathin tissue sections to localize intracellular receptors (Benhamou and Asselin, 1989) or on whole cells to detect cell-surface rearrangements of the cell wall during germination or adherence (Tronchin *et al.*, 1988) (Figure 2.5). In contrast to proteins, post-fixation with osmium tetroxide does not alter the lectin-binding sites. So, tissue processing is usually performed according to standard fixation and embedding procedures. The specificity of the labelling can also be assessed with appropriate controls.

Protocol 2.7
Detection of glycoconjugates by the lectin-gold technique

Equipment, materials and reagents

Note: The following protocol is designed for lectin labelling on whole cells. For labelling of ultrathin sections, grids are floated on drops of the different reagents.

15 ml Eppendorf-type tubes

Cells to be analysed

Glutaraldehyde

Osmium tetroxide

Cacodylate buffer 02 M, pH 74

70%, 95% and absolute ethanol

EPON 812® (Shell Chemicals Ltd., London, UK)

Phosphate buffer saline (PBS) 015 M, pH 74

Lectin-gold

Uranyl acetate

Nickel grids

Method

1. Wash cells (1 ml – 1×10^8 cells/ml) in distilled water.

2. Incubate the cells for 1 h with shaking in appropriate lectin-gold complex at 1:100 dilution in PBS.

3. Fix the cells for 1 h with 2.5% glutaraldehyde solution buffered at pH 7.4 with 0.1 M cacodylate.

4. Wash the cells overnight at 4 °C in cacodylate buffer.

5. Post-fix the cells for 30 min with 1% osmium tetroxide solution in 0.1 M cacodylate buffer pH 7.4.

6. Dehydrate the cells in a graded series of ethanol dilutions (70% for 10 min, 95% for 10 min and finally 100% for 2×10 min).

7. Embed material in EPON 812.

8. Contrast with uranyl acetate 5 min.

9. View under electron microscope.

Controls

1. Incubate with the lectin-gold complex in the presence of its specific sugar.

2. Incubate with the uncomplexed lectin, prior to incubation with the lectin-gold complex.

Note: For the indirect procedure, cells or ultrathin sections are successively incubated with the uncomplexed lectin diluted in PBS, washed in PBS and finally incubated with the gold-specific sugar complex diluted in PBS.

2.9 Immunogold detection of antigens on ultrathin sections of acrylic resin

One of the most useful applications of TEM is the identification of antigens and the study of their ultrastructural localization. In this context, the main objective of tissue fixation and embedding remains the preservation of both ultrastructure and immunoreactivity. Although there is no standard protocol for the preparation of tissue, the best approach is to use aldehyde fixation alone, since osmium tetroxide is known to denature protein antigens. Among resins, LR White® (Sigma-Aldrich, St. Louis, MO, USA) and Lowicryl® (Chemishe Werke Lowi, Waldkraiburg, Germany) are the most commonly used for immunoEM, both having advantages and disadvantages. LR White is provided in a ready-to-use form that does not require mixing. Its very low viscosity makes infiltration quite successful, but excessive heat produced by the reaction during polymerization may affect immunoreactivity. Hardening of Lowicryls requires exposure to ultraviolet. The hardening of both LR White and Lowicryls is inhibited by oxygen. For this reason, samples are embedded in gelatin capsules. In the post-embedding technique, thin sections of embedded tissue are incubated with the immune serum or purified antibodies. A two-step procedure is usual, with at first the fixation of the primary antibody to its corresponding epitope on the antigen, followed by its detection by a gold-labelled secondary antibody or using gold-labelled protein A. Several controls must be performed to assess the specificity of the immunogold labelling. The post-embedding technique has been widely applied for *in situ* localization of various intracellular and intracell-wall molecules in fungi (Mulholland *et al.*, 1994; Wright and Rine, 1989) (Figure 2.6). In addition, it has been used successfully to detect fungal adhesins using ligands such as fibrinogen or laminin and specific anti-ligand antibodies (Tronchin *et al.*, 1993).

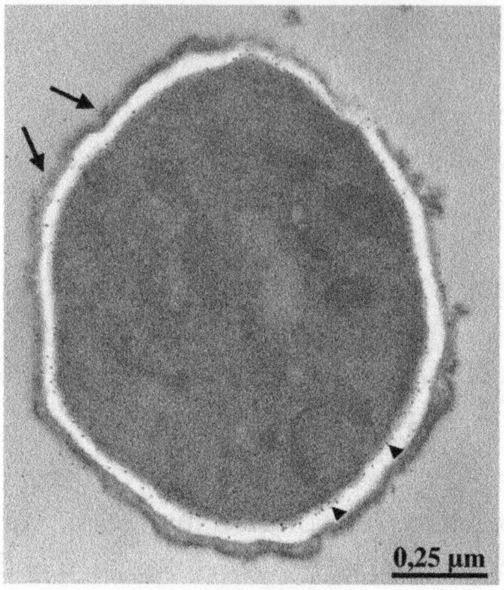

Figure 2.6 Immunolocalization of the 72 kDa laminin receptors in section of *A. fumigatus* conidia embedded in LR White resin using a specific antiserum and protein A-gold. Gold particles are distributed along the plasmalemma (arrowheads) and over the outer electron-dense layer of the cell wall (arrows), with permission of Elsevier SAS.

Protocol 2.8
Immunogold detection of antigens on ultrathin sections of acrylic resin

Equipment, materials and reagents

15 ml Eppendorf-type tubes

Cells to be analysed

Glutaraldehyde

Paraformaldehyde

Cacodylate buffer 02 M, pH 74

70%, 95% and absolute ethanol

Ammonium chloride

Phosphate buffered saline 015 M, pH 74 (PBS)

BSA

Primary antibody

Protein A-gold or gold-conjugated secondary antibody

LR White® (Sigma-Aldrich, St Louis, MO, USA) or Lowicryl® K4M (Chemishe Werke Lowi, Waldkraiburg, Germany)

Gelatin capsules

UV light (380 nm)

Uranyl acetate

Nickel grids

Method

1. Wash cells (1 ml – 1×10^8 cells/ml) in distilled water.

2. Fix the cells for 1 h in freshly prepared fixative (0.1% glutaraldehyde – 2% paraformaldehyde) buffered at pH 7.4 in 0.1 M sodium cacodylate.

3. Wash the cells in two changes of 0.1 M cacodylate buffer for 1 h each (or overnight at 4 °C).

4. Treat the cells with 50 mM ammonium chloride for 30 min to quench free aldehydes that might interact with antibodies.

5. Wash the cells in 0.1 M cacodylate buffer for 1 h.

6. Treat the cells for 30 min with 0.5% uranyl acetate solution.

7. Dehydrate in ethanol and embed in closed gelatin capsules with fresh resin at −18 °C.

8. Cut ultrathin sections and collect them on nickel grids with carbon-coated Formvar® film (Fondis Electronic, Guyancourt, France).

9. Rinse the grids in distilled water for 10 min.

10. Incubate the grids on drops of specific primary antibody at appropriate dilution in PBS (phosphate-buffered saline) pH 7.4 containing 1% BSA for 30 min at room temperature.

11. Wash the grids in three changes of PBS-1% BSA for 5 min each.

12. Incubate the grids with protein A-gold or gold-conjugated secondary antibody (10 or 20 nm; 1:100 dilution in PBS) for 30 min.

13. Wash the grids in two changes of distilled water for 5 min each.

14. Stain the grids with uranyl acetate 5 min.

15. View under electron microscope.

Controls

1. Omit the primary antibody.

2. Use a non-specific antibody instead of the primary antibody.

Acrylin resin

LR White Samples are dehydrated at −18 °C in 70% ethanol for 1 h, in 95% ethanol for 2 h and twice in 100% ethanol for 2 h each time. Dehydration is followed by treatment at −18 °C with mixtures of LR White/ethanol (1:2) overnight, LR White/ethanol (1:1) for 12 h and fresh LR White overnight. The samples are then placed in closed gelatin capsules with fresh LR White and polymerized for 2 days at −18 °C.

Lowicryl K4M Samples are dehydrated at 4 °C in 70% ethanol for 30 min, at −18 °C in 95% ethanol for 1 h and twice in 100% ethanol for 1 h each time. Dehydration is followed by treatment with mixtures of Lowicryl/ethanol (1:1) for 48 h, Lowicryl:ethanol (2:1) for 24 h and fresh Lowicryl overnight. The samples are then deposited in closed gelatin capsules with fresh Lowicryl, polymerized by indirect UV irradiation (360 nm) for 24 h at −18 °C and further polymerized at room temperature for 2 days.

Note: The treatment of fungal cells by uranyl acetate before embedding prevents the detachment of the resin from the fungal cell wall, which usually occurs during inclusion of fungal cells with acrylic resins.

2.10 Cryofixation and freeze substitution for ultrastructural analysis and immunodetection

Freeze substitution is a technique that combines the advantages of rapid freezing for immobilization of cell structures with those of conventional embedment and sectioning. Rapid cryofixation (within milliseconds) needed for adequate preservation can be obtained using specific equipment such as a high-pressure freezing machine or

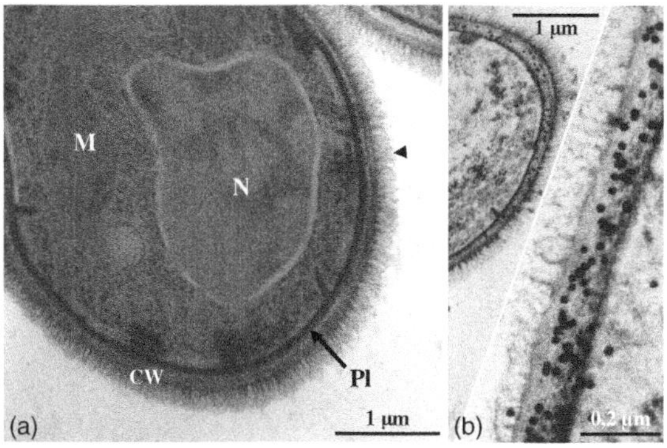

Figure 2.7 Cryofixation and cryosubstitution in osmium-acetone of *C. albicans* cells. (a) The ultrastructure of the cell wall and of the cytoplasmic organelles is well preserved (Pl = plasmalemma; N = nuclei; M = mitochondria). The external layer of the cell wall appears composed of fibrils densely arranged perpendicularly to the cell surface (arrowhead). (b) Wheat germ agglutinin-gold labelling (upper left, low magnification of a blastospore; lower right, large magnification of the cell wall). Gold particles are located in the intermediate and the inner layers of the cell wall by courtesy of Bobichon H.

by freezing in liquid propane. Then the ice in the frozen specimen is substituted at low temperatures (−80 °C to −90 °C) with organic solvents, like acetone or methanol, gradually warmed to room temperature. Depending on the information required (investigation of ultrastructure and/or preservation of antigenicity), chemical fixatives may be added. One of the most commonly used substitution fluids is osmium tetroxide in acetone. In the same way, infiltration and embedding can be obtained in epoxy resin and polymerization at 60 °C, or in acrylic resins and polymerization at low temperature. This procedure allows good preservation of the fungal cell wall and cytoplasmic organelles (Kurtz *et al.*, 1994) together with accessibility of fungal components to ligands conjugated to colloidal gold (Figure 2.7). For example, freeze-substituted material was successfully used to study the organization of tip-growth organelles and microtubules (Heath and Kaminskyj, 1989), to localize Concanavalin A- and wheat-germ agglutinin-binding sites in the endomembrane system and in the cell wall (Bourett *et al.*, 1993; Bobichon *et al.*, 1994) or to detect β-1,2-oligomannosides epitopes in the cell wall (Poulain *et al.*, 2002).

Protocol 2.9
Cryofixation and freeze substitution for ultrastructural analysis and immunodetection

Equipment, materials and reagents

Cells to be analysed

Osmium tetroxide

Acetone

Liquid propane

EPON 812® (Shell Chemicals Ltd., London, UK)

Uranyl acetate

Lead citrate

Copper grids

Method

1. Remove the cells and place them on squares of filter paper (or dialysis membrane or cellophane).

2. Freeze rapidly by plunging pieces of membrane bearing fungal cells into liquid propane.

3. Substitute in 2.5% osmium tetroxide in acetone for 1 to 2 days at –80 °C, then warm for 4 to 8 h at –20 °C and finally for 2 to 4 h at 4 °C.

4. Transfer to precooled (4 °C) pure acetone and then warm to room temperature.

5. Rinse twice in pure acetone.

6. Embed in EPON 812 (or in acrylic resin for immunological studies).

7. Stain sections with uranyl acetate and lead citrate.

8. View under electron microscope.

2.11 Overview

In this chapter, we have described various protocols which may constitute useful tools for ultrastructural, cytochemical and immunochemical studies of pathogenic fungi. During the last decades, these techniques have been widely used to localize and to identify various components and organelles of fungal cells. They allowed a better understanding of spore development and germination, of cell-wall organization and structure, and of host–pathogen interactions, as well as the characterization of cytoplasmic organelles and elucidation of their functions. In future research, these approaches have to be included in multidisciplinary programmes combining various experimental technologies, like biochemistry, immunology and molecular biology, to elucidate the different aspects of the highly complex strategy evolved by pathogenic fungi in host-tissue colonization, as well as in resistance to the host defence mechanisms or to antifungal drugs.

2.12 References

Bendayan, M. and Benhamou, N. (1987) Ultrastructural localization of glucoside residues on tissue sections by applying the enzyme-gold approach. *J. Histochem. Cytochem.* **35**: 1149–1155.

Benhamou, N. (1995) Immunocytochemistry of plant defense mechanisms induced upon microbial attack. *Microsc. Res. Tec.* **31**: 63–78.

Benhamou, N. and Asselin, A. (1989) Attempted localization of a substrate for chitinase in plant cells reveals abundant *N*-acetyl-D-glucosamine residues in secondary walls. *Biol. Cell* **67**: 341–350.

Benhamou, N. and Ouellette, G. B. (1987) Ultrastructural study and cytochemical investigation, by means of enzyme-gold complexes, of the fungus *Ascocalyx abietina. Can. J. Bot.* **65**: 168–181.

Benhamou, N., Chamberland, H., Ouellette, G. B. and Pauze, F. J. (1987) Ultrastructural localization of β-(1 $\rightarrow$4)-D-glucans in two pathogenic fungi and in their host tissues by means of an exoglucanase-gold complex. *Can. J. Microbiol.* **33**: 405–417.

Bobichon, H., Gache, D. and Bouchet, P. (1994) Ultrarapid cryofixation of *Candida albicans*: Evidence for a fibrillar reticulated external layer and mannans channels within the cell wall. *Cryo-Letters* **15**: 161–172.

Bouchara, J. P., Sanchez, M., Esnault, K. and Tronchin, G. (1999) Interactions between *Aspergillus fumigatus* and host proteins. *Contrib Microbiol.* **2**: 167–181.

Bourett, T. M., Picollelli, M. A. and Howard, R. J. (1993) Post-embedment labeling of intracellular concanavalin A-binding sites in freeze-substituted fungal cells. *Exp. Mycol.* **17**: 223–235.

Esnault, K., El Moudni, B., Bouchara, J. P., Chabasse, D. and G. Tronchin (1999) Association of a myosin immunoanalogue with cell envelopes of *Aspergillus fumigatus* conidia and its participation in swelling and germination. *Infect. Immun.* **67**: 1238–1244.

Gomori, G., (1952) *Microscopic Histochemistry: Principles and Practice* (2nd edn), Chicago, University of Chicago Press.

Heath, I. B. and Kaminskyj, S. G. W. (1989) The organization of tip-growth-related organelles and microtubules revealed by quantitative analysis of freeze-substituted oomycete hyphae. *J. Cell Sci.* **93**: 41–52.

Horisberger, M. and Clerc, M. F. (1988) Ultrastructural localization of anionic sites on the surface of yeasts, hyphal and germ-tube forming cells of *Candida albicans. Eur. J. Cell Biol.* **46**: 444–452.

Klomparens, K. L. (1990) The development and application of ultrastructural research in mycology. *Mycopathologia* **109**: 139–148.

Kurtz, M. B., Heath, I. B., Marrinan, J., Dreikorn, S. *et al.* (1994) Morphological effects of lipopeptides against *Aspergillus fumigatus* correlate with activities against (1,3)-β-D-glucan synthase. *Antimicrob. Agents Chemother.* **38**: 1480–1489.

Linnemans, W. A. M., Boer, P. and Elbers, P. F. (1977) Localization of acid phosphatase in *Saccharomyces cerevisiae*: a clue to cell wall formation. *J. Bacteriol.* 131: 638–644.

Mahvi, T., Spicer, S. S. and Wright, N. J. (1974) Cytochemistry of acid mucosubstance and acid phosphatase in *Cryptococcus neoformans*. *Can. J. Microbiol.* **20**: 833–838.

McKerracher, L. J. and Heath, I. B. (1985) Microtubules around migrating nuclei in conventionally-fixed and freeze-substituted cells. *Protoplasma* **125**: 162–172.

Mulholland, J., Preuss, D., Moon, A., Wong, A. *et al.* (1994) Ultrastructure of the yeast actin cytoskeleton and its association with plasma membrane. *J. Cell Biol.* **113**: 245–260.

Osumi, M. (1974) Ultrastructure of *Candida* yeasts grown on *n*-alkanes: Appearance of microbodies and its relationship to high catalase activity. *Arch. Microbiol.* **99**: 181–201.

Osumi, M. (1998) The ultrastructure of yeast: Cell wall structure and formation. *Micron.* **29**: 207–233.

Poulain D., Tronchin, G., Dubremetz, J. F. and Biguet, J. (1978) Ultrastructure of the cell wall of *Candida albicans* blastospores: Study of its constitutive layers by the use of a cytochemical technique revealing polysaccharides. *Ann. Microbiol. (Inst. Pasteur)* 129 **A**: 141–153.

Poulain, D., Slomianny, C., Jouault, T., Gomez, J. M. and Trinel, P. A. (2002) Contribution of phospholipomannan to the surface expression of β-1,2-oligomannosides in *Candida albicans* and its presence in cell wall extracts. *Infect. Immun.* **70**: 4323–4328.

Preuss, D., Mulholland, J., Franzusoff, A., Segev, N. and Botstein, D. (1992) Characterizarion of the *Saccharomyces* Golgi complex through the cell cycle by immunoelectron microscopy. *Mol. Biol. Cell.* **3**: 789–803.

Preuss, D., Mulholland, J., Kaiser, C. A., Orlean, P. *et al.* (1991) Structure of the yeast endoplasmic reticulum: Localization of ER proteins using immunofluorescence and immunoelectron microscopy. *Yeast* **7**: 891–911.

Rambourg, A., Clermont, Y., Ovtracht, L. and Képès, F. (1995) Three-dimensional structure networks, presumably Golgi in nature, in various yeast strains: A comparative study. *Anat. Rec.* **243**: 283–293.

Rambourg, A., Jackson, C. L. and Clermont, Y. (2001) Three-dimensional configuration of the secretory pathway and segregation of secretion granules in the yeast *Saccharomyces cerevisiae*. *J. Cell Sci.* **114**: 2231–2239.

Thiéry, J. P. (1967) Mise en évidence des polysaccharides sur coupes fines en microscopie electronique. *J. Microsc. (Paris)* **6**: 978–1018.

Tronchin, G., Bouchara, J. P. and Robert, R. (1989) Dynamic changes of the cell wall surface of *Candida albicans* associated with germination and adherence. *J. Cell. Biol.* **50**: 285–290.

Tronchin, G., Bouchara, J. P., Latgé, J. P. and Chabasse, D. (1993) Application of a Lowicryl K4M embedding technique for analysis of fungal adhesins. *J. Mycol. Med.* **3**: 74–78.

Tronchin, G., Bouchara, J. P., Robert, R. and Senet, J. M. (1988) Adherence of *Candida albicans* germ tubes to plastic: Ultrastructural and molecular studies of fibrillar adhesins. *Infect. Immun.* **56**: 1987–1983.

Tronchin, G., Poulain, D. and Biguet, J. (1979) Etudes cytochimiques et ultrastructurales de la paroi de *Candida albicans.* I. Localisation des mannanes par utilisation de concanavaline A sur coupes ultrafines. *Arch. Microbiol.* **123**: 245–249.

Tronchin, G., Poulain, D. and Biguet, J. (1980) Localization ultrastructurale de l'activité phosphatasique acide chez *Candida albicans*. *Biol. Cell.* **38**: 147–152.

Tronchin, G., Poulain, D. and Biguet, J. (1981) Cytochemical and ultrastructural studies of *Candida albicans*. II. Evidence for a cell wall coat using concanavalin A. *J. Ultrastruc. Res.* **75**: 50–59.

Tronchin, G., Poulain, D., Herbaut, J. and Biguet, J. (1981) Localization of chitin in the cell wall of *Candida albicans* by means of wheat germ agglutinin. Fluorescence and ultrastructural studies. *Eur. J. Cell Biol.* **26**: 121–128.

Tronchin, G., Poulain, D., Herbaut, J. and Biguet, J. (1982) Aspects ultrastructuraux et cytochimiques de la régénération des sphéro-protoplastes chez *Candida albicans*. *Ann. Microbiol. (Inst. Pasteur)* 133 **A**: 275–291.

Vorbrodt, A. W. (1987) Demonstration of anionic sites on the luminal and abluminal fronts of endothelial cells with poly-L-lysine-gold complex. *J. Histochem. Cytochem.* **35**: 1261–1267.

Wright, R. (2000) Transmission electron microscopy of yeast. *Microsc. Res. Tech.* **51**: 496–510.

Wright, R. and Rine, J. (1989) Transmission electron microscopy and immunocytochemical studies of yeast: analysis of HMG-CoA reductase overproduction by electron microscopy. *Methods Cell Biol.* **31**: 473–512.

3

Evaluation of molecular responses and antifungal activity of phagocytes to opportunistic fungi

Maria Simitsopoulou and **Emmanuel Roilides**

3.1 Introduction

The incidence of invasive fungal infections (IFIs) has dramatically increased during the last two decades as shown by recent epidemiological studies in the United States and Europe. The causes of this trend are not fully understood but are most probably related to the longer survival of immunocompromised, premature, aged or critically ill patients and the great expansion of the kinds and number of susceptible hosts in modern medicine. Examples of such susceptible hosts are patients with various malignancies, solid organ or haematopoietic stem cell transplants, acquired immunodeficiency syndrome, birth prematurity or those receiving immunosuppressive therapy. In addition, patients in the intensive care units as well as burn and surgical units frequently suffer from IFIs (Antachopoulos and Roilides, 2005).

While historically *Candida* spp. and *Aspergillus* spp. have caused the vast majority of the opportunistic fungal infections in immunocompromised patients, other less frequently encountered yeasts and filamentous fungi have recently emerged as important causes of IFIs. *Trichosporon* spp., zygomycetes, *Fusarium* spp. and *Scedosporium* spp. are such emerging fungi causing increased morbidity and mortality in immunocompromised patients.

The increase in the frequency of IFIs has been followed, after a latent period, by the rapid development of a number of novel antifungal agents and by the production of

Medical Mycology: Cellular and Molecular Techniques. Edited by Kevin Kavanagh

newer pharmaceutical formulations of older drugs like amphotericin B. In addition, our knowledge of the various steps of interaction of fungi with phagocytes has become much broader and deeper today, regarding molecular and cell functional aspects; it has thus led to insights into fungal pathogenesis and the improved management of IFIs.

Clinical and experimental data have demonstrated that innate immunity based on intact antifungal activity of phagocytes is critical to the outcome of invasive fungal infections by *Candida* spp. and *Aspergillus* spp. Quantitative and qualitative modulation of antifungal host defence using cytokines as adjuncts to antifungal drug therapy has been supported by extensive *in vitro* and *in vivo* preclinical data as well as some limited clinical results (Roilides and Walsh, 2004; Walsh *et al.*, 2005). Host innate immune response to *Candida* spp. is predominantly based on the antifungal efficacy of cytokine-activated tissue macrophages (i.e. peritoneal macrophages, Kupffer liver cells, splenic macrophages) and dendritic cells; circulating neutrophils (PMNs) as well as monocytes (MNCs) are also important in this challenge. By comparison, host innate immune response to *Aspergillus* spp. is predominantly based on the antifungal efficacy of pulmonary alveolar macrophages (PAMs) against conidia and of PMNs against hyphae. It is concluded that both circulating MNCs as well as tissue-residing mononuclear phagocytes are very important in the defence against medically important fungi.

Candida spp. enter the host usually after colonization of mucosal surfaces (i.e. gut, oral cavity and vagina), as well as through foreign bodies such as intravenous catheters. Airborne filamentous fungi reach the host by inhalation through the respiratory tract. The host responds to fungal challenge by recognizing the organisms via surface receptors, most frequent of which are Toll-like receptors (TLR) 1–12. Transduction of the recognition signal to the nucleus of phagocytes induces the gene expression and production of a series of cytokines and chemokines [e.g. tumour necrosis factor-α (TNF-α), interleukin (IL)-1, macrophage inflammatory protein-1] that have primarily pro-inflammatory and chemotactic properties. These agents act on the professional phagocytes including PMNs and mononuclear phagocytes and induce the activation of phagocytosis of small fungal elements and the activation of NADPH-oxidase with subsequent production of superoxide anion (O_2^-), hydrogen peroxide (H_2O_2) and other H_2O_2-dependent fungicidal cell metabolites. Secretion of these fungicidal oxidative products together with the secretion of a number of O_2-independent enzymes (i.e. defensins) leads to the destruction of the challenging fungus, either intracellularly (conidia, small elements) or extracellularly (hyphae, large elements).

A number of cytokines and chemokines are expressed and secreted upon infection with *Candida* or *Aspergillus*. Cytokines and chemokines modify the degree of the destructive capacity of the effector cells (Figure 3.1). Cytokines that are of most interest due to their ability to up-regulate the function of phagocytes are the haematopoietic growth factors granulocyte colony-stimulating factor (G-CSF), granulocyte-macrophage colony-stimulating factor (GM-CSF) and macrophage colony-stimulating factor (M-CSF). In addition, cytokines of T helper type-1 such as interferon-γ (IFN-γ), IL-12 and IL-15 as well as TNF-α are also of interest. Cytokines of T helper type-2

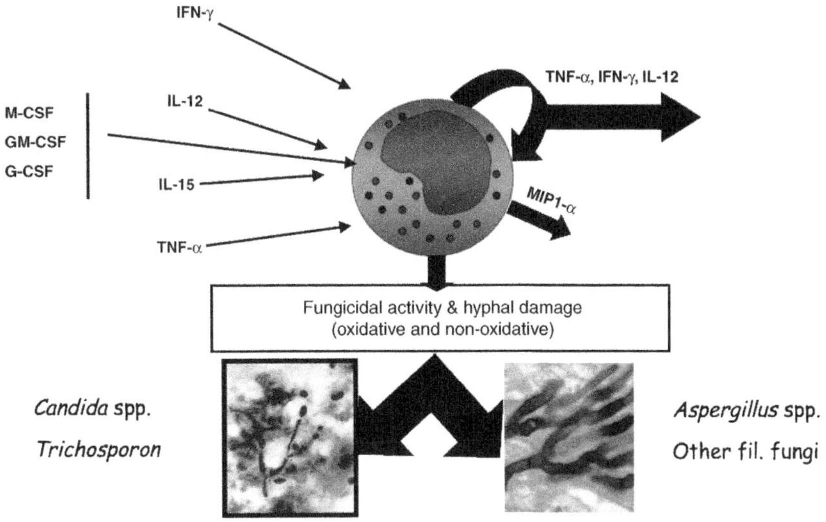

Figure 3.1 Augmentation of antifungal phagocytic host response by Th1 and proliferative cytokines

such as IL-4 and IL-10 exert an overall suppressive effect on the antifungal function of phagocytes against *Candida* spp. and *Aspergillus* spp.

In this chapter, we describe a series of protocols that help the scientists and the clinical researchers to evaluate gene expression and functional capacity of host phagocytes in response to fungal infection and after antifungal drug administration.

3.2 Preparation of conidia and hyphae of opportunistic fungi

The fungal strains used as targets or stimuli can be either primary isolates from patients with IFIs or purchased from culture maintenance banks. The strains are stored frozen at −20 °C in glycerol stocks or on agar slants until use. Although the optimal growth temperature for most fungi is 37 °C, some organisms prefer lower growth temperature than others. *Rhizopus* spp., for example, grow better at 25 °C than at 37 °C. For this reason, it is advisable to closely follow the optimal growth conditions for each organism as provided by the primary source.

Protocol 3.1
Preparation of conidia and hyphae of opportunistic fungi

Equipment, materials and reagents

Laminar-flow hood

Light microscope

Haemocytometer

CO_2 incubator

Centrifuge for 1.5 ml, 15 ml and 50 ml tubes and 96-well microtiter plates

Neubauer counting chamber

Sterile 50 ml polypropylene conical tubes (Costar®, Corning Incorporated, NY, USA)

Sterile 10 ml disposable pipettes

Clinical isolates or strains from a culture bank (i.e. American Tissue Culture Collection, Centraalbureau voor Schimmelcultures etc.)

Mycological media such as potato dextrose (PD) agar and Sabouraud dextrose (SD) agar

Glycerol

Peptone

Hanks' Balanced Salt Solution without calcium and magnesium ($HBSS^-$) (Gibco BRL, Life Technologies Ltd., Paisley, Scotland, UK)

10 x yeast nitrogen base (YNB) medium containing 20% glucose

Method

1. Retrieve stocks of appropriate fungi from freezer maintained either as solution containing 25% glycerol and 75% peptone, or on PD and SD agar slants.

2. Grow the fungal isolate by heavily streaking a PD or an SD agar plate and incubating at 25–37 °C, depending on the strain for 2 to 4 days.

3. Once the strain is fully conidiated, harvest conidia by adding 5–10 ml $HBSS^-$, scraping the agar surface with a sterile cotton swab and filtering the conidial suspension through sterile gauze.

   Note: This work should always take place under the hood, wearing a sterile mask and surgical gloves.

4. Repeat step 3 twice until the whole agar surface is clean of conidia and collect the conidial suspension in a 50 ml polypropylene conical tube.

5. Centrifuge the conidial suspension at 400 x *g* for 10 min and resuspend pellet in $HBSS^{-}$, count conidia on a Neubauer counting chamber under the microscope using a haemocytometer.

6. Standardize the fungal inoculum as needed for a particular experiment by CFU quantification on SD or PD plates after making serial dilutions of the stock solution.

7. Keep conidial stock fungal isolates up to 3 weeks in the refrigerator.

8. In order to obtain hyphae, seed the wells of microtiter plates, with the appropriate conidial suspension diluted in 1 x YNB media and incubate at 37 °C for 12–17 h.

Protocol 3.2
Preparation of hyphal fragments

In vitro experiments, where fungal organisms are incubated for long time periods with treated or untreated phagocytes to see an extended effect of fungus host-cell interaction, cannot be performed owing to hyphal overgrowth. This problem is overcome either by inactivating the fungal organisms with organic solvents or by preparing hyphal fragments. In our laboratory, in order to prepare hyphal fragments of *Aspergillus* spp. we use a procedure according to a modified version of Marshall *et al.* (1997) and Fontaine *et al.* (2000), as outlined in this protocol.

Equipment, materials and reagents

Ultrasonic processor equipped with a micro tip sonotrode (100 W, 30 kHz)

10 x YNB

Resuspension buffer (RB): 50 mM Tris-HCl, pH 7.5, containing 50 mM EDTA and 1 mM PMSF

Ice

Method

1. Harvest conidia from SD plates as detailed in Protocol 3.1.

2. Standardize the fungal inoculum as needed for a particular experiment by CFU quantification.

3. Transfer 10 ml of 1×10^6/ml *Aspergillus* conidia in a 50 ml tube and centrifuge at 3,700 x *g* for 10 min.

4. Resuspend the conidial pellet in 10 ml 1 x YNB and incubate at 37 °C, 5% CO_2 for 18 h to convert conidia to hyphae.

5. Centrifuge hyphae at 3,700 x *g* for 10 min and dissolve pellet in 5 ml of RB solution.

6. Sonicate the sample for 5 min total, in 10-sec bursts with 10-sec intervals (20 μm amplitude). Use ice for chilling sample and ear-protecting pads.

7. Repeat sonication process two more times.

8. Check for hyphal breakage under light microscope.

9. Store 100–120 μl aliquots of hyphal fragments corresponding to 10^5 conidia at −20 °C.

3.3 Isolation of human monocytes from whole blood

For the study of immunomodulatory gene expression and evaluation of cell metabolism in response to fungal or drug stimuli, we routinely use human peripheral blood mononuclear cells (hPBMCs) from healthy adult volunteers. Since continuous culturing of mononuclear cells may initiate genetic recombination events, which could potentially alter cell biology, we consider freshly isolated hPBMCs as a more appropriate cell population than human cell lines for immunological studies (Vonk *et al.*, 1998; Rogers *et al.*, 2002). In addition, cell-line maintenance and monitoring for contamination of cell lines are time-consuming but necessary routine procedures that require particular laboratory facilities for sterile handling and cell storage. In Protocol 3.3, we describe an isolation procedure of MNCs based on ficoll separation of blood cell populations. Enrichment of the cell population in MNCs by cell attachment on 12-well cell culture plates is also described.

Protocol 3.3
Isolation of human MNCs from whole blood

Equipment, materials and reagents

Equipment as for Protocol 3.1

Sterile 21-gauge needle

Heparin (1000 U ml^{-1})

Hanks' Balanced Salt Solution without calcium or magnesium ($HBSS^-$)

3% (w/v) Dextran T500 (Sigma-Aldrich Chemical Co, St Louis, MO, USA; cat. no. D5251) in $HBSS^-$. Dissolve dextran in $HBSS^-$ by swirling on a warm plate

Ficoll Histopaque-1077: lymphocyte separation medium from Gibco BRL

Trypan blue (Sigma-Aldrich)

May–Grunwald/Giemsa stain

Tissue culture plates with 6, 12 or 24 wells (Costar®, Corning Incorporated, NY, USA)

18 mm round glass coverslips (Thomas Scientific; cat. no. 6662-Q43)

Cell scrapers (Costar®, Corning Incorporated, NY, USA)

Complete medium (CM) with Fetal Calf Serum (FCS): RPMI-1640 without glutamine supplemented with 10% FCS, 100 U ml^{-1} of penicillin, and 100 μg ml^{-1} of streptomycin

Method

Note: If you start from whole blood, follow the isolation procedure of Protocol 3.2 up to step 4. If you start from Buffy coat, dilute it with equal volume of $HBSS^-$ and go directly to step 4 of Protocol 3.2.

1. In a 50 ml polypropylene conical tube, place heparinized (10 U ml^{-1} heparin) blood and add half volume of 3% Dextran T500 solution. Mix carefully by gentle inversion. Take care not to introduce bubbles during mixing to avoid cell-membrane damage.

2. Incubate the homogeneous solution at room temperature for 20 min to allow the formation of two layers. The lower layer contains the sedimentary erythrocytes and the upper layer the leukocytes.

3. Transfer the top layer to a new 50 ml tube and under layer 13 ml of Ficoll (Histopaque), inserting the tip of the pipette at the bottom of the tube and slowly release the solution to create a distinct interface between the two layers. Alternatively, in a 50 ml tube, add 13 ml of Ficoll first and overlay the leukocyte suspension of step 2. This can be facilitated by tilting the tube that contains Ficoll so that the leukocyte suspension trickles down the side of the tube. At the end,

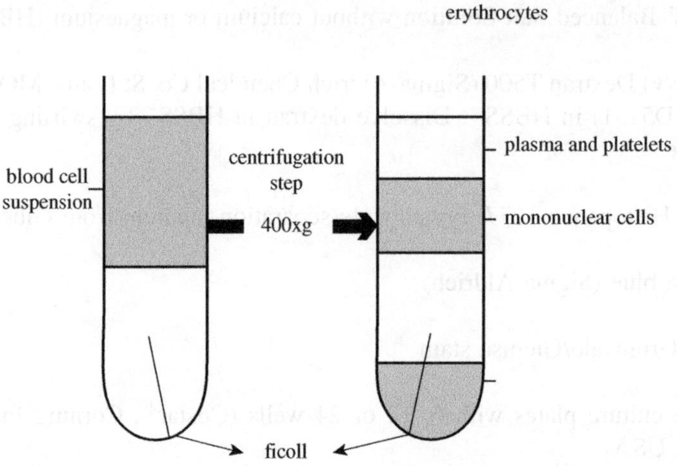

Figure 3.2 Leukocyte separation using ficoll gradient centrifugation

Ficoll is at the bottom and the leukocyte at the top. It is essential that both the Ficoll and leukocyte suspension are at room temperature and there is a distinct interface between Ficoll and leukocyte suspension.

4. Centrifuge at 400 x *g* at room temperature for 20 min. After centrifugation, PMNs and remaining erythrocytes are in the pellet and MNCs are collected on a cloudy layer above the Ficoll layer. Above MNC layer there is a supernatant containing plasma and platelets (Figure 3.2).

5. Aspirate and discard the top layer (plasma).

6. Transfer the MNC layer to a clean 50 ml tube with a sterile plastic pipette and add up to 40 ml of $HBSS^-$.

7. Centrifuge at 400 x *g* for 10 min to pellet the cells.

8. Aspirate the supernatant and resuspend pellet in $HBSS^-$.

9. Determine the cell concentration by trypan blue staining.

10. Estimate the percentage of MNCs within the MNC population by modified May–Grunwald/Giemsa stain procedure as follows:

    - Smear the cell suspension onto a microscopic slide.

    - Immerse slide in May–Grunwald stain for 10 min and in Giemsa stain for 20 min.

- Rinse with water and dry. Calculate the percentage of MNCs out of a total 100 cells.

For cell enrichment of MNCs by adherence

1. Place the MNCs in a 12-well culture plate at an estimated concentration of $1–2 \times 10^6$ MNCs per ml of CM. Use appropriate number of cells and volumes of medium if you use culture plates with different sizes of wells. For phagocytosis assays, place alcohol-sterilized 18 mm round glass coverslips in the centre of each well of a 12-well plate. Carefully place 0.2 ml of a MNC suspension with 5×10^6 MNCs per ml on each coverslip.

2. Incubate at 37 °C for 2 h (MNCs adhere on plastic or glass surfaces).

3. Wash twice with warm $HBSS^-$. Non-adherent T lymphocytes and many B lymphocytes are removed; the remaining adherent cells are enriched MNCs.

4. Gently scrape MNCs off the bottom of wells with cell scrapers. To facilitate cell dissociation place the 12-well culture plate on an ice-cold surface. Centrifuge at 400 x *g* for 10 min, aspirate the supernatant and resuspend the pellet of enriched MNCs at desired cell density in appropriate medium (CM, $HBSS^-$, RPMI).

3.4 Analysis of human MNC function in response to fungal infection

In this section, we describe methods (superoxide anion assay and hydrogen peroxide-rhodamine assay) for assessing the generation of toxic by-products in phagocytes in response to fungal stimulation. In addition, we describe methods for assessing phagocytosis of conidia and hyphal damage by XTT metabolic conversion.

XTT is a tetrazolium dye used to quantify phagocyte-mediated damage to hyphae. XTT diffuses into intact fungal cells, where it is reduced by the dehydrogenase enzyme system to an orange-coloured formazan derivative via the electron-coupling agent Coenzyme Q. Disruption of the dehydrogenase activity is affected by hyphal damage after incubation with phagocytes. Percent hyphal damage induced is assessed colorimetrically. The O_2^- production by phagocytes in response to fungal stimulation is assessed as reduction of ferricytochrome c. The rate of reduction is expressed in nM O_2^- per 10^6 phagocytes per 1 h using $29.5 \times 10^4 M^{-1} cm^{-1}$ as the extinction coefficient for reduced cytochrome c at 550 nm. The bacterial chemoattractive tripeptide FMLP or PMA (see Protocol 3.5) can be incubated with PMNs as soluble stimuli of O_2^- production. The enzyme SOD converts O_2^- into H_2O_2 at neutral or alkaline pH and thus the phagocyte nicotinamide adenine dinucleotide phosphate-specific reduction of ferricytochrome c is prevented. Stimulated production of reactive oxygen species is a process named 'respiratory burst', owing to the rapid consumption of oxygen, and is performed by the NADPH oxidase enzyme system of phagocytes (Thannical and Fanburg, 2000; Forman and Torres, 2002). In this protocol, generation of H_2O_2 and H_2O_2-dependent intracellular intermediates is assessed via the oxidation of dihydrorhodamine 1,2,3 (DHR-123) to rhodamine 1,2,3 (R-123) by flow cytometry.

Protocol 3.4 XTT microassay

Equipment, materials and reagents

Equipment as for Protocol 3.1

Microtiter plate reader equipped with 450 nm, 550 nm and 690 nm optical filters

Polystyrene, flat-bottomed, sterile 96-well microtiter plates (Costar®, Corning Incorporated, NY, USA)

Multichannel pipette

10 x YNB media

RPMI-1640 with phenol red (Biochrom KG., Berlin, Germany)

Pooled human serum (HS)

HBSS with calcium and magnesium ($HBSS^+$)

1 x PBS: 10 mM phosphate buffer, 2.7 mM KCl, 137 mM NaCl, pH 7.4

XTT (2,3-bis[2-methoxy-4-nitro-5-sulfophenyl]2H-tetrazolium-5-carboxanilide sodium salt; Sigma-Aldrich). Dissolve 0.25 mg XTT per ml of 1 x PBS solution and incubate at 37 °C for 15–20 min until solution becomes clear

Coenzyme Q_0 (2,3-dimethoxy-5-methyl-1,4-benzoquinone, Sigma-Aldrich). Prepare 5 mg ml^{-1} stock solution and use it at a final concentration of 40 µg ml^{-1}. Mix XTT and coenzyme Q_0 only when needed

Method

1. Seed the wells of a microtiter plate with 0.2 ml of a suspension containing 2.5×10^4–7.5×10^4 conidia per ml in 1 x YNB and incubate at appropriate temperature for 14 to 16 h to generate hyphae.

2. Centrifuge the plate at 800 x *g* for 20 to 30 min at ambient temperature.

3. Aspirate 1 row at a time and immediately add either $HBSS^+$, if the incubation period with the effector cells will be 2 h, or RPMI with 10% HS for longer incubation times. Since hyphae of zygomycetes adhere to the plastic surface of the wells very loosely, a centrifugation step always must precede an aspiration step in this case.

4. Add the required volume of effector cells (MNCs) at desired effector cell to target (E:T) ratios.

5. Incubate at 37 °C and 5% CO_2 for 2 h.

6. Centrifuge the plate at 800 x *g* 15 °C for 20–30 min.

7. Aspirate contents from each well 1 row at a time, taking care not to disturb the lawn of hyphae covering the bottom of each well and add 200 µl of sterile water.

8. Centrifuge microplate at 800 x *g* and 15 °C for 20–30 min.

9. Repeat the washing step twice.

10. Aspirate 1 row at a time and immediately add 150 µl XTT with Coenzyme Q_0. Add 150 µl XTT with Coenzyme Q_0 to a blank well for subtracting background values from XTT solution.

11. Incubate at 37 °C and 5% CO_2 for 1 h.

12. At the end of the incubation period, transfer 100 µl from each well to a new microtiter plate and read Absorbance at 450 nm with a reference wavelength set at 690 nm.

13. The % hyphal damage $= (1 - X/C) \times 100$, where X is the absorbance of test wells and C is the absorbance of control wells with hyphae only.

Protocol 3.5
Superoxide anion assay in 96-well plate

Equipment, materials and reagents

Equipment and materials as for Protocol 3.3

HBSS with calcium and magnesium without phenol red ($HBSS^+$)

1.2 mM cytochrome c (horse heart; Sigma-Aldrich cat. no. C4186)

1 mM *N*-formyl-methionyl-leucyl-phenylalanine (FMLP; Sigma-Aldrich cat. no. F3506)

10 μg ml^{-1} phorbol myristate acetate (PMA; Sigma-Aldrich cat. no. P-8139)

1 mg ml^{-1} superoxide dismutase (SOD; Sigma-Aldrich cat. no. S-8524)

Method

Note: Hanks' balanced solution used to perform the superoxide anion assay should not contain phenol red because it adds to colour intensity of ferrocytochrome c produced and gives erroneously high Absorbance values. Cytochrome c can be stored in stock solution frozen at −70 °C but must be freshly diluted for each assay.

1. Dilute suspension of conidia in 1 x YNB to a final concentration of 10^6 ml^{-1}. Add 200 μl of conidial suspension per well in a 96-well flat-bottomed plate and incubate at 37 °C and 5% CO_2 for appropriate times depending on the fungal organism to allow germination of conidia into hyphae.

2. In wells where O_2^- production in response to non-opsonized hyphae is assessed, aspirate and replace RPMI-1640 with 100 μl $HBSS^+$.

3. In order to opsonize hyphae, add 100 μl 50% pooled HS in wells. Incubate at 37 °C for 30 min.

4. Wash twice with $HBSS^+$. Add 50 μM ferricytochrome c and phagocytes at required E:T ratio with opsonized or non-opsonized hyphae in wells. An E:T ratio of 1:1 or 2:1 is frequently used and is sufficient to detect differences in the amounts of reduced cytochrome c in response to opsonized vs. non-opsonized hyphae.

5. Assess O_2^- production by MNCs to soluble stimuli adding 0.5 or 1 μM FMLP or 50 ng ml^{-1} PMA and cytochrome c in wells with MNCs alone. Wells with MNCs and cytochrome c in $HBSS^+$ are used to determine baseline O_2^- production.

6. Incubate at 37 °C on a rotating rack for 1 h. Transfer 100 μl to a new 96-well plate and read Absorbance at 550 nm with a reference wavelength set at 690 nm. Take absorbance values using as blank 100 μl cytochrome c. The extinction coefficient of reduced cytochrome c used to calculate the O_2^- production is $29.5 \times 10^4\ M^{-1}cm^{-1}$.

7. In some wells, add 50 μg ml^{-1} SOD, phagocytes, ferricytochrome c, and a stimulus. Any absorbance read in the SOD wells is non-specific and should be subtracted from the values obtained in the assay wells.

Protocol 3.6
Hydrogen peroxide-rhodamine assay

Equipment, materials and reagents

Equipment as for Protocol 3.1

EPICS XL Flow Cytometer Coulter Beckman

2 mM DHR-123 solution

10 μg ml^{-1} PMA

10 x YNB

Method

1. Inoculate 1×10^7 conidia ml^{-1} in 1 x YNB broth at final volume of 2 ml and incubate at appropriate temperature with 5% CO_2 for a 7-day growth period.

2. Centrifuge the broth solution at 400 x *g* for 15 min and recover the fungal supernatant.

3. Incubate 1×10^6 MNCs with 100 μl (i.e. 1×10^6 conidia) fungal supernatant for 2 h at appropriate temperature. Add 50 ng ml^{-1} PMA to 1×10^6 MNCs to maximally stimulate them (positive control).

4. Add 20 μM DHR-123 solution to the samples 1 h prior to termination of the 2-h incubation period with fungal supernatant.

5. Assess the oxidation of DHR-123 to R-123 by flow cytometry using an argon laser emitting 15 mV at 488 nm. The percentage of DHR-123 positive cells is calculated by flow cytometry.

Protocol 3.7
Phagocytosis assay

Equipment, materials and reagents

Equipment as for Protocol 3.1

18 mm round glass coverslips

Pooled HS(human serum) for opsonization

CM-HS: CM with pooled human serum (HS): 25% v/v HS in RPMI-1640 containing $100\ \mathrm{U\,ml^{-1}}$ of penicillin and $100\ \mu\mathrm{g\,ml^{-1}}$ of streptomycin

$HBSS^+$

May–Grunwald/Giemsa stain

75% ethanol

Method

1. Keep the coverslips in 75% ethanol and flame briefly before use. Place sterile round coverslips into the wells of a 12-well tissue culture plate.

2. On each coverslip, distribute 200 µl of a suspension of 5×10^6 MNCs $\mathrm{ml^{-1}}$ in CM-HS evenly. Avoid spillage on the surface of the well.

3. Incubate at 37 °C 5% CO_2 for 2 h to allow MNC adherence on the glass coverslips.

4. Wash each coverslip with 1 ml of warm $HBSS^-$ twice. Adherent MNCs remain, whereas non-adherent lymphocytes are washed off.

5. Add 1×10^6 opsonized conidia per millilitre in CM to each coverslip and incubate at 37 °C 5% CO_2 for 2 h, 4 h, 6 h and 12 h to obtain different snapshots of the phagocytosis process.

6. Wash with 1 ml of warm $HBSS^-$, remove, and air-dry coverslips.

7. Fix and stain the coverslips according to May–Grunwald/Giemsa stain procedure (Protocol 3.2, step 10).

8. Mount coverslips on microscopic slides with Permount.

9. Observe and count phagocytosed conidia under a light microscope. Distinguish between those that have a complete phagosome membrane around them or are ingested by more than 50% in a deepening of phagocytic membrane. Do not count conidia that are attached or ingested by less than 50%.

10. Calculate the number of conidia phagocytosed/100 MNCs as follows:

    - Percentage of phagocytosis $= 100 \times (B + C + D + E + F)/(A + B + C + D + E + F)$

- Phagocytic index = (B + 2C + 3D + 4E + 5F)/(B + C + D + E + F)

where *A* is the number of cells that phagocytosed no conidia, *B* is the number that phagocytosed one conidium, *C* is the number that phagocytosed two conidia, *D* is the number that phagocytosed three conidia, *E* is the number that phagocytosed four conidia, and *F* is the number that phagocytosed more than four conidia.

3.5 Evaluation of immunomodulators in response to fungal infection

In this section, we describe protocols for evaluating gene expression of cytokines and chemokines in phagocytes in response to fungal infection or drug cell-signalling analysed by RT-PCR, low-density microarray technology, or ELISA (Figure 3.3). RT-PCR combines both complementary deoxyribonucleic acid (cDNA) synthesis and PCR in a single tube using gene-specific primers and the polymerase mixture: Superscript™ II (Invitrogen Corporation, Carlsbad, CA, USA) H–RT and Platinum Taq polymerase. PCR products are quantitated including a second set of internal control primers, which amplify a housekeeping gene in the RT-PCR (Marone *et al.*, 2001; Selvey *et al.*, 2001).

The amount of immunomodulatory mRNAs are expressed as a ratio between the sample mRNA and the internal control mRNA (Reischl and Kochanowski, 1995).

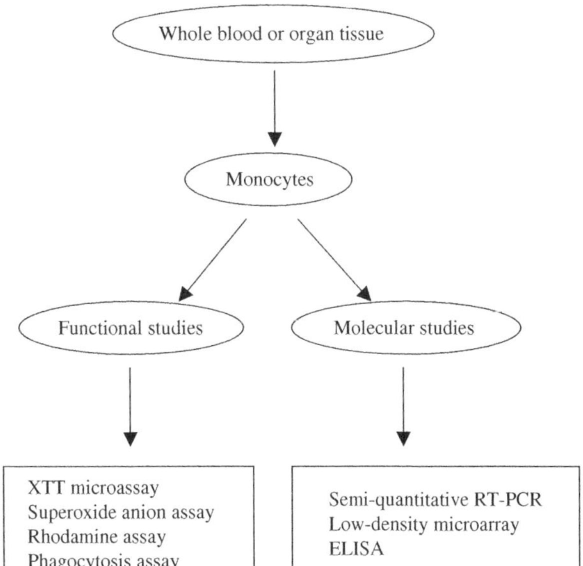

Figure 3.3 Flow chart for functional and molecular analyses of immunomodulatory proteins expressed in phagocytes in response to opportunistic fungi

Low-density microarrays are designed to study the gene expression profile of well-defined biological pathways under different experimental conditions. Ninety-six to 440 cDNA fragments are arrayed on a nylon membrane by non-contact printing technology, hybridized to biotinylated RNA samples and detected by chemiluminescense. The array image can be acquired for data analysis either by a CCD camera benchtop station or recorded using X-ray film and a desktop scanner in TIFF or JPEG format. Data analysis and interpretation of results are performed by a Web-based software program. The advantages that this system offers are cost-effectiveness, sensitivity and performance with standard molecular biology laboratory equipment. In Protocol 3.10 we describe the sandwich-type method of ELISA that requires a monoclonal antibody specific for the protein to be detected and a polyclonal antibody conjugated to horseradish peroxidase. The test sample is sandwiched between the antibodies bound by non-covalent interactions. The protein is detected by the addition of equal amounts of hydrogen peroxide and tetramethylbenzidine, the substrate on which the enzyme acts to produce a colour change (Desphande, 1996). The absorbance of each sample is determined at 450 nm.

Protocol 3.8
Analysis of gene expression by RT-PCR

Equipment, materials and reagents

Equipment as for Protocol 3.1

Thermal cycler

Horizontal gel electrophoresis

Power supply

UV transilluminator

Filter and spin columns

Sterile 1.5 ml snap-cap tubes

Sterile thin-walled 0.2 ml polymerase chain reaction (PCR) tubes

Lysis and binding buffer solution

Desalting buffer

Washing buffer

Absolute ethanol

Diethylpyrocarbonate (DEPC)-treated sterile water

TE buffer: 50 mM Tris-HCl, 10 mM ethylenediaminetetraacetic acid (EDTA), pH 7.5

Target and internal control primers (TIB MOLBIOL, Berlin, Germany)

Reverse transcriptase (RT)/Platinum Taq mix (Invitrogen Corporation, Carlsbad, CA, USA)

2 x reaction mix, containing 0.4 mM of each deoxynucleotide 5′-triphosphatases

2.4 mM $MgSO_4$ (Invitrogen)

1 x PBS: 10 mM phosphate buffer, 2.7 mM KCl, 137 mM NaCl, pH 7.4

Agarose gel, MB grade

0.5 M 3-[N-morpholino]propanesulfonic acid, free acid (MOPS)

0.5 M sodium acetate

1 M EDTA

Formamide

37% formaldehyde

Bromophenol blue

10 $\mu g\, ml^{-1}$ ethidium bromide

Method

RNA isolation by silica membrane spin column purification

Note: The following protocol can be used with any isolation kit using silica membrane spin columns for RNA isolation and purification. This method yields up to 30–40 μg RNA from 5×10^6 cultured cells.

1. Isolate MNCs by plastic adherence (follow Protocol 3.2).

2. Discard culture supernatant and add 1 ml of 1 x PBS.

3. Scrape cells off the wells of culture plate and transfer solution to 1.5 ml snap-cap tubes.

4. Pellet cells by centrifugation at 400 x *g* for 5 min.

5. Loosen cell pellet by gently flicking the tube and lyse cells (5×10^6 MNCs) with 400 μl of lysis and binding buffer solution. The lysis buffer contains chaotropic components, which dissolve cell-membrane constituents, prevent RNA degradation and maximize retention of the RNA on the spin column. Before proceeding to the next step, make sure to pipette or gently vortex the cell lysate to avoid transferring cell clumps to the filter column. If more MNCs are processed, they will not be fully lysed and contaminants will not be completely removed.

6. Add the entire volume of cell lysate on a filter column and collect the filtrate in a 1.5 ml tube after centrifugation at 11 000 x *g* for 1 min. This step reduces the viscosity of the cell lysate and creates a homogeneous solution. If too many cells have been used, the homogenized lysate will be too viscous to pipette.

7. Add 1 volume of 70% ethanol to the homogenized lysate, mix well by pipetting and add entire volume of sample to a spin column.

8. Centrifuge at 8000 x *g* for 1 min. Discard the flow-through. Add 1 volume of desalting buffer to the spin column, centrifuge at 11 000 x *g* for 1 min. Discard the flow-through. If the flow-through will come into contact with the spin column, in order to eliminate possible buffer carryover, repeat centrifugation step 8.

9. Wash 3 times with washing buffer. Discard the flow-through each time. Steps 6 and 7 remove salts, metabolites and proteins.

10. Elute total RNA from the spin column by adding 30–50 μl of DEPC-treated deionized water to the spin column. Collect the filtrate in a 1.5 ml tube.

11. Determine sample purity and quantify the RNA yield, using a UV spectrophotometer.

    - Read and record the absorbance at 260 and 280 nm.

    - Calculate the concentration:

      A260 × 40 × Dilution Factor = Sample Conc. μg/ml

    - Determine sample purity by calculating 260/280 ratio. Clean samples have a ratio within 1.8 to 2.2. A low 260/280 ratio indicates possible protein contamination.

Electrophoresis of RNA on a denaturing agarose gel

The overall quality of the RNA preparation should be assessed by electrophoresis on a denaturing agarose gel. An RNA sample known to be intact can be used as positive control.

1. Prepare 10 x formaldehyde gel buffer
   - 200 mM 3-[N-morpholino]propanesulfonic acid, free acid (MOPS)
   - 50 mM sodium acetate
   - 10 mM EDTA, pH to 7.0 with NaOH
2. Prepare a 1.2% formaldehyde agarose gel
   - 1.2 gram agarose
   - 10 ml 10 x formaldehyde gel buffer
   - RNase-free water to 100 ml
3. Melt the agarose, cool to 45 °C, add 1.8 ml of 37% formaldehyde and 1 μl of ethidium bromide (10 mg ml^{-1}). Mix thoroughly and pour onto the gel casting unit.
4. Add 1 x FA gel running buffer to the gel tank.
   - 100 ml 10 x FA gel buffer
   - 20 ml 37% formaldehyde
   - 880 ml RNase-free water
5. Add 1 volume of 5 x RNA loading buffer and 1 volume of RNA sample (1.5 μg) and load the sample. (5 x RNA loading buffer: 20–30 mg bromophenol blue; 80 μl 500 mM EDTA, pH 8.0; 720 μl formaldehyde; 2.0 ml glycerol, 3.1 ml formamide, 4.0 ml 10 x FA gel buffer and RNase-free water to 10 ml).
6. Run gel at 70 V.
7. Visualize total RNA on UV transilluminator. The 28S rRNA band should be approximately twice as intense as the 18S rRNA band (Figure 3.4).

cDNA synthesis and RT-PCR

Note: Suitable internal control RNAs to be used in RT-PCR are glyceraldehyde-3-phosphate dehydrogenase or aldolase A. The first RNA transcript is used for normalization of highly expressed transcripts, whereas aldolase A mRNA can be used to quantitate low-abundant mRNAs. If there is competition observed between the target PCR product and the co-amplified internal control, add $MgSO_4$ to a final concentration of 3 mM.

1. Use thin-walled 0.2 ml PCR tubes and add the following reagents:

2. To rule out contamination of genomic DNA in RNA preparation, include a negative RT-PCR control tube containing all reagents except the polymerase mix. Replace the RT/platinum Taq polymerase mix with 2 U of Taq DNA polymerase.

3. The following cycling conditions for cDNA and double-stranded DNA synthesis are applicable for Perkin-Elmer Thermal Cycler 9600.

   a. 1 cycle: 48 °C for 30 min (cDNA synthesis)

   b. 94 °C for 2 min (inactivation of RT, reactivation of Taq polymerase)

Reagents	Volume	Final concentration
2x reaction mix	25 µl	1 ×
Forward primer: 10 µM	× µl	0.2 µM
Reverse primer: 10 µM	× µl	0.2 µM
Internal control forward primer: 5 µM	× µl	0.1 µM
Internal control reverse primer: 5 µM	× µl	0.1 µM
RT/platinum Taq polymerase mix	1 unit	
Template: 0.1–1 µg total RNA	× µl	
DEPC-treated water	to 50 µl	

   c. 35 cycles: denaturation step: 94 °C for 15 sec

   d. Annealing step: 50–60 °C

   e. Extension step: 68 °C for 1 min

   f. 1 cycle: 68 °C for 7 min (final extension)

4. Detect PCR products on 1.2% agarose gel by electrophoresis and ethidium bromide staining (0.5 µg ml^{-1}).

5. Quantitate PCR products using a CCD camera to capture the DNA image and a computer software program to express the band intensity into ng ml^{-1}.

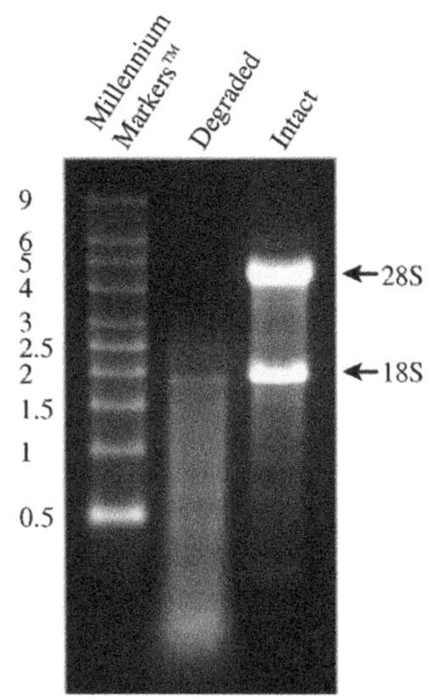

Figure 3.4 Profile of intact and degraded RNA on a denaturing RNA gel. 2 μg of degraded total RNA and intact total RNA were run beside Ambion's RNA size marker on a 1.2% denaturing agarose gel

Protocol 3.9 Analysis of pathway-specific gene expression by microarray technology

Note: The following protocol is based on the GEArray kit supplied by Superarray Bioscience Corporation (http://www.superarray.com).

Equipment, materials and reagents

Hybridization oven

CCD camera image station or X-ray film and scanner

Primer mix solution

50 nmol biotin-16-dUTP

Sheared salmon sperm DNA

20 x SSC: 3 M NaCl, 0.3 M sodium citrate

20% SDS

Superscript II™ reverse transcriptase (Invitrogen Corporation, Carlsbad, CA, USA)

5 x reverse transcriptase (RT) buffer (Invitrogen)

Hybridization solution

Wash solution 1 for hybridization: 2 x SSC, 1% SDS

Wash solution 2 for hybridization: 0.1 x SSC, 0.5% SDS

Blocking solution

5 x AP (alkaline-phosphatase-conjugated streptavidin) binding buffer

Alkaline phosphatase (AP)

Wash solution for detection

CDP-Star chemiluminescent substrate

Method

Follow the RNA isolation method and electrophoresis of total RNA on a denaturing gel as outlined in Protocol 3.7.

Probe synthesis

1. Prepare the annealing mixture, by adding 4–5 μg of total RNA (1 $\mu g\ ml^{-1}$) to 3 μl of a primer mix solution, containing oligonucleotides (18- to 21-mers) complementary to an internal region of each of the cDNA fragments arrayed on the membrane. Add RNase-free H_2O to 10 μl final volume.

2. Denature the reaction mixture in a thermal cycler at 70 °C for 3 min, cool to 42 °C for 5 min.

3. Prepare the reverse transcriptase (RT) reaction: 50 μM each dATP, dCTP, and dGTP, 5 μM dTTP, 10 μM DTT, 1 x RT buffer, 200 U of Superscript II reverse transcriptase, and 1 x RT buffer in a final volume of 10 μl.

4. Transfer 10 μl of the RT reaction to the 10 μl annealing mixture, mix and incubate at 42 °C, 90 min for cDNA synthesis.

5. Denature the cDNA probe at 94 °C for 5 min and quickly chill on ice. The labelled cDNA probe mixture is now ready for hybridization.

Hybridization

1. Heat the sheared salmon sperm DNA at 100 °C for 5 min and immediately chill on ice.
2. Mix 2 ml of hybridization buffer with denatured salmon sperm DNA (100 $\mu g\,ml^{-1}$) and pre-hybridize in hybridization oven at 60 °C for 1 to 2 h with mild agitation.
3. Prepare GEAhyb: Add the entire volume of denatured cDNA probe to 0.75 ml of pre-warmed GEAprehyb. Mix well, and keep the GEAhyb at 60 °C.
4. Replace the pre-hybridization solution with 2 ml of fresh hybridization solution, containing the entire volume of denatured cDNA probe and hybridize overnight at 60 °C with continuous agitation.
5. Wash the membrane twice with 5 ml Wash Solution 1 at 60 °C with agitation for 15 min each.
6. Wash the membrane twice with 5 ml Wash Solution 2 at 60 °C with agitation for 15 min each.

Chemiluminescent detection

1. Discard the last wash and add 2 ml blocking solution. Incubate for 40 min with agitation.
2. Discard the blocking solution from the tube and add 2 ml 1 x AP binding buffer, containing AP diluted 1:8000. Incubate for 10 min with agitation.
3. Wash the membrane 4 times with 4 ml of 1 x washing buffer for detection for 5 min with gentle agitation.
4. Add 1.0 ml CDP-Star chemiluminescent substrate to the hybridization tube. Incubate at room temperature for 2 to 10 min.
5. Capture the image with a CCD camera or record array image using X-ray film and scanner.
6. Save array images in TIFF format and use the Web-based software program provided by Superarray Bioscience Corporation for data analysis. After setting

background correction values and normalizing values relative to housekeeping gene spots, data analysis is performed by scatter plot, clustergram or K-means clustering.

Protocol 3.10
Assessment of cytokines and chemokines by ELISA

Equipment

Equipment as for Protocols 3.1 and 3.3

Lysis buffer: 25 mM HEPES, pH 7.8; 0.5% Nonident-P40; 0.1% sodium dodecyl sulphate (SDS); 0.5 M NaCl; 5 mM EDTA; 0.1 mM sodium deoxycholate; 1 mM phenylmethyl-sulphonyl fluoride; 0.1 mg ml^{-1} aprotinin; 0.1 mg ml^{-1} leupeptin; 0.1 mg ml^{-1} pepstatin

Murine monoclonal antibody specific for the protein of interest

25 x wash buffer solution of buffered surfactant (R&D Systems)

Polyclonal antibody specific for the protein of interest conjugated to horseradish peroxidase

Chromogenic substrate solution: hydrogen peroxide and tetramethylbenzidine

Stop solution: 2N H_2SO_4

Method

Isolation of intracellular extract

1. Isolate MNCs by plastic adherence (follow Protocol 3.2).

2. Save culture supernatant to measure immunomodulatory proteins secreted extracellularly in response to fungal infection.

3. Wash once with 1 ml of 1 x PBS and lyse attached MNCs by adding 1 ml of freshly prepared lysis buffer to each well of culture plate.

4. Detach remaining MNCs by scraping the wells of culture plate and transfer solution to 1.5 ml snap-cap tubes. Keep samples on ice.

5. Vortex samples vigorously for 30 sec and pellet solubilized cell fragments by centrifugation at 12 000 x *g*, 4 °C for 10 min.

6. Combine culture supernatants with the corresponding intracellular extracts to measure the total amount of specific protein expressed. Alternatively, one can treat culture supernatants and intracellular extracts separately for protein quantitation.

7. Keep samples at −20 °C until processed by ELISA using the appropriate Quantikine kits from R&D Systems, depending on the cytokines under investigation.

Enzyme-linked immunosorbent assay (ELISA)

General guidelines and important points to consider for quantitation of proteins using the Quantikine kits supplied by R&D Systems (http://www.R&Dsystems.com) are listed:

1. To quantitate the expression levels of a specific protein, the measured amount should fall within the concentration range of the standards provided in the kit. Reserve two ELISA strips for a trial experiment where the wells of one of the strips will contain various dilutions of the sample and the second strip will be used for the standards.

2. Dilutions of standards and samples should be made with utmost accuracy and consistency.

3. Before reading the microplate, make sure the colour change appears uniform in all wells by passing the contents of each well several times through a multichannel pipette.

4. Convert the absorbance readings into picograms per millilitre of measured protein using a computer software program.

3.6 Overview

The above techniques are a portion of assays that one can apply to study antifungal responses of human phagocytes upon challenge by opportunistic fungi. Since innate immunity is critical in host response to common fungi, methods of evaluation of its level of activity are both important and helpful. While most of these techniques have been developed and applied in relation to *C. albicans* and *A. fumigatus*, they are easily adapted to be used (and they have been used) in the study of host responses to other more rare fungi. The fact that they are not very sophisticated and they do not require expensive equipment permits their application to a wide range of laboratories with the scope of antifungal host response.

3.7 References

Antachopoulos, C. and Roilides, E. (2005) Cytokines and fungal infections. *Br. J. Haematol.* **129**: 583–596.

Desphande, S. S. (1996) *Enzyme Immunoassays: From Concept to Product Development*, New York, Chapman & Hall.

Fontaine, T., Simenel, C., Dubreucq, G., Adam, O. *et al.* (2000) Molecular organization of the alkali-insoluble fraction of *Aspergillus fumigatus* cell wall. *J. Biol. Chem.* **275**: 27594–27607.

Forman, H. J. and Torres, M. (2002) Reactive oxygen species and cell signalling. Respiratory burst in macrophage signaling. *Am. J. Respir. Crit. Care. Med.* **166**: 54–58.

Marone, M., Mozzetti, S., De Ritis, D., Pierelli, L. and Scambia, G. (2001) Semi-quantitative RT-PCR analysis to assess the expression levels of multiple transcripts from the same sample. *Biol. Proced.* **3**: 19–25.

Marshall, M., Gull, K. and Jeffries, P. (1997) Monoclonal antibodies as probes for fungal wall structure during morphogenesis. *Microbiology* **143**: 2255–2265.

Reischl, U. and Kochanowski, B. (1995) Quantitative PCR: A survey of the present technology. *Mol. Biotechnol.* **3**: 55–71.

Rogers, P. D., Pearson, M. M., Cleary, J. D., Sullivan, D. C. and Chapman, S. W. (2002) Differential expression of genes encoding immunomodulatory proteins in response to amphotericin B in human mononuclear cells identified by cDNA microarray analysis. *J. Antimicrob. Chemother.* **50**: 811–817.

Roilides, E. and Walsh, T. J. (2004) Recombinant cytokines in augmentation and immunomodulation of host defenses against *Candida* spp. *Med. Mycol.* **42**: 1–13.

Selvey, S., Thompson, E. W., Matthaei, K., Lea, R. A. *et al.* (2001) Beta-actin: An unsuitable internal control for RT-PCR. *Mol. Cell. Probes.* **15**: 307–311.

Thannical, V. J. and Fanburg, B. L. (2000) Reactive oxygen species in cell signalling. *Am. J. Physiol. Lung Cell Mol. Physiol.* **279**: 1005–1028.

Vonk, A. G., Netea, M. G., Denecker, N. E., Verschueren, I. C. *et al.* (1998) Modulation of the pro- and anti-inflammatory cytokine balance by amphotericin B. *J. Antimicrob. Chemother.* **42**: 469–74.

Walsh, T. J., Roilides, E., Cortez, K., Kottilil, S. *et al.* (2005) Control, immunoregulation, and expression of innate pulmonary host defenses against *Aspergillus fumigatus*. *Med. Mycol.* **43**: S165–S172.

4

Determination of the virulence factors of *Candida albicans* and related yeast species

Khaled H. Abu-Elteen and **Mawieh Hamad**

4.1 Introduction

Candida albicans is arguably the most important causative agent of human fungal infections. It is a major opportunistic infection in AIDS patients, patients undergoing chemotherapy and radiotherapy, transplant recipients and other classes of immunocompromised patients. *C. albicans* ranks fourth among the leading causes of nosocomial infections (Sullivan *et al.*, 2005). Studies indicate that up to 90% of AIDS patients suffer from oropharyngeal candidosis mainly due to *C. albicans* infection (Akpan and Morgan, 2002). In transplant patients, patients undergoing chemotherapy or radiotherapy and those with AIDS, *C. albicans* and other *Candida* species are common causes of invasive disease (Akpan and Morgan, 2002). Paradoxically, advanced medical care that tends to prolong the lives of such patients is causally associated with increased rates of fungal infections (Abu-Elteen and Hamad, 2005), hence the growing interest and importance of applying better diagnosis and treatment approaches to fungal infections. The capacity of yeast to cause disease depends (a) on its ability to survive and thrive in special microenvironments within the host including the mucosa and the bloodstream and (b) on a number of virulence factors that aid in the pathogen's adherence to and invasion of host tissues and cells (Calderone and Fonzi, 2001). The fungus can exist as a commensal on the mucosa of the oral cavity, oesophagus, vagina and the gastrointestinal tract. While localized *C. albicans* (oropharyngeal, oesophageal, vulvovaginal and cutaneous) infections generally lead to significant morbidity, systemic infections especially in the

Medical Mycology: Cellular and Molecular Techniques. Edited by Kevin Kavanagh

Table 4.1 Parameters affecting adhesion of *Candida* species to epithelial cells, plastic surfaces and denture-base resins

Yeast	Epithelial cells	Inanimate surfaces	Environment
Species, strains and phenotype	Cell type	Hydrophobic forces (London–van der Waals forces)	Hydrogen-ion concentration and CO_2 level
Culture medium	Variability and viability	Electrostatic forces (surface charge)	Temperature and time of contact
Culture age and viability		Surface-free energy	Buffer
Culture concentration		Surface roughness and physical presentation	Serum and saliva
Germination		Wettability	Presence of commensal bacteria
Cell-surface hydrophobicity			Presence of antibodies
			Age-related alterations

Adapted from Ghannoum and Radwan (1990), with permission from © CRC Press, Boca Raton, Chapter 5

immunocompromised can be lethal (Sullivan *et al.*, 2005). Systemic infections may develop following adhesion of *C. albicans* to epithelial cells and colonization of mucosal surfaces. Subsequently, *C. albicans* cells penetrate epithelial and endothelial cell barriers and disseminate throughout the system (Odds, 1998). Accordingly, vascular endothelium plays a significant role in the dissemination and establishment of systemic candidosis. The endothelium is sensitive to invasion and damage by *C. albicans* through the attachment of germinated *C. albicans* to endothelial cells, yeast-cell phagocytosis and activation of hypha-specific genes.

It is well established that adhesion represents the initial intimate contact between the pathogen and the host (Sturtevant and Calderone, 1997). Therefore, thorough evaluation of the adhesion potential of *C. albicans* isolates is of significant scientific and clinical value. A number of factors can influence the association between *Candida* species and various contact surfaces (Table 4.1) (Ghannoum and Radwan, 1990). Numerous *in vitro* and *in vivo* models have been developed to quantify and characterize *Candida* adhesion to both animate and inanimate surfaces (Table 4.2) (Kennedy, 1988).

4.2 Measurement of *Candida* species adhesion *in vitro*

A number of *in vitro* methods have been developed and tested for the evaluation of *Candida* species adhesion to freshly isolated cells, cultured cell lines and inanimate surfaces (Douglas, 1987).

Protocol 4.1
The visual assessment of candidal adhesion to BECs

Light microscopy, immunofluorescent microscopy and electron microscopy can be used to ascertain adhesion of *Candida* to cells or surfaces. Quantitative determination

Table 4.2 Various models for studying *C. albicans* adhesion and associated mechanisms

In vitro models	
Exfoliated epithelia cells	Buccal epithelial cells
	Corneocytes
	Urogenital epithelial cells
	Vaginal epithelial cells
Tissue culture	HeLa cells
	Cervical epithelial cells
	Vaginal epithelial cells
Tissue slices, disks or explants	Endothelial tissue
	Oral mucosa
	Gastrointestinal mucosa
	Firbrin-platelet matrices
	Intestinal mucus gel
Non-biological surfaces	Acrylic
	Glass
	Plastic
	Denture-based resins
Hydrophobicity testing	Phase-partition test
	Hydrophobic interaction
	Chromatography (HIC)
	Modified HIC
	Contact angle measurements
	Salting-out procedures
	Replica plate test
In vivo models	
Gastrointestinal mucosa	Infant mice
	Adult euthymic mice
	Adult germ-free mice
	Adult conventional and antibiotic-treated hamsters and mice Adult conventional rats
	Dogs
Heart	Rabbits
Oral mucosa	Mice
	Hamsters
	Rats
	Monkeys
Renal endothelium	Rabbits
Vaginal mucosa	Mice
	Rats

Adapted from Kennedy (1988) In: *Current Topics in Medical Mycology*, Vol. 2, Chapter 4, with permission from © 1988, Academic Press, New York.

of *C. albicans* adhesion to buccal epithelial cells (BECs) is also possible. This involves the incubation of equal volumes of standardized suspensions of yeast and epithelial cells followed by the removal of unattached yeast cells by repeated washing of the mixture over a 12–14 μm polycarbonate membrane filter. Polycarbonate filters can also be used for recovery of epithelial cells from the incubation mixture (Kimura and Pearsall, 1978). Polycarbonate filters with a pore size that allows passage of unattached yeast cells but not epithelial cells permits for yeast cells adhering to BECs to be fixed, stained and counted directly on the transparent filter. The procedure has been used by several groups to study the adhesion of *C. albicans* and other *Candida* species to swabbed or scrapped human buccal or vaginal epithelial cells (Douglas, 1987).

Equipment, materials and reagents

Sterile tongue depressor

Rotary shaker

C. albicans ATCC reference strain (recommended strain: *C. albicans* ATCC 10231) and *C. albicans* isolates to be tested

PBS containing 0.15 M NaCl; 0.01 M phosphate, (pH 7.2)

Yeast Nitrogen Base containing amino acids and ammonium sulphate and supplemented with 500 mM galactose

Preparation of BECs

1. Collect and pool BEC from at least 10 healthy adults who are not on antimicrobial therapy, by gently rubbing the mucosal surface of the cheek with a sterile tongue depressor.

2. Suspend the epithelial cells in 50 ml PBS. Ideally, cells should be used shortly after collection, but to avoid problems associated with day-to-day or donor-to-donor variations, the same cell pool may be stored at 4 °C for up to 2 weeks before use.

Preparation of *C. albicans* cells

1. It is highly recommended to use a well-established strain of *C. albicans* for these kinds of studies. Based on our own experience, *C. albicans* ATCC 10231 strain yields consistent adherence results.

2. Inoculate 1 ml from an overnight culture of *C. albicans* into 50 ml YNBG medium (galactose at 500 mM) in a 250 ml Erlenmeyer flask.

3. Incubate in a rotary shaker running at 100 rpm and 37 °C for 12 h (early stationary phase).

Yeast cells prepared in this way should be standardized and used immediately.

Method

1. Take 2 ml of the yeast culture containing 4×10^7 cells, centrifuge (4 °C, 3300 x *g*, 10 min) and resuspend in 2 ml of assay medium.

2. Take 2 ml of the BEC suspension containing 4×10^5 cells, centrifuge at 4 °C, 1400 x *g* for 10 min and suspend in 2 ml of the same assay medium.

3. Mix the two suspensions in a McCartney bottle and incubate at 37 °C with shaking water bath at 100 rpm for 1 hour.

4. Immediately filter through a 20 μm (pore size) filter to remove non-adhering yeast cells.

5. Wash the cells on the filter twice with 5 ml portions of HBSS and suspend them by gently swirling with 1 ml of HBSS.

6. Mount a drop of the suspension on a glass slide, air dry, heat fix by passing the slide quickly through a Bunsen burner flame, and stain with 0.5% (w/v) crystal violet for 1 min.

7. Determine adherence microscopically by counting the mean number of yeast cells adhering to 100 BECs. A yeast/epithelial cell ratio of 50–100:1 is commonly employed. Only yeast cells in direct contact with the BEC plasma membrane are counted.

8. The assay should be performed in duplicate or triplicate. Using the Student t-test to evaluate adherence values (average number of yeast cells / 1 epithelial cell). A *P* value of <0.05 should indicate a significant difference between the results of two determinations.

Protocol 4.2
The radiometric measurement of candidal adhesion

The radiometric adhesion method is used to measure the adherence of several *Candida* species to buccal and vaginal epithelial cells and to fibrin-platelet matrix (King *et al.*, 1980; Maisch and Calderone, 1980).

Equipment, materials and reagents

Orbital water bath shaker

Polypropylene filters (10 μm)

Liquid scintillation counter (to measure *β* emission)

BECs (see Protocol 4.1 for preparation)

C. albicans isolates to be tested (see Protocol 4.1 for preparation)

PBS (pH 7.2)

Solubilizing agent (Protosol)

Premixed scintillation solution (Econofluor)

D-[U-^{14}C] glucose

Method

1. The *C. albicans* inoculate is grown in 1% phytone-peptone broth supplemented with a final concentration of 0.1 μCi of D-[U-^{14}C] glucose/ml of medium and incubated in an orbital water bath shaker at 25 °C for 24 h, 120 rpm. Cells are washed 3 times with PBS.

2. Add 0.5 ml from a standardized suspension of labelled yeast cells (final concentration: 10^8 cells/ml PBS) to 0.5 ml suspension of epithelial cells (final concentration: 10^5 cells/ml PBS) and incubated at 37 °C for 30 min, 200 rpm.

3. Epithelial cells with adherent yeast cells are collected on a 10 μm polypropylene filter.

4. The filter is washed 5 times with PBS, treated with a 0.5 ml of Protosol and incubated at 50 °C for 2 h.

5. The vials are then chilled to 4 °C and each receives 5 ml Econofluor.

6. Measure radioactivity using a liquid scintillation counter.

7. Mean number of yeast cells/epithelial cell = [{adherence disintegration per minute − background disintegration per minute}/disintegration per minute per yeast]/total number of epithelial cells (5×10^4).

It has been shown that *C. albicans* adsorbs in higher numbers to polypropylene filters than to polycarbonate filters. Differential centrifugation can be used to separate epithelial cells from unattached yeast cells as an alternative to filtration. A similar radiometric protocol has been used in studies of adherence of *C. pintolopesii* (formerly *Torulopsis pintolopesii*) to murine gastrointestinal epithelial cells using ^{3}H-methionine (Suegara *et al.*, 1979). The method is rapid and considerably less laborious than the visual method.

4.3 Adhesion to inanimate surfaces

Various approaches have been employed to study the adherence of yeast to catheters, glass plates, clay minerals, metal hydroxides and various substances present in the denture-based resin material. The significance of these studies stems from the fact that such surfaces represent important inanimate surfaces closely associated with several forms of candidosis like chronic atrophic candidosis or denture stomatitis. Here we describe the method used to assess adsorption of *Candida* to silica. The method relies on measuring the number of yeasts remaining in suspension, using a coulter counter, or by determining the nitrogen content of the silica-yeast complex. Techniques for measuring attachment of *Saccharomyces cerevisiae* to glass slides and to cellulose or ECTEOLA-cellulose films have been reported (Douglas, 1987); in the latter investigation, adsorption of ^{35}S-labelled yeast cells to weighed papers was determined using a liquid scintillation counter. In the case of denture stomatitis, *C. albicans* is usually recovered from the fitting surfaces of the denture; this pattern suggests that the denture acrylic functions as a reservoir for the fungus (Samaranayake and MacFarlane, 1980).

Protocol 4.3
Assessment of candidal adhesion to denture acrylic material

Equipment, materials and reagents

Yeast cell isolate to be tested

Small transparent acrylic strips

PBS (pH 7.2)

Clean glass slides and a light microscope

Method

1. Acrylic strips are incubated with yeast-cell suspensions (final concentration of the yeast-cell suspension must be predetermined).

2. Non-adherent yeast cells are removed by rinsing with PBS.
3. Adherent yeast cells are counted under a microscope. Adhesion is directly proportional to yeast concentration within the range tested.

A similar experimental protocol can be used to measure yeast adhesion to a variety of denture-based resin material.

Protocol 4.4
Adherence of *Candida* to plastic catheter surfaces

In this assay, it is important to know that *Candida* species attach more readily to polyvinyl chloride catheters than to Teflon catheters (Rotrosen *et al.*, 1983).

Equipment, materials and reagents

Yeast cell isolate to be tested (*C. albicans* or *C. tropicalis*)

[^{14}C] glucose

Intravenous Catheters (length, 4 cm of 16 gauge)

PBS (pH 7.2)

10 ml Petri dishes

Roller drum

HBSS

Scintillation liquid (AQUASOL; obtained from New England Nuclear, Boston, USA)

Liquid scintillation counter (to measure β emission)

Method

C. albicans or *C. tropicalis* are labelled with [^{14}C] glucose and standardized to 10^7 CFU/ml.

1. Catheters are immersed in the yeast suspension and incubated for 60 min on a roller drum. Adherence must be restricted to the external surface of the catheters via sealing of catheter ends.

2. After 60 min, catheters are washed 3 times with HBSS, the ends are removed and the remaining 4 cm sections are counted in AQUASOL.

3. Values for yeast adherence in CFUs/catheter section are obtained by comparing the radioactivity of suspensions obtained from the catheter section versus those obtained from standardized labelled yeast-cell suspensions.

Adherence of *Candida* species to catheters can also be tested by submerging 2.5 cm^2 pieces of polytetrafluorethylene; polyethylene terephthalate; polymethylmerthacrylate and polystyrene Petri dishes at equal depth in a normal saline bath containing 5×10^6 yeasts/ml. Following agitation at 50 rpm for 20 h, submerged pieces are removed, dried and stained with crystal violet. Adherent yeast cells are then counted from photomicrographs of stained preparation.

4.4 *C. albicans* strain differentiation

In order to establish disease, the pathogen must enter the host, replicate within it, evade the host's defences and damage cells and tissues. Given that each of these steps entails a complex series of sequential events, several pathogenicity determinants are usually involved to reach the overall effect. Lack of one member of the full complement of determinants may result in considerable attenuation of the pathogen. Several groups have tried to relate the differences in virulence between strains of *C. albicans* to differences in pathogenicity determinants, namely adherence, proteinase and phospholipase production and germ tube formation. Several methods have been devised to classify *C. albicans* strains into subgroups below the species level. Although such subdivision schemes may seem of limited relevance, strain differentiation and typing procedures still contribute to our ever-growing knowledge of the epidemiology of *C. albicans* infections.

Protocol 4.5
Resistogram typing

This method is based on differences in the resistance of *C. albicans* to five different chemicals incorporated into the agar media at known concentrations. Each chemical is tested at a number of different concentrations and allotted a letter of the alphabet as shown in Table 4.3 (McCreight and Warnock, 1982).

Equipment, materials and reagents

Yeast isolate to be tested

Molten YMA or SDA

Table 4.3 Chemicals selected for differentiation of *C. albicans* strains

Chemical compound	Letter code	Concentration of stock solution (g/L)	Volume added to resistogram plates (ml/20 ml agar)			
Sodium selenite	A	20.0	0.5	0.6	0.7	0.8
Boric acid	B	20.0	1.2	1.3	1.4	1.5
Cetrimide	C	20.0	0.05	0.1	0.15	0.2
Sodium periodate	D	20.0	0.3	0.4	0.5	0.6
Silver nitrate	E	2.0	0.25	0.5	1.0	2.0

Adapted from McCreight and Warnock (1982) with permission from © *Mykosen*, **25**: 589–598.

Sodium selenite (designated as A), boric acid (B), cetrimide (C), sodium periodate (D)and silver nitrate (E); a stock solution for each of the 5 chemicals is prepared as indicated in Table 4.3

Multipoint inoculator

5 ml Petri dishes

Method

1. Stock solutions of the 5 chemicals are prepared in distilled water and added separately at specific concentration to 20 ml of molten YMA or SDA.
2. Plates are dried at 40 °C for 15 min and stored overnight at 4 °C before inoculation.
3. Plates are inoculated with 10^6 yeast cells/ml with a multi-inoculator, dried at room temperature and incubated at 37 °C for 40 h.
4. Based on the growth, or the lack of it, using different concentrations of the 5 chemicals, a resistogram of a specific *C. albicans* strain can be formulated. For example, the resistogram AB — — E signifies that the tested strain is resistant to sodium selenite (A), boric acid (B) and silver nitrate (E), but sensitive to cetrimide (–) and sodium periodate (–). A strain resistant to all 5 chemicals is designated as ABCDE, while a strain susceptible to all 5 chemicals is designated as — — — — —.

The resistogram method permits the discrimination between 32 different resistotype strains of *C. albicans*. This method was successfully used to differentiate oral isolates of *C. albicans* from normal subjects, patients with denture-induced stomatitis, and those with oral and laryngeal cancer (Nakamura *et al.*, 1998; Abu-Elteen, 2000; Abu-Elteen *et al.*, 2001). Variations in phospholipase and proteinase production as well as adherence to BEC are not necessarily related to resistogram-based strains (Abu-Elteen *et al.*, 2001).

4.4.1 Serotyping

Hasenclever and Mitchell were first to demonstrate the presence of serotypes among *C. albicans* isolates; namely serotypes A and B (Hasenclever and Mitchell, 1961). A third *C. albicans* serotype (C) was proposed by Müller and Kirchoff (1969). Serotype differences among *C. albicans* isolates are based on differences in cell-wall mannan branches. Antigenic differences result from variations in the bonding positions between mannose residues and the number of residues in mannan chains. Methods commonly used for serotyping of *C. albicans* isolates include the use of the Hasenclever tube agglutination test or the slide agglutination test (Stiller *et al.*, 1982). These tests are considered positive when *C. albicans* cells agglutinate following the addition of rabbit polyclonal antibodies specific for *C. albicans* antigens. A commercial slide agglutination test, which uses monospecific rabbit antiserum against a polysaccharide antigen, Iatron factor 6 (IF6) of *C. albicans* serotype A cells (Shinoda *et al.*, 1981) is available. Additionally, *C. albicans* serotyping has been carried out by immunofluorescence assays using either serotype A-specific rabbit antisera or monoclonal antibodies (H9) against extracted mannans of *C. albicans* serotype A (Miyakawa *et al.*, 1986). The expression of Hasenclever and IF6 antigens on *C. albicans* cells was also demonstrated by flow cytometric analysis (Mercure *et al.*, 1996).

It is not unusual for serotyping to give different results for the same *C. albicans* strain either due to a lack of standardized methodology or due to the use of different antigen-antibody detection systems. Slide agglutination tests using the H9 monoclonal antibody have matched poorly with those using the anti-IF6 antibody. This is mostly due to the fact that the two antibodies recognize different *C. albicans* antigens with variable patterns of expression. In the Hasenclever tube agglutination test, lots of antisera obtained from different animals produced different serotype results. Again, variations in the polyclonal set of antibodies may be the prime cause behind such inconsistencies. Even with the introduction of flow cytometry, an objective and quantitative tool, serotyping inconsistencies are bound to persist as long as immune reagents are not standardized.

Protocol 4.6
The slide agglutination technique

Equipment, materials and reagents

Yeast isolates to be analysed

Reagent number 6 from Candida Check Kit (Iatron Laboratories, Inc., Higashi-Kanda, Chiyoda, Tokyo, Japan)

PBS with 0.15 M NaCl; 0.01 M phosphate; pH 7.2

SDA plates

Glass slides, tubes and other plastic wear

Light microscope

Method

1. Following the culture of *C. albicans* isolates on SDA plates for 48 h at 25 °C, small amounts of specimen yeast cells are first tested in PBS to exclude the presence of auto-agglutination, then inoculated onto a Candida Check test tray. If auto-agglutination occurs, cells are washed with PBS and then resuspended and tested to ensure that auto-agglutination is no longer present.

2. Approximately 50 µl of anti-IF6 specific serum is added for test tubes and 50 µl PBS is added to control tubes.

3. Glass test tray is stirred for 1 to 2 min.

4. The presence of aggregates is indicative of *C. albicans* serotype A (positive agglutination reaction), while absence of aggregates is indicative of *C. albicans* serotype B (negative agglutination reaction). Yeast isolates that auto-agglutinate in PBS after repeated testing are characterized as non-serotypeable.

Protocol 4.7
Serotyping of *C. albicans* by flow cytomerty

Equipment, materials and reagents

C. albicans isolates to be analysed

Yeast nitrogen base media

Centrifuge, vortex and shaking water bath

Phosphate buffer saline

Multiparameter/multipurpose-type flow cytometer

2% paraformaldehyde

Anti-IF6 antibodies and goat anti-rabbit IgG coupled to DTAF fluorochrome

Method

1. Grow *C. albicans* cells to the stationary phase in 10 ml yeast nitrogen base in shaking water bath at 30 °C.

2. Yeast cells are pelleted by centrifugation at 1000 x *g* for 10 min, cells are then washed with 10 ml PBS, recentrifuged and resuspended in PBS to approximately 5×10^6 cells/ml.

3. Aliquots of 100 µl of the diluted *Candida* cells (5×10^5) are transferred to flow cytometry tubes prior to the addition of the primary antiserum (5 µl of the Ca/A-B-antiserum or 1 µl of the undiluted IF6 antiserum as provided by the manufacturer). The mixture is gently shaken and incubated on ice for 1 h.

4. After incubation, cells are washed by the addition of 3 ml of PBS to each sample, tubes are centrifuged for 10 min at 1000 x *g*, the supernatants are carefully discarded, leaving the pellets in approximately 200 µl of PBS. This washing procedure is repeated prior to the addition of 4 µl of purified goat anti-rabbit IgG coupled to DTAF to the yeast cells. Tubes are gently shaken and incubated on ice for 30 min.

5. Cells are washed twice in PBS and fixed using 0.5 ml of 2% paraformaldehyde/ tube. Tubes are sealed from light and kept refrigerated until time of analysis. It is always best if flow cytometric analysis is carried out on the same day of staining; however, stained and fixed samples can be analysed within a week of preparation if refrigerated.

6. The analysis is performed using a 488 nm argon laser beam; the fluidics system on the flow must be equipped with a 75 µm diameter nozzle tip. Note that DATF emits a green colour fluorescence; therefore, the instrument must be set to collect data for green colour emission.

4.5 Phenotypic switching in *C. albicans*

In what resembles dimorphism, *C. albicans* can undergo a process called phenotypic switching or the generation of different antigenic variants, in which different cellular morphologies are generated spontaneously. Switching can easily be distinguished from bud-hypha transition as it occurs spontaneously and at much lower frequency than bud-hypha transition. The phenomenon is not limited to fungi; prokaryotes like *Salmonella typhimurium*, *Borrelia hermsii* and *Niesseria gonorrhoeae* are known to produce multiantigenic variants as means of protection against the immune responses of the host (Soll, 1992). Different strains of *C. albicans* reversibly switch, often at high frequency, between different colony morphologies (Slutsky *et al.*, 1985;

Pomes *et al.*, 1985). One switching system of *C. albicans* consists of a reversible transition between two phases: a white hemispherical colony morphology referred to as the white phase and a grey flat colony morphology referred to as the opaque phase. This white-opaque transition system was initially identified in strain WO-1 isolated in 1986 from the bloodstream of a bone-marrow transplant patient (Soll, 1992; Soll *et al.*, 2003). Microscopically, opaque-phase cells of strain WO-1 are morphologically unique while white-phase cells of the same strain are round to ovoid and budded, just like most other strains of *C. albicans.* Opaque-phase cells are larger in size than white-phase cells, asymmetrical, elongated and bean-shaped. In the case of the common laboratory strain 3153A, switching occurs among at least seven colony phenotypes, each dictated by differences in the temporal dynamics and spatial distributions of budding cells, pseudohyphae and hyphae in the colony dome. Several reports have also demonstrated reversible, high-frequency switching in *C. tropicalis, C. parapsilosis, C. glabrata, C. lusitaniae,* and even *Cryptococcus neoformans* (Soll *et al.*, 2003; Yoon *et al.*, 1999; Lachke *et al.*, 2002).

Protocol 4.8
Evaluation of phenotype switching in *C. albicans*

Equipment, materials and reagents

C. albicans strain to be tested

Lee's medium (Lee *et al.*, 1975) supplemented with 70 μg of arginine per ml and 0.1 μM zinc sulphate ($ZnSO_4$).

Water bath shaker

2% Bacto^{TM} Agar (Difco Laboratories, Detroit, MI, USA)

Double-distilled water

Method

1. *C. albicans* cells from agar streaks are diluted into 125 ml Erlenmeyer flask containing 25 ml of the defined Lee's nutrient medium.
2. Culture flasks are rotated at 200 rpm at 25 °C, cells are diluted into fresh Lee's nutrient medium and grown for two more rounds of growth, then used in mid-log phase.
3. Cells are diluted in double-distilled water at a concentration of 500–1000 CFU/ml. An 0.1 ml aliquot is spread on 2% Bacto^{TM} Agar containing the nutrient components of Lee's medium.

4. Plates are incubated at 24 °C for at least 6 days before being scored for colony phenotype, each plate should contain about 50–100 colonies. Colonies of white and opaque phase cells can also be visualized on media containing 5 μg/ml of phloxine B.

5. Colonies exhibiting the original phenotype and those exhibiting the variant phenotype are counted; the total number of colonies is considered as 100% and number of colonies of the original phenotype versus those of the variant phenotype are converted to percentages of the total number. The probability of a mother cell producing a daughter cell of opposite phenotype can also be calculated if a clonal (1 cell) plating (on a glass slide containing a thin agar film) strategy is followed. In such a case, a high-resolution inverted microscope equipped with a long-working lens is used to count and score the different microcolonies over time. The probability of a mother cell producing a daughter cell of opposite phenotype is then calculated according to the formula: $\alpha = 2 - ([T - S]/M)^{1/k}$ where T is the final number of cells in the two phases, S is the final number of switched cells, M is the number of mother cells and k is the calculated number of generations. For full details of this procedure, consult Bergen *et al.* (1990).

4.6 Extracellular enzymes secreted by *C. albicans*

To aid in the invasion of host tissues, microbial cells possess constitutive and inducible hydrolytic enzymes, which destroy or derange constituents of membranes leading to membrane dysfunction and physical disruption (Ghannoum and Radwan, 1990). Since membranes are made of lipids and proteins, membranes constitute targets for microbial degrading enzymes. The three most significant extracellular hydrolytic enzymes produced by *C. albicans* are the secreted aspartyl proteinases (Saps), phospholipase B enzymes and lipases (Ghannoum and Radwan, 1990). Saps have been comprehensively studied as key virulence determinants of *C. albicans* (Hube, 1996, 2000; Na *et al.*, 1997; Martinez *et al.*, 1998).

All proteinases catalyze the hydrolysis of peptide bonds (CO-NH); they differ markedly, however, in specificity and mechanism of action. Four classes of proteinase have been described: serine proteinase, cystine proteinases, aspartyl proteinases and the metalloproteinases. *C. albicans*, *C. dubliniensis*, *C. tropicalis* and *C. parapsilosis* produce active extracellular proteinases when grown in media-containing proteins as the sole nitrogen source (Staib, 1965). The ability of *C. albicans* to produce extracellular proteinases was first described by Staib and colleagues. Table 4.4 summarizes some of the physical and chemical properties of this family of enzymes. The role of proteinases in virulence and pathogenesis has been amply studied; the reader is advised to refer to Chaffin *et al.* (1998) and Naglik *et al.* (2003) for detailed reviews.

The secretion of extracellular phospholipases (PL) by *C. albicans* was first reported in the 1960s (Werner, 1966). Yeast cells were first grown on a solid medium containing egg yolk or lecithin, then the formation of lipid breakdown products

Table 4.4 Characteristics of proteinases isolated from *Candida* spp. that may contribute to pathogenicity

Isolated from	Group	Mr	Isoelectric point	Denaturation pH	Optimum pH	Type of activity	Temp. of inactivation
C. albicans and *C. tropicalis*	Carboxyl proteinase	45 000	4.45	>8.4	2.5–3.9	Proteolytic	45 °C
C. albicans	Carboxyl proteinase	42 000	4.0	<2.5 & >6.0	4.0	Keratinolytic	>50 °C
C. albicans	Neutral proteinase	—	—	—	6–7.5	Proteolytic	
C. albicans	—	46 000	4.2	<6.0	3.5–4.0	Collagenolytic	70 °C
C. parapsilosis	Aspartic proteinase	33 000	5.3	>7.0	4.3	Proteolytic	—

Adapted from Ghannoum and Radwan (1990) with permission from © CRC Press, Boca.

was tested. Subsequent to these earlier reports, PL activity was reported in many pathogenic *C. albicans* strains grown on media containing blood serum and sheep erythrocytes (Costa *et al.*, 1968). The types of PL produced by pathogenic fungi include PLA, PLB and PLC, lysophospholipase and lysophospholipase transacylase (Ghannoum, 2000). The observation that *C. albicans* secretes PL prompted the development of a lecithin-based cytochemical method to detect this enzyme (Pugh and Cawson, 1975, 1977). Subsequent studies were directed at developing simpler methods to detect candidal PLs. Odds and Abbott described a biochemical assay to measure intracellular PLA and Lyso-PL activity in *C. albicans.* Using this assay, it was found that phosphatidylcholine (PC, lecithin) is hydrolyzed to yield Lyso-PC (Odds and Abbott, 1980). The disadvantage of this assay is that it is time-consuming and hence not suitable for testing large numbers of isolates. Price *et al.* (1982) described a plate method for the detection of PL activity in *C. albicans*. Since egg yolk contains large amounts of phospholipids, predominantly PC and phosphatidylethanolamine (PE), it was incorporated into SDA-based media. When grown on this medium, phospholipase-positive candidal isolates produce a distinct, well-defined, dense white zone of precipitation around the colony. The white zone forms due to the formation of complexes of calcium and fatty acids released from yolk phospholipids by the action of PL (MacFarlane and Knight, 1941). In this assay, PL activity (*Pz*) is defined as the ratio of the diameter of the colony to that of the dense white zone of precipitation around phospholipase-positive colonies. However, because the yolk contains substrates for both PL (phospholipids) and lipases (triglycerides), the egg-yolk-based assay is not specific, and therefore its use should be limited to initial screens. The assay is not suitable for screening of fungal isolates that produce low levels of PL. Confirmation of PL activity necessitates the use of specific radiometric or colorimetric assays on concentrated culture filtrate, particularly in poorly PL-producing strains.

Protocol 4.9
Measurement of extracellular proteinase production by *C. albicans*

Equipment, materials and reagents

C. albicans isolates to be tested

1.2% yeast carbon base medium adjusted to pH 4 supplemented with filter sterilized BSA at a final concentration of 0.2%

Sterile 0.22 μm membrane filter

Centrifuge and orbital shaker

15 mM sodium citrate buffer (pH 5.6) or 50 mM sodium phosphate buffer (pH 6.5), Ammonium sulphate (75%)

DEAE-Sephadex A-25 elution column (1.6 by 13 cm) equilibrated with the buffer to be used

NaCl stock solutions (0.1, 0.2, 0.3 and 0.5 M)

Spectrophotometer

1% w/v BSA in 50 mM KCl-HCl buffer (pH 2.5)

Ice-cold 10% w/v TCA.

Method

1. *C. albicans* isolates to be tested are maintained on SDA. Single colonies from plates are inoculated into 1.2% yeast carbon base supplemented with filter-sterilized BSA at a final concentration of 0.2%. Incubate at 30 °C in a rotary shaker for 2 to 3 days.

2. *C. albicans* cells are removed by centrifugation at 8400 x *g* for 10 min, the culture filtrate is passed through a 0.22 μm Millipore filter to remove any remaining yeast cells. Culture filtrate is precipitated with 10 ml of ammonium sulphate (75%) and centrifuged at 8400 x *g* for 10 min.

3. The culture filtrate is dialyzed against 15 mM sodium citrate buffer or 50 mM sodium phosphate buffer. The dialysate is applied to a DEAE-Sephadex A-25

column equilibrated with the same buffer and eluted with the same buffer at a flow rate of 40 ml/hour.

4. Adsorbed proteins are eluted with 0.1, 0.2, 0.3 and 0.5 M NaCl in a stepwise gradient. Eluant (2.5 ml/fraction) is collected for measurement of enzyme activity at 280 nm. Fractions containing maximal enzyme activity are pooled, dialyzed against distilled water at 4 °C.

5. Enzyme activity is determined spectrophotometrically following the digestion of BSA. To 30 µl enzyme solution, 270 µl 1% w/v BSA in 50 mM KCl-HCl buffer is added. The mixture is incubated at 37 °C for 2 h, the reaction is stopped by adding 700 µl ice-cold 10% w/v TCA.

6. Precipitated protein is removed by centrifugation at 8400 x *g* for 5 min and the amount of proteolysis is determined by measuring the absorbance of the supernatant at 280 nm. Protein concentration is measured by the Lowry method (Lowry *et al.*, 1951) using BSA as the standard.

Protocol 4.10
Measurement of extracellular proteinase produced by *C. albicans* (staib method)

Equipment, materials and reagents

C. albicans isolates to be tested

Test medium consisting of 1% w/v Bacto™ Agar, 0.1% w/v KH_2PO_4, 0.5% w/v $MgSO_4$, and 1% w/v glucose, pH 5.0, containing 0.16% w/v BSA as the sole source of nitrogen

Phosphate buffer saline (0.15 M NaCl; 0.01 M phosphate, pH 7.2)

Centrifuge and orbital shaker

20% TCA

1.25% amido black (90% methanol + 10% acetic acid)

15% acetic acid

Schnelltaster caliper

Method

1. *C. albicans* isolates to be tested are maintained on SDA. Three flasks, each containing 50 ml SDB per isolate are incubated at 26 °C in an orbital shaker for 24 h.

2. Cells are harvested by centrifugation at 5000 x *g* for 30 min, washed with PBS and resuspended at a density of 10^8 cells/ml.

3. 10 µl of the cell suspension are placed onto test medium containing BSA and incubated for 5 days at 37 °C. The diameter of the colony is measured by a Schnelltaster caliper.

4. De-staining is performed with 15% acetic acid; the clear zones around each colony are measured by a Schnelltaster caliper and used to determine the precipitation zone (*Pz*) according to the formula:

$$Pz = 1 - A/B$$

where *A* is the diameter of the colony + the proteolysis zone and *B* is the diameter of the colony only.

Protocol 4.11
Measurement of extracellular phospholipases of *C. albicans*

Equipment, materials and reagents

C. albicans isolates to be tested

SDA: (13 gram; NaCl, 11.7 gram and $CaCl_2$, 0.111 gram) supplemented with 10% (w/v) sterile egg yolk emulsion and dissolved in 184 ml of distilled water. The egg yolk is centrifuged at 500 x *g* for 10 min at room temperature, and 20 ml of the supernatant is added to sterilized medium.

Phosphate buffer saline

Centrifuge and orbital shaker

Schnelltaster caliper

Method

5. *C. albicans* isolates are cultured for 18 h at 30 °C in Sabouraud broth containing 2% glucose in an orbital shaker.

6. Cells are harvested by centrifugation (5000 x *g*, 30 min), washed twice with PBS and resuspended at a density of 1×10^8 cells/ml.

7. 10 microlitres of the cell suspension are spotted onto test medium containing egg yolk emulsion and left to dry at room temperature. The plates are incubated at 37 °C for 48–72 h after which the plates are flooded with staining solution (amido black) for 15 min and decolourized with 10% glacial acetic acid for 36 h.

8. The production of phospholipase is measured by measuring the diameter of the precipitation zone around the colony by Schnelltaster caliper, using the following equation:

$$Pz = 1 - A/B$$

where A is the diameter of colony + diameter of the white precipitation zone and B is the diameter of the colony.

4.7 Germ-tube formation in *C. albicans*

One of the most important characteristics of *C. albicans* is its ability to exist as ovoid blastospores and elongated hyphal filaments (i.e. dimorphism). The growth phenotype depends on environmental conditions (i.e. serum, growth temperature greater than 35 °C, nutrient starvation and pH greater than 6.5). Budding cells are round to ellipsoidal and have a diameter between 2 and 5 μm. The pseudohyphal morphology of *C. albicans* is an intermediate between the ellipsoidal blastospore and elongated hypha. An individual pseudohypha is typically 2 to 5 μm in diameter and 26 to 30 μm in length. True hyphal cells of *C. albicans* are compartmentalized tubes or filaments with diameters measuring typically 0.8–1.3 μm, which is about one-third the diameter of a comparable blastospore. Formation of hyphae from blastospores is initiated by the emergence of germ tubes which grow continuously by extension. Until 1995, *C. albicans* was the only species known to produce germ tubes and chlamydospores (i.e. the two 'gold standard' tests commonly used for the identification of *C. albicans*). However, in 1995, a novel, closely related species, *C. dubliniensis*, was identified in HIV-infected individuals, which can also exhibit these two morphological characteristics (Sullivan *et al.*, 2005).

Protocol 4.12
Germ-tube formation assay

Equipment, materials and reagents

BSA; working solution may be stored at 2–8 °C; stock solution can be dispensed into small tubes and stored at −20 °C.

Clean glass microscope slides

Glass cover slips

Glass tubes (13 × 100 mm)

Pasteur pipettes

Method

1. Place 3–4 drops of serum into a small glass test tube.
2. Using a Pasteur pipette, scratch a colony from the plate including the cultured yeast isolate in question and gently emulsify it in the test tube containing the serum.
3. Incubate the test tube at 35 °C to 37 °C for no more than 3 h.
4. Transfer a drop of the serum to a slide for examination.
5. Cover-slip and examine microscopically using 40 x objective.

Germ tubes are appendages half the width and 3–4 times the length of a yeast cell from which they arise. The presence of yeast cells with single short lateral filaments is indicative of germ-tube formation.

4.8 References

Abu-Elteen, K. H. (2000) *Candida albicans* strain differentiation in complete denture wearers. *New Microbiol.* **23**: 329–337.

Abu-Elteen, K. H., Elkarmi, A. Z. and Hamad, M. (2001) Characterization of phenotype-based pathogenic determinants of various *Candida albicans* strains in Jordan. *Jpn. J. Infect. Dis.* **54**: 229–236.

Abu-Elteen, K. H. and Hamad, M. (2005) Antifungal agents for use in human therapy. In: *Fungi: Biology and Applications* (ed. K. Kavanagh), John Wiley & Sons, London, pp 191–217.

Akpan, A. and Morgan, R. (2002) Oral candidiasis: A review. *Postgrad. Med. J.* **78**: 455–459.

Bergen, M., Voss, E. and Soll, D. J. (1990) Switching at the cellular level in the white-opaque transition of *Candida albicans. J. Gen. Microbiol.* **136**: 1925–1936.

Calderone, R. A. and Fonzi, W. A. (2001) Virulence factors of *Candida albicans. Trends in Microbiol.* **9**: 327–335.

Chaffin, W. L., Lopez-Ribot, J. L., Casanova, M., Gozalbo, D. and Martinez, J. P. (1998) Cell wall and secreted proteins of *Candida albicans*: Identification, function, and expression. *Microbiol. Mol. Biol. Rev.* **62**: 130–180.

Costa, A., Costa C., Misefari, A. and Amato, A. (1968) On the enzymatic activity of certain fungi. VII: Phosphatidase activity on media containing sheep's blood of pathogenic strains of *Candida albicans. Atti. Soc. Sci. Fis. Mat. Nat.* **XIV**: 93–101.

Douglas, L. J. (1987) Adhesion to surfaces. In: *The Yeasts: Volume 2* (2nd edn.) (eds A. H. Rose, and J. S. Harrison), Academic Press, London, pp 239–280.

Ghannoum, M. A. (2000) Potential role of phospholipases in virulence and fungal pathogenesis. *Clin. Microbiol. Rev.* **13**: 122–143.

Ghannoum, M. A. and Radwan, S. S. (1990) *Candida Adherence to Epithelial Cells*. CRC Press. Inc., Boca Raton, Florida.

Hasenclever, H. F. and Mitchell, W. O. (1961) Antigenic studies of *Candida*. I. Observation of two antigenic groups in *Candida albicans*. *J. Bacteriol.* **82**: 570–573.

Hube, B. (1996) *Candida albicans* secreted aspartyl proteinases. *Curr. Top. Med. Mycol.* **7**: 55–69.

Hube, B. (2000) Extracellular proteinases of human pathogenic fungi. *Contrib. Microbiol.* **5**: 126–137.

Kennedy, M. J. (1988) Adhesion and association mechanisms of *Candida albicans*. In: *Current Topics in Medical Mycology*, (ed M. R. McYinnis), Academic Press, New York, pp 73–169.

Kimura, L. H. and Pearsall, N. N. (1978) Adherence of *Candida albicans* to human buccal epithelial cells. *Infect. Immun.* **21**: 64–68.

King, R. D., Lee, J. C. and Morris, A. L. (1980) Adherence of *Candida albicans* and other *Candida* species to mucosal epithelial cells. *Infect. Immun.* **27**: 667–674.

Lachke, S. A., Joly, S., Daniels, K. and Soll, D. R. (2002) Phenotypic switching and filamentation in *Candida glabrata*. *Microbiol.* **148**: 2661–2674.

Lee, K. L., Buckley, H. R. and Campbell, C. C. (1975) An amino acid liquid synthetic medium for development of mycelial and yeast forms of *Candida albicans*. *Sabouraudia* **13**: 148–153.

Lowry, O. H., Rosebrugh, N. J., Farr, A. L. and Randall, R. J. (1951) Protein measurement with the folin phenol reagent. *J. Biol. Chem.* **193**: 265–275.

MacFarlane, M. G. and Knight, B. C. J. G. (1941) The biochemistry of bacterial toxins. I: Lecithinase activity of *Cl. welchii* toxins. *Biochem. J.* **35**: 884–902.

Maisch, P. A. and Calderone, R. A. (1980) Adherence of *Candida albicans* to a fibrin-platelet matrix formed *in vitro*. *Infect. Immun.* **27**: 650–656.

Martinez, J. P., Gil, M. L., Lopez-Ribot, J. L. and Chaffin, W. L. (1998) Serologic response to cell wall mannoproteins and proteins of *Candida albicans*. *Clin. Microbiol. Rev.* **11**: 121–141.

McCreight, M. C. and Warnock, D. W. (1982) Enhanced differentiation of isolates of *Candida albicans* using a modified resistogram method. *Mykosen* **25**: 589–598.

Mercure, S., Senechal, S., Auger, P., Lemay, G. and Montplaisir, S. (1996) *Candida albicans* serotype analysis by flow cytometry. *J. Clin. Microbiol.* **34**: 2106–2112.

Miyakawa, Y., Kagaya, K., Kukazawa, Y. and Soe, G. (1986) Production and characterization of agglutinating monoclonal antibodies against predominant antigenic factors for *Candida albicans*. *J. Clin. Microbiol.* **23**: 881–886.

Müller, H. I. and Kirchoff, G. (1969) Serologische type von *Candida albicans*. *Zent. Bakteriol. Parasit. Infek. Hygiene.* **210**: 114–121.

Na, B-K., Lee, S., Kim, S-O., Park, Y-K. *et al.* (1997) Purification and characterization of extracellular aspartic proteinase of *Candida albicans*. *J. Microbiol.* **35**: 109–116.

Naglik, J. R., Challacombe, S. J. and Hube B. (2003) *Candida albicans* secreted aspartyl proteinases in virulence and pathogenesis. *Microbiol. Molec. Biol. Rev.* **67**: 400–428.

Nakamura, K., Ito-Kuwa, S., Nakamura Y., Aoki, S. *et al.* (1998) Resistogram typing of oral *Candida albicans* isolates from normal subjects in three successive trials. *Rev. Iberoam. Micol.* **15**: 19–21.

Odds, F. C. (1998) *Candida and Candidosis* (2nd edn.), Bailliere Tindall, London.

Odds, F. and Abbott, A. (1980) A simple system for presumptive identification of *Candida albicans* and differentiation of strains within the species. *Sabouraudia* **18**: 301–307.

Pomes, R., Gil, C. and Nombela, C. (1985) Genetic analysis of *Candida albicans* morphological mutants. *J. Gen. Microbiol.* **131**: 2107–2113.

Price, M. F., Wilkinson, I. D. and Gentry, L. O. (1982) Plate method for detection of phospholipase activity in *Candida albicans*. *Sabouraudia* **20**: 7–14.

Pugh, D. and Cawson, R. A. (1975) The cytochemical localization of phospholipase A and lysophospholipase in *Candida albicans*. *Sabouraudia* **13**: 110–115.

Pugh, D. and Cawson, R. A. (1977) The cytochemical localization of phospholipase in *Candida albicans* infecting the chick chorio-allantoic membrane. *Sabouraudia* **15**: 29–35.

Rotrosen, D., Gibson, T. R. and Edwards J. E. Jr. (1983) Adherence of *Candida* species to intravenous catheters. *J. Infect. Dis.* **147**: 594.

Samaranayake L. P. and MacFarlane, T. W. (1980) An *in vitro* study of the adherence of *Candida albicans* to acrylic surfaces. *Archs. Oral Biol.* **25**: 603–609.

Samaranayake L. P. and MacFarlane, T. W. (1981) The adhesion of the yeast *Candida albicans* to epithelial cells of human origin *in vitro*. *Archs. Oral Biol.* **26**: 815–820.

Shinoda, T., Kaufman, L. and Padhye, A. A. (1981) Comparative evaluation of the Iatron serological *Candida* Check kit and the API 20C kit for identification of medically important *Candida* species. *J. Clin. Microbiol.* **13**: 513–518.

Slutsky, B., Buffo, J. and Soll, D. R. (1985) High-frequency switching of colony morphology in *Candida albicans*. *Science* **230**: 666–669.

Sobel, J. D., Myers, P., Levison, M. E. and Kaye, D. (1982) Comparison of bacterial and fungal adherence to vaginal exfoliated epithelial cells and human vaginal epithelial tissue culture cells. *Infect. Immun.* **35**: 697–701.

Soll, D. R. (1992) High-frequency switching in *Candida albicans*. *Clin. Microbiol. Rev.* **5**: 183–203.

Soll, D. R., Lockhart, S. R. and Zhao, R. (2003) Relationship between switching and mating in *Candida albicans*. *Eukaryotic Cell.* **2**: 390–397.

Staib, F. (1965) Serum-proteins as nitrogen source for yeast-like fungi. *Sabouraudia* **4**: 187–193.

Stiller, R. L., Bennett, J. E., Scholer, H. J., Wall. M. *et al.* (1982) Susceptibility to 5-fluorocytosine and prevalence of serotype in 402 *Candida albicans* isolates from the United States. *Antimicrob. Agents Chemother.* **22**: 482–487.

Sturtevant, J. and Calderone, R. (1997) *Candida albicans* adhesions: Biochemical aspects and virulence. *Rev. Iberoam. Micol.* **14**: 90–97.

Suegara N., Siegel, J. E. and Savage D. C. (1979) Ecological determinants in microbial colonization of the murine gastrointestinal tract: Adherence of *Torulopsis pintolopesii* to epithelial surfaces. *Infect. Immun.* **25**: 139–145.

Sullivan, D., Moran, G. and Coleman, D. (2005) Fungal diseases of humans. In: *Fungi: Biology and Applications*, (ed. K. Kavanagh), John Wiley & Sons, London, pp 171–190.

Werner, H. (1966) Untersuchungen uber die lipase-Aktivitate bei Hefen und Hefeahnlichen Pilzen. *Zentbl. Bakteriol. Micrrobiol. Hyg. I Abt. Orig. A.* **200**: 113.

Yoon, S. A., Vazquez, J. A., Steffan, P. E., Soll, J. D. and Akins, R. A. (1999) High-frequency, *in vitro* reversible switching of *Candida lusitaniae* clinical isolates from amphotericin B susceptibility to resistance. *Antimicrob. Agents Chemother.* **43**: 836–845.

5

Analysis of drug resistance in pathogenic fungi

Gary P. Moran, Emmanuelle Pinjon, David C. Coleman and **Derek J. Sullivan**

5.1 Introduction

The prevalence of yeast and fungal infections has increased dramatically during the last two decades, mainly due to the AIDS pandemic and the increased use of invasive and immunosuppressive medical procedures (Fridkin and Jarvis, 1996; VandenBergh *et al.*, 1999; Kaplan *et al.*, 2000). The increased diagnosis of fungal infection has been followed by more widespread use of antifungal agents both therapeutically and prophylactically. In the early 1990s, a new generation of azole antifungal drugs, the triazoles, was introduced to combat the rising incidence of fungal infection (Saag and Dismukes, 1988). In particular, the triazole drug fluconazole changed the face of antifungal chemotherapy as it is well tolerated, relatively non-toxic and can penetrate most tissues (Pfaller and Diekema, 2004). However, since the introduction and widespread use of fluconazole for the treatment of *Candida* infections, there have been increasing numbers of reports of clinical resistance to azole antifungal drugs, particularly in HIV-infected patients with oral candidosis (Willocks *et al.*, 1991; Johnson *et al.*, 1995; Xu *et al.*, 2000). The increasing frequency of reports of therapeutic failure with antifungal agents during the 1990s identified an important need for the development of new methods for accurately measuring antifungal drug susceptibility. This culminated in the publication of the National Committee for Clinical Laboratory Standards (NCCLS, now known as the Clinical and Laboratory Standards Institute (CLSI)) Protocol M27-A to allow conformity in the antifungal susceptibility testing of yeasts around the world. A protocol for susceptibility testing moulds and filamentous fungi (M38-A) has also recently been approved by the CLSI.

Medical Mycology: Cellular and Molecular Techniques. Edited by Kevin Kavanagh

It is now widely accepted that resistance to fluconazole is relatively common in *Candida* species. Since its introduction, there have been many reports concerning the emergence of species such as *Candida krusei* and *Candida glabrata* which exhibit intrinsic or primary resistance to this agent (Powderly, 1992; Krcmery and Barnes, 2002; Sanguinetti *et al.*, 2005). There have also been reports of developed, or secondary, resistance in isolates of normally susceptible species such as *C. albicans* and *Candida dubliniensis* and reports of cross-resistance to other azole drugs such as ketoconazole and itraconazole (Moran *et al.*, 1997; Muller *et al.*, 2000). Azole resistance is not limited to *Candida* yeasts. Fluconazole resistance has been reported in isolates of *Cryptococcus neoformans* recovered from AIDS patients with recurrent cryptococcal meningitis (Sullivan *et al.*, 1996). Resistance to azoles has also been reported in filamentous fungi of agricultural and clinical significance. *Aspergillus fumigatus*, the most important of the medically significant filamentous fungi, although inherently resistant to fluconazole, is susceptible to itraconazole. However, secondary resistance to itraconazole has been reported in isolates of *A. fumigatus* (Mosquera and Denning, 2002). In order to improve therapy of infections caused by azole-resistant fungi, new azole drugs have been developed with a wider spectrum of antifungal activity. Three of these new azoles, ravuconazole, voriconazole and posaconazole have been shown to have activity against fungi resistant to fluconazole (Pfaller *et al.*, 2003a).

Resistance to other classes of antifungal agents has also been reported. Flucytosine (5-fluorocytosine), which shows activity against *Candida* spp., is now only used in combination with other drugs, such is the incidence of primary resistance to this agent (Hope *et al.*, 2004). In *C. albicans*, resistance to flucytosine is mainly associated with a genetic subgroup of the species identified by DNA fingerprint analysis with the species-specific Ca3 DNA-fingerprinting probe. Over 70% of isolates whose Ca3 fingerprint profiles are grouped in *C. albicans* clade I are flucytosine-resistant, whereas the incidence of resistance outside clade I is only 2% (Dodgson *et al.*, 2004). A similar association with flucytosine resistance and DNA fingerprint group has recently been shown in *C. dubliniensis* (Al Mosaid *et al.*, 2005). In contrast to flucytosine, the polyene antibiotic amphotericin B has been in use for over fifty years, yet the incidence of resistance to this agent has remained low (Leenders *et al.*, 1996). However, a small number of species, such as *A. terreus* and *A. nidulans*, may be intrinsically resistant, while other species such as *C. lusitaniae* have a significant tendency towards secondary resistance (White *et al.*, 1998; Moore *et al.*, 2000). As well as the newer azole drugs, a new class of echinocandin antifungal drug has become available. Caspofungin is the first of these to come into clinical practice; however, this drug has not been in use long enough to determine whether resistance will become a clinical problem (Gil-Lamaignere and Muller, 2004).

The molecular biology of antifungal-agent action and resistance has been intensely studied. The azole drugs are generally fungistatic and their major mode of action is the inhibition of the cytochrome P-450 enzyme lanosterol demethylase, that is involved in the production of ergosterol, the main sterol component of fungal membranes (White *et al.*, 1998). Several studies have described point mutations in this enzyme which reduce the affinity of the active site for azole drugs (White, 1997; Sanglard *et al.*, 1998). Point mutations in genes encoding lanosterol demethylase have been described in *Candida* species (encoded by the *ERG11* gene), *A. fumigatus* (encoded by *cyp51A*) and *C. neoformans* (White *et al.*, 1998; Diaz-Guerra *et al.*, 2003; Rodero *et al.*, 2003; Mellado *et al.*, 2004). Azole resistance can also develop through the activity of multidrug transport systems, including ABC-transporters and major facilitator transporters. Increased expression of these proteins has been well documented during the development of azole resistance in *C. albicans*, *C. glabrata* and *C. dubliniensis* (Sanglard *et al.*, 1995. 1999; Moran *et al.*, 1998; White *et al.*, 1998). At least two ABC-transporters, Cdr1p and Cdr2p, encoded by the genes *CDR1* and *CDR2*, can be involved in azole resistance in *C. albicans.* These membrane proteins act as general transporters of small hydrophobic molecules including azole drugs and actively remove them from the cell envelope to the external medium in an ATP-dependent process. Similarly, Mdr1p, a protein of the major facilitator class encoded by the *MDR1* gene, can efflux fluconazole and other toxic molecules by utilizing the electrochemical proton gradient of the cell membrane. Azole resistance in *C. albicans* is often multifactorial, sometimes involving more than one pump in conjunction with point mutations in the *ERG11* gene (White *et al.*, 1998). Increased expression of ABC-transporter-encoding genes has also been implicated as a mechanism of itraconazole resistance in *A. fumigatus* (Nascimento *et al.*, 2003). As occurs in *Candida* species, this can also occur in conjunction with point mutations in the *Cyp51A* gene encoding lanosterol demethylase (Nascimento *et al.*, 2003).

Unlike the azoles, amphotericin B is fungicidal in action. The mode of action of this drug is complex and it is believed to bind to ergosterol in the cell membrane resulting in the formation of pores in the cell envelope, leakage of cellular components and severe oxidative damage. Resistance to amphotericin B is poorly understood and most resistant fungi are believed to harbour nonsense mutations in the gene encoding the sterol C5,6 desaturase enzyme. Inactivation of this enzyme in the ergosterol biosynthesis pathway prevents the formation of ergosterol but leads to the accumulation of ergosterol precursors, which can replace ergosterol in the membrane and permit growth. These mutants are also resistant to azole drugs as inactivation of the sterol C5,6 desaturase circumvents azole-mediated growth arrest.

Resistance to flucytosine is very well understood in *Candida* species. Flucytosine enters the cell by active transport and is then metabolised via the pyrimidine salvage pathway to produce toxic nucleotides that inhibit DNA and protein synthesis. Within the yeast cell, 5-fluorocytosine is converted to 5-fluorouracil by the enzyme cytosine deaminase (encoded by *FCA1)* and is further metabolized by uracil phosphoribosyltransferase (encoded by *FUR1*) (White *et al.*, 1998). In *C. albicans*, a single point mutation in the *FUR1* gene leading to an arginine to cysteine substitution at position 101 seems to be the most common mechanism of resistance to flucytosine (Hope *et al.*, 2004). Heterozygous mutants are more resistant than homozygous mutants. Mutations in *FCA1* and alterations in flucytosine uptake may also be involved in resistance (Dodgson *et al.*, 2004; Hope *et al.*, 2004).

Caspofungin is the first antifungal drug in the echinocandin class to be introduced into clinical practice. Caspofungin represents a huge breakthrough in antifungal chemotherapy as it allows clinicians to prescribe a new class of drugs with a novel mode of action that is effective against azole-resistant organisms. Caspofungin inhibits the production of glucan, one of the major components of the fungal cell wall, by inhibiting the enzyme 1,3-beta-D-glucan synthase, encoded by *FKS1* (Groll *et al.*, 1998). Caspofungin is effective in many cases against fluconazole-resistant yeasts, but has not been in clinical use for sufficient time to assess if resistance will be a significant clinical problem (Pfaller *et al.*, 2003b).

The introduction of new antifungal agents and the continued emergence of novel pathogenic fungi emphasize the need for continued surveillance of fungal drug susceptibility. In this chapter, we present some basic methods for analysing the susceptibility of pathogenic yeasts to antifungal agents. Although these are essentially simple procedures, great care should be taken to ensure that the susceptibility tests performed are standardized in accordance with the methods used in other laboratories and that well-established control isolates are used to allow accurate comparisons to be made. Reliable and reproducible susceptibility testing procedures are the cornerstone for success of any study of antifungal resistance, whether in clinical surveillance projects or projects involving the molecular biology of antifungal action.

5.2 Method for the determination of minimum inhibitory concentrations (MICs) of antifungal agents for yeasts

In the 1990s, the CLSI began a rigorous assessment of antifungal susceptibility testing methodologies in order to establish a standardized method that would allow interlaboratory comparison of MICs and the establishment of defined breakpoints for resistance. This culminated in the establishment of the M27-A procedure for the susceptibility testing of fermentative yeasts such as *Candida* spp. to antifungal drugs (Canton *et al.*, 1999). The CLSI have recently approved methods for the susceptibility testing of conidia-forming filamentous fungi, including *Aspergillus* spp. (Protocol M38-A) (Espinel-Ingroff *et al.*, 2002). The method presented here is a modified procedure developed by the European Committee for Antifungal Drug Susceptibility Testing (EUCAST), which utilizes a modified glucose-rich medium to improve growth and shorten incubations times (Cuenca-Estrella *et al.*, 2003). The method also entails the use of a 96-well format that allows for the spectrophotometric analysis of results. The method is especially useful for testing susceptibility to azoles and flucytosine. For amphotericin B susceptibility testing, a different growth medium is recommended (See Note on amphotericin B susceptibility testing on p. 101). The basic method yields minimum inhibitory concentrations (MICs) for antifungal agents, but can also be used to determine minimum fungicidal concentrations (MFCs).

Protocol 5.1
Method for the determination of minimum inhibitory concentrations (MICs) of antifungal agents for yeasts

Equipment, materials and reagents

Yeast strains to be analysed (24 to 48 h-old plate cultures)

Reference strain of known antifungal susceptibility

0.5 McFarland standard

Antifungal powder

Dimethyl sulphoxide

Sterile distilled water

RPMI 1640 powdered medium (with glutamine and pH indicator, without bicarbonate)

Morpholinopropanesulfonic acid

Glucose

Flat-bottomed, clear-plastic, 96-well plates

Spectrophotometric plate reader (recommended)

Method

1. Prepare the 2 x RPMI-2% glucose growth medium as described in Table 5.1. Add the components to 900 ml water, dissolve by stirring and adjust to pH 7.0 with 10 M NaOH. Adjust the volume to 1 litre and filter sterilize with a 0.22 μm filter. Note that 2 x RPMI-2% glucose can be stored at 4 °C for several weeks.

2. Prepare stocks of antifungal powders. Powders should ideally be obtained directly from the manufacturer or reliable commercial sources. Fluconazole and flucytosine can be dissolved in water, whereas most other antifungal drugs require DMSO as a solvent at the concentrations recommended in Table 5.2. Drugs should be stored in aliquots at −70 °C for no longer than 6 months.

Table 5.1 Recommended growth medium for antifungal susceptibility tests

		Per litre	
		(1 x concentration)	(2 x concentration)
RPMI-2%G	RPMI-1640 powder	10.40 gram	20.80 gram
	MOPS[a]	34.53 gram	69.06 gram
	Glucose	18.00 gram	36.00 gram
Buffer to pH 7.0 with 10 M NaOH			
AM3-2%G	AM3 powder (Difco)[b]	17.50 gram	35.00 gram
	Glucose	19.00 gram	38.00 gram
Buffer to pH 7.0 with 10 mM phosphate			

[a]3-(N-morpholino)propanesulfonic acid (0.165 M)
[b]Some investigators have reported differences in performance of AM3 from different manufacturers and different lots. In our laboratory, Difco™ medium (BD, Franklin Lakes, NJ, USA) is used and the above recipe is based on their instructions

3. Upon thawing, a working stock of drug should be prepared in 2 x assay medium. The concentration of this stock should be twice that of the highest concentration in the assay range (128 μg/ml for fluconazole and flucytosine, 16 μg/ml for other azole and amphotericin B). Unused drug should be discarded at the end of each day.

Table 5.2 Antifungal drugs: stock solutions, assay medium and assay endpoints

Drug	Stock μg/ml	Solvent[a]	Assay range μg/ml	Assay medium[b]	Endpoint[c]
Fluconazole	5000	10% (v/v) DMSO	0.125–64	2 x RPMI-2%G	50%
Flucytosine	5000	Water	0.125–64	2 x RPMI-2%G	50%
Itraconazole	1000	DMSO	0.015–8	2 x RPMI-2%G	50%
Voriconazole	1000	DMSO	0.015–8	2 x RPMI-2%G	50%
Amphotericin B	1000	DMSO	0.015–8	2 x AM3-2%G	100%

[a]DMSO: dimethylsulfoxide
[b]2 x RPMI-2% G containing 0.165 M Morpholinopropane sulfonic acid pH 7.0 and antibiotic medium 3–2% pH 7.0 (See Table 5.1)
[c]Endpoint refers to the percentage reduction in growth relative to the growth control at which the MIC for that drug is defined

4. In transparent 96-well flat-bottomed plates, 200 μl of this drug solution should be added to wells in column 1 (Figure 5.1). Columns 2 to 12 should be dispensed with 100 μl of the assay medium.

5. In order to prepare a concentration gradient of drug, 100 μl of the solution in column 1 is transferred to column 2 and mixed, thereby diluting the drug two-fold.

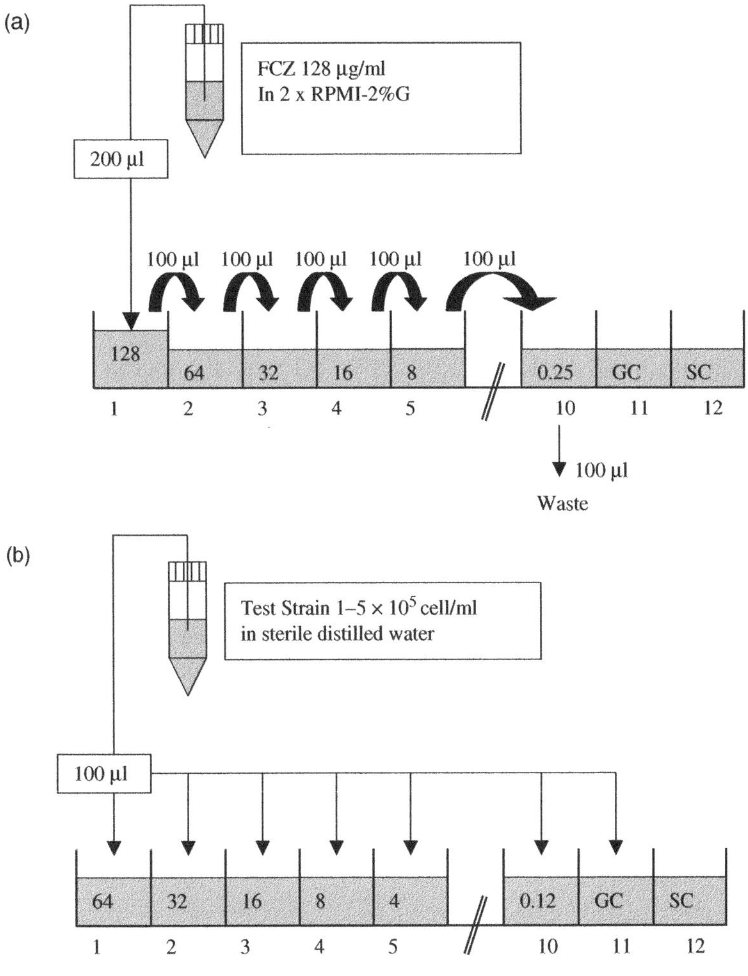

Figure 5.1 Outline of the dilution scheme in protocol 5.1 for preparing 96-well dishes for fluconazole susceptibility testing. (a) First, 200 μl of drug at twice the highest assay concentration is added to column 1 (for fluconazole, 128 μg/ml). Wells in columns 2–12 are dispensed with 100 μl of drug-free 2 x RPMI-2% glucose. From column 1, 100 μl of the 128 μg/ml solution should be serially diluted from column 2 to column 10 to yield a two-fold range of concentrations from 128-0.25 μg/ml. Wells 11 and 12 are kept drug-free to act as growth controls (GC) and sterility controls (SC). (b) Each well in columns 1–11 is dispensed with 100 μl of cell suspension to yield a final cell density of 0.5–2.5 × 10^5 cells/ml in a range of fluconazole concentrations from 64 to 0.125 μg/ml

Continue this series of two-fold dilutions from column 2 through to column 10 (Figure 5.1). Remove 100 μl from column 10 to leave 100 μl of drug solution. Wells 11 and 12 are kept drug-free. Plates prepared in this way can be sealed in plastic or foil and frozen at −70 °C for up to 6 months.

6. To prepare the inoculum, resuspend 5 yeast colonies of ∼1 mm diameter in 5 ml sterile distilled water. The suspension should be resuspended thoroughly by vortexing and the turbidity adjusted to that of a 0.5 McFarland standard by comparing the absorbance at 530 nm and adjusting the volume with sterile distilled water if necessary. The suspension should then be diluted 1/10 in sterile distilled water. This will yield a suspension of $1–5 \times 10^5$ CFU/ml.

7. For each strain to be tested, duplicate rows of wells should be inoculated, allowing four strains to be tested per plate. To wells 1 to 11, 100 μl of the suspension should be added to yield a final concentration of $0.5–2.5 \times 10^5$ CFU/ml. Column 11 will serve as a drug-free growth control. Column 12 should not be inoculated and acts as a sterility control.

8. The plate should be incubated at 35–37 °C for 24 h. Reading of the MIC can be done visually; however, spectrophotometric readings are more reliable. The optical density of each well should be measured in a plate reader at 540 nm (other wavelengths such as 405 nm or 450 nm can also be used). Column 12 should be used to blank measurements against the growth medium. The MIC is defined as the lowest concentration of drug to reduce growth by 50% or more compared to the drug-free growth control in well 11 (Figure 5.2). This value is termed the MIC or MIC-50% (Figure 5.2). If the strain has grown poorly after 24 h incubation (i.e. the absorbance at 540 nm < 0.5) the plate should be reincubated for a further 24 h. If the absorbance value is still <0.5, the test has failed and should be repeated. Interpretive breakpoints for antifungal drug resistance in *Candida* species are shown in Table 5.3.

9. Minimum fungicidal concentrations can be determined by resuspending the medium in each well and plating out 100 μl of this suspension onto a routine mycological medium (e.g. Sabouraud's dextrose agar). The lowest concentration of drug which yields no viable growth is deemed the minimum fungicidal concentration (MFC).

Note on quality control: Quality control procedures ensure consistency and reproducibility of results. The routine quality of test results are monitored by the use of control strains of known susceptibility. If no such strains exist in the laboratory, suitable controls strains should be acquired from a reliable source such as the American Type Culture Collection (ATCC) or other international culture collections. Control strains should ideally be included in each batch of tests and should be used to assess each new batch of media or drug stocks. The CLSI

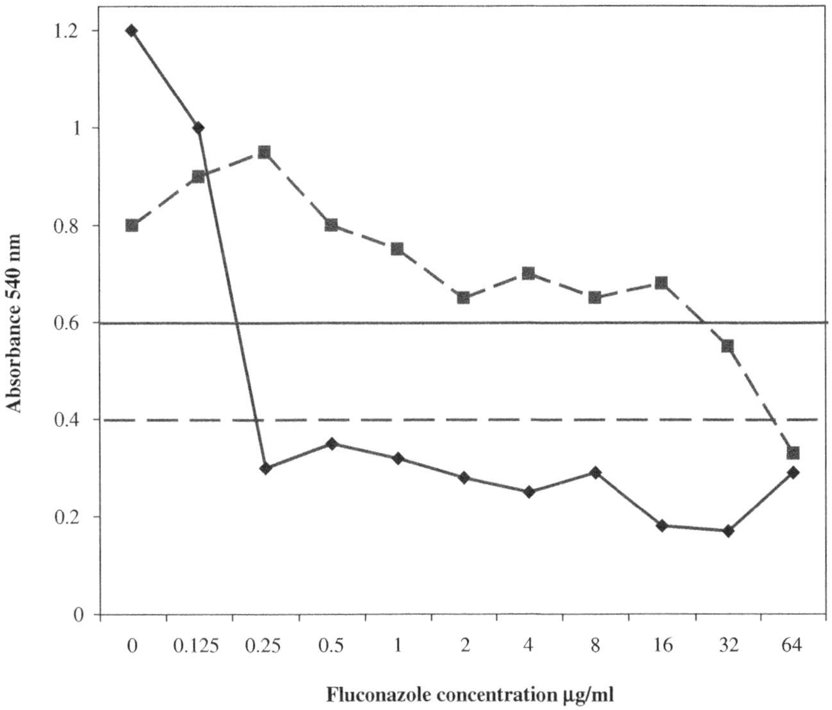

Figure 5.2 Typical dose responsive curves to fluconazole for fluconazole-susceptible (solid line with diamonds) and -resistant (broken line with squares) *C. albicans* isolates. Growth at each fluconazole concentration was determined by measuring absorbance at 540 nm using the MIC method in Protocol 5.1. The MIC-50% for each isolate is the lowest fluconazole concentration to reduce growth to below 50% of that in the in drug free medium (0 concentration). The MIC-50% cut off for each isolate is represented by the horizontal lines. The MIC-50% of the fluconazole susceptible isolate (solid straight line) is 0.25 µg/ml whereas the resistant isolate (dashed straight line) has an MIC-50% of 64 µg/ml

recommend several strains such as *Candida krusei* ATCC 6258 and *Candida parapsilosis* ATCC 22019.

Note on amphotericin B susceptibility testing: Amphotericin B susceptibility testing offers a novel challenge as studies have shown that RPMI-medium-based assays are unable to detect resistance. The above protocol can be modified to detect resistance by substituting 2 x RPMI-2% glucose with 2 x Antibiotic medium 3

Table 5.3 Interpretive breakpoints for antifungal drug resistance (µg/ml)

Drug	Susceptible	Susceptible-dose dependant	Resistant
Fluconazole	≤8	16–32	≥64
Itraconazole	≤0.125	0.25–0.5	≥1
Flucytosine	≤4	n/a	≥32

supplemented with 2% w/v glucose, pH 7. Amphotericin B should be diluted in an 8 to 0.015 μg/ml range (Rex *et al.*, 1995). The inoculum should also be diluted in sterile distilled water so that the final concentration of organism in each well is $0.5–2.5 \times 10^3$ CFU/ml. As amphotericin B is fungicidal, the MIC is taken as the lowest concentration of drug to completely inhibit the growth of the organism (MIC-100%).

5.3 Measurement of rhodamine 6G uptake and glucose-induced efflux by ABC transporters

Reduced accumulation of drug due to increased efflux is one of the major mechanisms of azole resistance in fungi. Two types of efflux transporters can contribute to azole resistance in fungal cells: the ABC transporters (e.g. *C. albicans* Cdr1p and Cdr2p) and major facilitator proteins (e.g. *C. albicans* Mdr1p). The following protocol is a modification of the method of Maesaki *et al.* (1999), which aims to detect increased energy-dependent efflux in yeast. Using the fluorescent compound Rhodamine 6G (R6G), this method aims to detect the upregulation of the ABC transporter genes *CDR1* and *CDR2* since Rhodamine 6G is a substrate of the Cdr1p and Cdr2p multidrug transporters but not of the MF transporter Mdr1p. Most methods used to investigate the activity of multidrug transporters use the intracellular accumulation of a compound as an indirect measurement of efflux and do not take into account the contribution of drug import by diffusion. In contrast, the following method uses energy-starved cells to study efflux directly in order to minimize the contribution of drug import. Energy-starved cells are incubated in the presence of R6G, which rapidly diffuses into the cells. Efflux is then initiated by energy supplementation (addition of glucose). Efflux is measured by comparing the concentration of R6G in the medium of cells supplemented with glucose to that present in the medium of control cells (without any addition of glucose).

Protocol 5.2 Measurement of rhodamine 6G uptake and glucose-induced efflux

Equipment, materials and reagents

YPD broth

Phosphate-buffered saline (PBS) buffer

Rhodamine 6G (10 mM stock)

Glucose (0.1 M stock)

250 ml culture flasks

Haemocytometer

50 ml conical polypropylene tubes (e.g. Falcon tubes, BD Biosciences)

1.5 ml microfuge tubes

Cuvettes for spectrophotometer

Shaking incubator

Benchtop centrifuge

Microcentrifuge

Spectrophotometer

Graph paper

Method

1. Grow yeast cells overnight in YPD broth in culture flasks at 37 °C with shaking (Fig. 5.3).

2. Count cells using a haemocytometer and transfer a total of 1×10^8 cells to 100 ml fresh YPD broth in a culture flask.

3. Incubate for 4 to 5 h at 37 °C with shaking.

4. Harvest the cells by centrifugation at 3000 x *g* for 5 min in 50 ml tubes.

5. Wash the pellets twice in PBS with centrifugation as before.

6. Resuspend pellets in PBS at a density of 1×10^8 cells/ml in 10 ml PBS.

7. Incubate for 1 h at 37 °C with shaking.

8. Add 10 µl of R6G stock (final concentration: 10 µM).

9. Incubate at 37 °C for 30 min with shaking.

10. Transfer 1 ml of cell suspension to microfuge tube. Spin at 10 000 x *g* for 1 min. Transfer supernatant to a fresh tube and keep for later use.

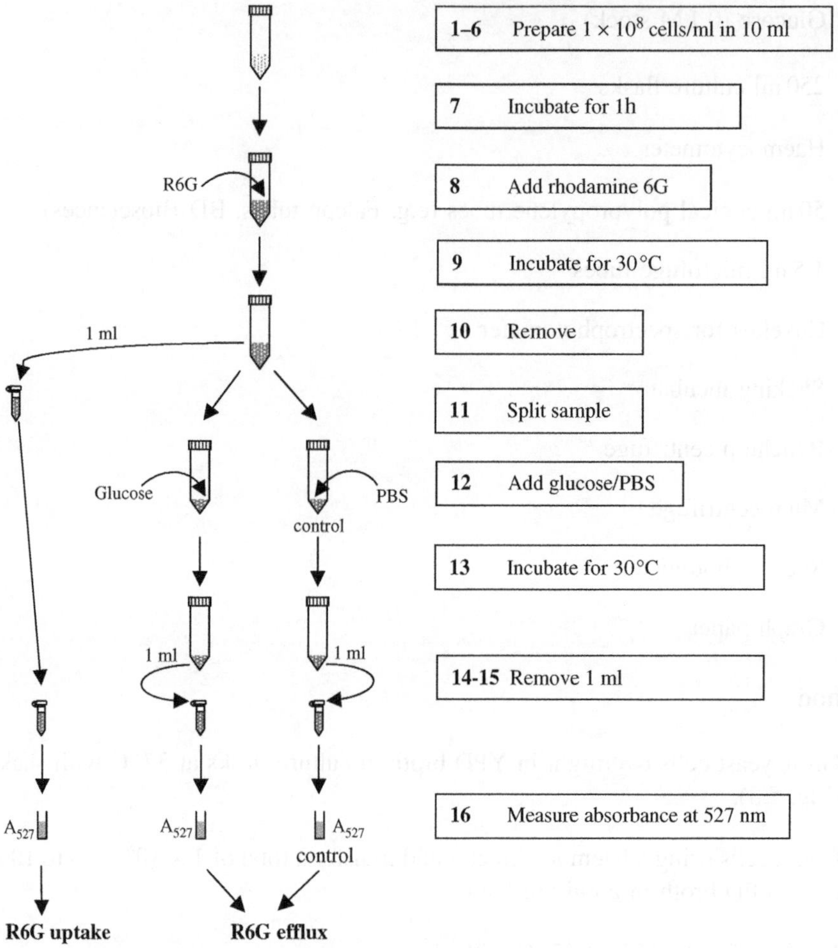

Figure 5.3 Protocol for the measurement of rhodamine 6G uptake and glucose induced efflux

11. Split the remaining 9 ml of cell suspension into 2 fresh 50 ml tubes.

12. Add 500 μl PBS to the control tube and add 500 μl glucose stock (to a final concentration of 10 mM) to the other Falcon tube.

13. Incubate at 37 °C for a further 30 min.

14. Remove 1 ml from each Falcon tube and transfer to microfuge tube. Spin at 10 000 x *g* for 1 min.

15. Immediately transfer supernatants to fresh microfuge tubes or cuvettes.

16. Measure the absorbance at 527 nm of the three supernatants.

17. Calculate the concentration of R6G using a standard concentration curve of R6G. This can be generated by measuring the absorbance at 527 nm of known concentrations of R6G (0-5 μM) diluted in PBS buffer.

18. Uptake of R6G (before glucose supplementation) is calculated by substracting the value of R6G extracellular concentration after 30 min incubation in glucose-free PBS (step 10) from the initial value of R6G concentration added to the medium. Glucose-induced efflux of R6G is calculated by subtracting the value of R6G extracellular concentration in the absence of glucose (control sample) from the value of R6G extracellular concentration of the glucose-supplemented sample.

5.4 Analysis of expression of multidrug transporters in pathogenic fungi

Reduced accumulation of azole drugs due to energy-dependent efflux is a common mechanism of resistance in pathogenic fungi. As described in section 5.3, Rhodamine 6G is an easily detected substrate for ABC-transporters that can be used to measure the activity of some efflux mechanisms. Increased levels of efflux in resistant cells is in many cases associated with increased expression of genes encoding ABC-transporters or major-facilitator transporters. In *C. albicans*, deletion of these genes in resistant strains has directly shown that they are responsible for the drug-resistance phenotype. Increased expression of transporter encoding genes can be detected in several ways, including reverse-transcriptase PCR analysis, Real-time PCR analysis or Northern blot analysis. Itraconazole resistance in *A. fumigatus* has been associated with increased expression of the transporters *Afumdr3*, *Afumdr4* and *AtrF* (Sanglard *et al.*, 1995, Nascimento *et al.*, 2003). In *Candida* spp. fluconazole resistance is associated with increased expression of the ABC-transporters *CDR1* and *CDR2* and the major-facilitator *MDR1*. Northern blot analysis is probably the most commonly used method to measure the relative expression levels of drug transporter-encoding mRNAs in fungi. Northern blot analysis is described in detail in Chapter 8 of this book. However, the following details should be considered when measuring drug transporter mRNA expression levels:

1. Expression levels of drug transporter-encoding genes in resistant strains should, if at all possible, be compared to those in isogenic drug susceptible strains. Laboratory-derived drug-resistant mutants should therefore be compared to their drug-susceptible parental strains. Drug-resistant clinical isolates should ideally be compared to drug susceptible strains that have been recovered from the same patient and have been demonstrated by DNA fingerprint analysis to be clonally related.

2. Expression levels of drug transporters should be measured *in vitro* with cells grown in the presence and absence of the antifungal agent in question. This will indicate whether expression of the transporter is inducible by the drug.

3. Probes for Northern blot hybridization should be carefully designed if they are to be specific for the gene of interest. ABC-transporter encoding genes from the same species can share >80% identity at the nucleotide sequence level. For example, in the case of *C. albicans*, the *CDR1* and *CDR2* genes are highly homologous and most probes derived from their sequences will hybridize to both mRNA molecules, even at high stringencies. However, Sanglard *et al.* (1997) show that probes derived from first ~250 bp of the coding sequence exhibit enough divergence to be able to discriminate between the two sequences. Alternatively, some authors have used generic CDR probes to detect changes in expression of both genes simultaneously.

4. Always normalize expression levels of transporter encoding genes to that of a suitable housekeeping gene (e.g. *ACT1*, *TEF3*) before comparing expression levels between strains.

5.5 Analysis of point mutations in genes encoding cytochrome P-450 lanosterol demethylase

A major mechanism of resistance to azole antifungal drugs in pathogenic fungi is the acquisition of point mutations in the protein encoding the target of azole drugs, the cytochrome P-450 lanosterol demethylase. Mutations affecting the binding of azole drugs to the enzyme have been described in *Candida* spp., *C. neoformans* and *A. fumigatus* (Sanglard *et al.*, 1998; Rodero *et al.*, 2003; Mellado *et al.*, 2004). These mutations mediate resistance without significantly affecting the catalytic activity of the enzyme. PCR amplification is commonly used to isolate the *ERG11* genes from resistant strains of *Candida* spp. using high-fidelity thermostable polymerases such as *Pfu* (Promega Corp., Madison, Wisconsin, USA; Table 5.4).

Similarly, the *Cyp51A* and *Cyp51B* genes can be amplified from *A. fumigatus* (Diaz-Guerra *et al.*, 2003; Table 5.4). As with Northern blot analysis, it is preferable that genes amplified from resistant strains are compared directly with genes amplified from isogenic-susceptible strains (see section 5.4). Sequence analysis of the amplified genes will quickly determine if point mutations are present in the coding sequence. PCR-amplified genes can be sequenced directly or cloned into a suitable plasmid vector for analysis. If the amplimers are cloned into a plasmid vector prior to sequence analysis, it is recommended that several clones are sequenced as even high-fidelity thermostable polymerases can introduce errors. It is also worth noting that in diploid organisms some mutations may be heterozygous (White *et al.*, 1998).

In *C. albicans*, at least 83 point mutations have been identified in the *ERG11* gene. The substitutions R467K, G464S and Y132H have been identified in fluconazole-resistant

Table 5.4 Primers used to PCR amplify cytochrome P-450 lanosterol demethylase encoding genes from pathogenic fungi

Species	Gene	Primers	Notes	Reference
C. albicans	*CaERG11*	5′-GCGGATCCTTAA AACATACAAGTTTC TCTTTT-3′	*Bam*HI site (underlined)	Sanglard *et al., 1998*
		5′-ACGCGTCGACAATA TGGCTATTGTTGAC-3′	*Sal*I site (underlined)	
C. dubliniensis	*CdERG11*	5′-GCTCGAGCGCAATAT GGCTATTGTTGAAAC TGTC-3′	*Xho*I site (underlined)	Pinjon *et al.*, 2003
		5′-GCTCTAGAGCTTAA AACATACAAGTTTCT CTTTT-3′	*Xba*I site (underlined)	
C. glabrata	*CgERG11*	5′-ATGTCCACTGAAAACA CTTCTTTGG-3′	–	Sanguinetti *et al.*, 2005
		5′-GTACTTTTGTTCTGGA TGTCTCTTTTC-3′	–	
A. fumigatus	*Cyp51A*	5′-ATGGTGCCGATGCTATGG-3′	–	Diaz-Guerra *et al.*, 2003
		5′-CTGTC-TCACTTGGATGTG-3′	–	
	Cyp51B	5′-ATGGGTCTCATCGCGTTC-3′	–	Diaz-Guerra *et al.*, 2003
		5′-TCAGGCTTTGGTAGCGG-3′	–	
C. neoformans	ERG11	5′-TCGTCGAACCATCTTTCG-3′	–	Rodero *et al.*, 2003
		5′-CGTCTATGACTTCATGACC-3′	–	

isolates by several authors (White *et al.*, 1998). In *A. fumigatus*, substitution of G54 or M220 of *Cyp51A* have been associated with itraconazole resistance (Diaz-Guerra *et al.*, 2003; Mellado *et al.*, 2004). The role of such point mutations in the resistance phenotype can only be fully assessed by a functional analysis of the mutant allele. This can be achieved by either purification and biochemical analysis of the mutant protein or by heterologous expression of wild-type and mutant forms of the protein in a drug-susceptible organism. The latter strategy has been used to characterize mutations in the *A. fumigatus Cyp51A* gene by expression of mutant alleles in a wild-type susceptible strain (Diaz-Guerra *et al.*, 2003; Mellado *et al.*, 2004). Introduction of mutant *Cyp51A* alleles into a wild-type *A. fumigatus* strain can render the transformed derivatives resistant to itraconazole. Similarly, Sanglard *et al.* (1998) have used *Saccharomyces cerevisiae* as a tool to characterize heterologously expressed mutant alleles of *ERG11* from *C. albicans* (Sanglard *et al.*, 1998). Unfortunately, a thorough description of these heterologous expression methodologies is beyond the scope of this chapter. However, the authors of these methods have described them in detail and the reader is urged to consult the original publications for further details.

5.6 Qualitative detection of alterations in membrane sterol contents

Alterations in the ergosterol biosynthetic pathway such as the inactivation of the C5,6-desaturase enzyme can lead to azole resistance by avoiding the accumulation of the toxic sterol metabolite, which normally accumulates in the membrane in the presence of azoles (Kelly *et al.*, 1995; Pinjon *et al.*, 2003). In these mutants, growth is supported by 14α-methylfecosterol and ergosterol is not synthesized. Since ergosterol is the target of amphotericin B, cross-resistance to amphotericin B can also be a consequence of these alterations. The following protocol is a rapid method (just over an hour) allowing qualitative detection of such alterations of membrane sterol contents. This method takes advantage of the unique spectral absorption profile produced by extracted sterols between 240 and 330 nm, which is indicative of the presence of ergosterol (and of another late sterol pathway intermediate, 24(28)dehydroergosterol). A typical wild-type profile shows peaks at 261, 273, 282 and 294 nm while the profile of a C5,6-desaturation defective mutant lacks these peaks. Figure 5.4 shows typical wild-type and mutant sterol profiles.

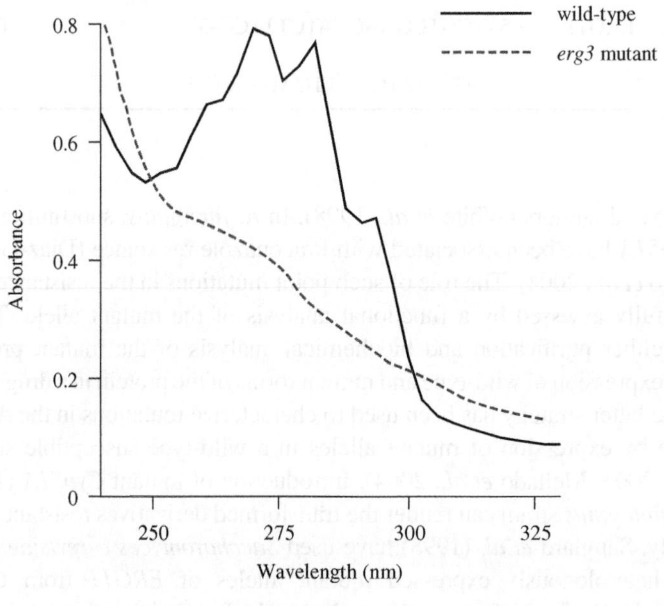

Figure 5.4 Typical UV spectrophotometric profiles of non-saponifiable sterols obtained from wild-type and *erg3* mutant *C. albicans* strains

Protocol 5.3
Qualitative detection of alterations in membrane sterol contents

Equipment, materials and reagents

YPD broth

Screw-cap microfuge tubes

20% (w/v) KOH, 60% (v/v) EtOH solution

Hexane

Quartz cuvettes for the spectrophotometer

Shaking incubator

Microcentrifuge

Heating block

Spectrophotometer with scanning capability

CAUTION! The KOH/EtOH solution is very corrosive. Wear adequate protective clothing and eye protection and use screw-cap microfuge tubes for the incubation step at 95 °C.

Method

1. Grow cells overnight in YPD broth at 37 °C with shaking.

2. Put 1 ml of overnight culture into a screw-cap microfuge tube and centrifuge for 1 min at top speed to pellet the cells.

3. Resuspend pellet in 0.5 ml of 20% (w/v) KOH, 60% (v/v) EtOH solution.

4. Incubate at 95 °C for 1 h (with shaking if possible, but not strictly necessary).

5. Following incubation, add 0.6 ml hexane and vortex for a few sec.

6. Centrifuge at 10 000 x *g* for 1 min.

7. Remove the top hexane layer, which contains the sterols, and transfer to a fresh tube.

8. Add 0.6 ml hexane and mix well.

9. Using a quartz cuvette, measure the absorption between 240 and 330 nm using the scanning function of the spectrophotometer (see Figure 5.4 for an example of a typical spectrum).

5.7 Overview

During the last 10 years, there have been dramatic improvements in our knowledge of antifungal agent mode of action and resistance. Although much of this work has been carried out in *Candida* species, our understanding of resistance mechanisms in other pathogenic fungi is now also improving. An important lesson from these studies is that antifungal drug resistance is multifactorial in nature, often involving mutation of target enzymes and active efflux mechanisms. However, a full understanding of antifungal resistance will require further study. Many researchers are now using the new 'omic' technologies such as transcriptomics and proteomics (both detailed elsewhere in this book) to study molecular mechanisms of antifungal resistance. These new technologies should greatly improve our knowledge of antifungal resistance in the coming years.

5.8 References

Al Mosaid, A., Sullivan, D. J., Polacheck, I., Shaheen, F. A. *et al.* (2005) Novel 5-Flucytosine-Resistant Clade of *Candida dubliniensis* from Saudi Arabia and Egypt identified by Cd25 fingerprinting. *J. Clin. Microbiol.* **43**: 4026–4036.

Canton, E., Peman, J., Carrillo-Munoz, A., Orero, A. *et al.* (1999) Fluconazole susceptibilities of bloodstream *Candida* sp. isolates as determined by National Committee for Clinical Laboratory Standards method M27-A and two other methods. *J. Clin. Microbiol.* **37**: 2197–2200.

Cuenca-Estrella, M., Moore, C. B., Barchiesi, F. *et al.* (2003) Multicenter evaluation of the reproducibility of the proposed antifungal susceptibility testing method for fermentative yeasts of the Antifungal Susceptibility Testing Subcommittee of the European Committee on Antimicrobial Susceptibility Testing (AFST-EUCAST). *Clin. Infect. Dis.* **9**: 467–474.

Diaz-Guerra, T. M., Mellado, E., Cuenta-Estrella, M. and Rodriguez-Tudela, J. L. (2003) A point mutation in the 14alpha-sterol demethylase gene *Cyp51A* contributes to itraconazole resistance in *Aspergillus fumigatus*. *Antimicrob. Agents Chemother.* **47**: 1120–1124.

Dodgson, A. R., Dodgson, K. J., Pujol, C., Pfaller, M. A. and Soll, D. R. (2004) Clade-specific flucytosine resistance is due to a single nucleotide change in the FUR1 gene of *Candida albicans*. *Antimicrob. Agents Chemother.* **48**: 2223–2227.

Espinel-Ingroff, A., Chaturvedi, V., Fothergill, A. W. and Rinaldi, M. G. (2002) Optimal testing conditions for determining MICs and minimum fungicidal concentrations of new and established antifungal agents for uncommon molds: NCCLS collaborative study. *J. Clin. Microbiol.* **40**: 3776–3781.

Fridkin, S. K. and Jarvis, W. R. (1996) Epidemiology of nosocomial fungal infections. *Clin. Microbiol. Rev.* **9**: 499–511.

Gil-Lamaignere, C. and Muller, F. M. (2004) Differential effects of the combination of caspofungin and terbinafine against *Candida albicans*, *Candida dubliniensis* and *Candida kefyr*. *Int. J. Antimicrob. Agents* **23**: 520–523.

Groll, A. H., De Lucca, A. J. and Walsh, T. J. (1998) Emerging targets for the development of novel antifungal therapeutics. *Trends Microbiol.* **6**: 117–124.

Hope, W. W., Tabernero, L., Denning, D. W. and Anderson, M. J. (2004) Molecular mechanisms of primary resistance to flucytosine in *Candida albicans*. *Antimicrob. Agents Chemother.* **48**: 4377–4386.

Johnson, E. M., Warnock, D. W., Luker, J., Porter, S. R. and Scully, C. (1995) Emergence of azole drug resistance in Candida species from HIV-infected patients receiving prolonged fluconazole therapy for oral candidosis. *Antimicrob. Agents Chemother.* **35**: 103–114.

Kaplan, J. E., Hanson, D., Dworkin, M. S., Frederick, T. *et al.* (2000) Epidemiology of human immunodeficiency virus-associated opportunistic infections in the United States in the era of highly active antiretroviral therapy. *Clin. Infect. Dis.* **30**(suppl 1), S5–14.

Kelly, S. L., Lamb, D. C., Corran, A. J., Baldwin, B. C. and Kelly, D. E. (1995) Mode of action and resistance to azole antifungals associated with the formation of 14a-methylergosta-8,24(28)-dien-3B,6a-diol *Biochem. Biophys. Res. Commun.* **207**: 910–915.

Krcmery, V. and Barnes, A. J. (2002) Non-albicans *Candida* spp. causing fungaemia: pathogenicity and antifungal resistance. *J. Hosp. Infect.* **50**: 243–260.

Leenders, A. C., de Marie, S., ten Kate, M. T., Bakker-Woudenberg, I. A. and Verbrugh, H. A. (1996) Liposomal amphotericin B (AmBisome) reduces dissemination of infection as compared with amphotericin B deoxycholate (Fungizone) in a rate model of pulmonary aspergillosis. *J. Antimicrob. Chemother.* **38**: 215–225.

Maesaki, S., Marichal, P., Vanden Bossche, H., Sanglard, D. and Kohno, S. (1999) Rhodamine 6G efflux for the detection of *CDR1*-overexpressing azole-resistant *Candida albicans* strains. *J. Antimicrob. Chemother.* **44**: 27–31.

Mellado, E., Garcia-Effron, G., Alcazar-Fuoli, L., Cuenta-Estrella, M. and Rodriguez-Tudela, J. L. (2004) Substitutions at methionine 220 in the 14alpha-sterol demethylase (*Cyp51A*) of *Aspergillus fumigatus* are responsible for resistance *in vitro* to azole antifungal drugs. *Antimicrob. Agents Chemother.* **48**: 2747–2750.

Moore, C. B., Sayers, N., Mosquera, J., Slaven, J. and Denning, D. W. (2000) Antifungal drug resistance in *Aspergillus*. *J. Infect.* **41**: 203–220.

Moran, G. P., Sullivan, D. J., Henman, M. C., McCreary, C. E. *et al.* (1997) Antifungal drug susceptibilities of oral *Candida dubliniensis* isolates from human immunodeficiency virus (HIV)-infected and non-HIV-infected subjects and generation of stable fluconazole-resistant derivatives *in vitro*. *Antimicrob. Agents Chemother.* **41**: 617–623.

Moran, G. P., Sanglard, D., Donnelly, S. M., Shanley, D. B. *et al.* (1998) Identification and expression of multidrug transporters responsible for fluconazole resistance in *Candida dubliniensis*. *Antimicrob. Agents Chemother.* **42**: 1819–1830.

Mosquera, J. and Denning, D. W. (2002) Azole cross-resistance in *Aspergillus fumigatus*. *Antimicrob. Agents Chemother.* **46**: 556–557.

Muller, F-M. C., Weig, M., Peter, J. and Walsh, T. J. (2000) Azole cross-resistance to ketoconazole, fluconazole, itraconazole and voriconazole in clinical *Candida albicans* isolates from HIV-infected children with oropharyngeal candidosis. *J. Antimicrob. Chemother.* **46**: 323–342.

Nascimento, A. M., Goldman, G. H., Park, S. *et al.* (2003) Multiple resistance mechanisms among *Aspergillus fumigatus* mutants with high-level resistance to itraconazole. *Antimicrob. Agents Chemother.* **47**: 1719–1726.

Pfaller, M. A., Diekema, D. J., Messer, S. A., Boyken, L. *et al.* (2003a) *In vitro* activities of voriconazole, posaconazole, and four licensed systemic antifungal agents against *Candida* species infrequently isolated from blood. *J. Clin. Microbiol.* **41**: 78–83.

Pfaller, M. A., Messer, S. A., Boyken, L., Rice, C. *et al.* (2003b) Caspofungin activity against clinical isolates of fluconazole-resistant *Candida. J. Clin. Microbiol.* **41**: 5729–5731.

Pfaller, M. A. and Diekema, D. J. (2004) Twelve years of fluconazole in clinical practice: Global trends in species distribution and fluconazole susceptibility of bloodstream isolates of *Candida. Clin. Microbiol. Infect.* **10** Suppl 1, 11–23.

Pinjon, E., Moran, G. P., Jackson, C. J., Kelly, S. L. *et al.* (2003) Molecular mechanisms of itraconazole resistance in *Candida dubliniensis. Antimicrob. Agents Chemother.* **47**: 2424–2437.

Powderly, W. G. (1992) Mucosal candidiasis caused by non-*albicans* species of *Candida* in HIV-positive patients *AIDS* **6**: 604–605.

Rex, J. H., Cooper, C. R., Merz, W. G., Galgiani, J. N. and Anaissie, E. J. (1995) Detection of amphotericin B-resistant *Candida* isolates in a broth based system. *Antimicrob. Agents Chemother.* **39**: 906–909.

Rodero, L., Mellado, E., Rodriguez, A. C., Salve, A. *et al.* (2003) G484S amino acid substitution in lanosterol 14-alpha demethylase (ERG11) is related to fluconazole resistance in a recurrent *Cryptococcus neoformans* clinical isolate. *Antimicrob. Agents Chemother.* **47**: 3653–3656.

Saag, M. S. and Dismukes, W. E. (1988) Azole antifungal agents: Emphasis on new triazoles. *Antimicrob. Agents Chemother.* **23**: 1–8.

Sanglard, D., Kuchler, K., Ischer, F., Pagani, J.-L. *et al.* (1995) Mechanisms of resistance to azole antifungal agents in *Candida albicans* isolates from AIDS patients involve specific multidrug transporters. *Antimicrob. Agents Chemother.* **39**: 2378–2386.

Sanglard, D., Ischer, F., Monod, M. and Bille, J. (1997) Cloning of *Candida albicans* genes conferring resistance to azole antifungal agents: Characterisation of *CDR2*, a new multidrug ABC transporter gene *Microbiol.* **143**: 405–416.

Sanglard, D., Ischer, F., Koymans, L. and Bille, J. (1998) Amino acid substitutions in the cytochrome P-450 lanosterol 14a-demethylase (CYP51A1) from azole-resistant *Candida albicans* clinical isolates contribute to resistance to azole antifungal agents. *Antimicrob. Agents Chemother.* **42**: 241–253.

Sanglard, D., Ischer, F., Calabrese, D., Majcherczyk, P. A. and Bille, J. (1999) The ATP binding cassette transporter gene *CgCDR1* from *Candida glabrata* is involved in the resistance of clinical isolates to azole antifungal agents. *Antimicrob. Agents Chemother.* **43**: 2753–2765.

Sanguinetti, M., Posteraro, B., Fiori, B., Ranno, S. *et al.* (2005) Mechanisms of azole resistance in clinical isolates of *Candida glabrata* collected during a hospital survey of antifungal resistance. *Antimicrob. Agents Chemother.* **49**: 668–679.

Sullivan, D., Haynes, K., Moran, G., Shanley, D. and Coleman, D. (1996) Persistence, replacement, and microevolution of *Cryptococcus neoformans* strains in recurrent meningitis in AIDS patients. *J. Clin. Microbiol.* **34**: 1739–1744.

VandenBergh, M. F., Verweij, P. E. and Voss, A. (1999) Epidemiology of nosocomial fungal infections: invasive aspergillosis and the environment. *Diagn. Microbiol. Infect. Dis.* **34**: 221–227.

White, T. C. (1997) The presence of an R467K amino acid substitution and loss of allelic variation correlate with an azole-resistant lanosterol 14a demethylase in *Candida albicans. Antimicrob. Agents Chemother.* **41**: 1488–1494.

White, T. C., Marr, K. A. and Bowden, R. A. (1998) Clinical, cellular, and molecular factors that contribute to antifungal drug resistance *Clin. Microbiol. Rev.* **11**: 382–402.

Willocks, L., Leen, C. L. S., Brettle, R. P., Urquhart, D. *et al.* (1991) Fluconazole resistance in AIDS patients. *J. Antimicrob. Chemother.* **28**: 937–939.

Xu, J., Ramos, A. R., Vilgalys, R. and Mitchell, T. G. (2000) Clonal and spontaneous origins of fluconazole resistance in *Candida albicans. J. Clin. Microbiol.* **38**: 1214–1220.

6

Animal models for evaluation of antifungal efficacy against filamentous fungi

Eric Dannaoui

6.1 Introduction

Human invasive infections due to filamentous fungi are life-threatening diseases occurring primarily in immunocompromised patients. *Aspergillus* species are among the most common filamentous fungi responsible for these infections and *A. fumigatus* is the aetiologic agent in more than 90% of the cases of aspergillosis (Latge, 1999). Neutropoenia, seen frequently in leukaemic and bone marrow transplant patients as well as in patients treated with anti-neoplastic chemotherapy, is the major predisposing factor for invasive aspergillosis.

Amphotericin B and itraconazole were until recently the drugs of choice for these infections (Stevens *et al.*, 2000). The mortality was about 85% in invasive pulmonary aspergillosis and >95% in disseminated infections. Moreover, recent reports of *de novo* and acquired resistance of *A. fumigatus* clinical isolates to itraconazole (Denning *et al.*, 1997; Dannaoui *et al.*, 1999, 2001) as well as *in vivo* resistance to amphotericin B (Johnson *et al.*, 2000) may complicate the management of invasive aspergillosis.

New antifungal drugs with activity against *Aspergillus* spp. have been recently marketed or are in development. These new drugs include azoles such as posaconazole and voriconazole and echinocandins such as caspofungin. Nevertheless, therapeutic efficacy remains limited with response rate of about 40–50% for caspofungin (Keating and Jarvis, 2001) or for voriconazole (Herbrecht *et al.*, 2002). It is therefore mandatory to find new approaches to antifungal therapy of aspergillosis.

Medical Mycology: Cellular and Molecular Techniques. Edited by Kevin Kavanagh

Less common filamentous fungi are also emerging and are often responsible for devastating infections in patients with haematological malignancies and in organ transplant recipients (Marr *et al.*, 2002; Husain *et al.*, 2003). These emerging fungi include *Fusarium* spp., *Scedosporium* spp. and zygomycetes. Invasive infection due to these fungi often have a rapidly fatal course with a mortality rate ranging between 50% and 100% despite aggressive antifungal therapy.

Zygomycoses occur particularly in patients with diabetes mellitus with or without ketoacidosis, neutropoenia, iron overload or in patients receiving prolonged corticosteroid treatment. The most common clinical forms are rhinocerebral, pulmonary, gastrointestinal or disseminated infections. Species of *Rhizopus*, *Mucor*, *Rhizomucor* and *Absidia* are the most frequent aetiologic agents of zygomycosis in humans. Zygomycetes are resistant to most of the antifungal drugs currently available and recent reports have documented breakthrough zygomycosis in patients treated with voriconazole (Kontoyiannis *et al.*, 2005).

A better knowledge of the efficacy of antifungal drugs, either alone or in combination, against these filamentous fungi is critical for a better management of the patients. *In vitro* antifungal susceptibility testing of filamentous fungi remains difficult to perform and to interpret. The CLSI (Clinical and Laboratory Standards Institute), formerly NCCLS (National Committee for Clinical Laboratory Standards) has recently proposed a methodology that showed good intra- and interlaboratory reproducibility (CLSI, 2002). Nevertheless, correlation between *in vitro* results and therapeutic *in vivo* efficacy remains very limited (Odds *et al.*, 1998; Johnson *et al.*, 2000). For these reasons, animal models that take into account host factors and the pharmacokinetics of the drugs are crucial for the evaluation of antifungal efficacy.

Different types of animal models can be used. On the one hand, there are 'infection' models in which animals, generally not immunocompromised, develop an infection that does not fully mimic the clinical infection and are therefore partially relevant. This is the case, for example, of animal models of disseminated aspergillosis obtained by intravenous injection of a spore suspension into immunocompetent mice. These 'infection' models are generally easy to use and highly reproducible. On the other hand, 'infectious diseases' models closely mimic the human infection with similar means of infection, host factors (immunosuppression) and target organs. This is the case of models of invasive pulmonary aspergillosis obtained by intranasal inoculation in neutropoenic mice. The choice of the model to be used essentially depends upon the question that has to be answered.

The most common parameters for evaluation of antifungal therapy are the mortality rate and the fungal burden in target organs (determined by CFU counting or by quantitative PCR), but other endpoints could also be used such as monitoring of antigenaemia, scoring tissue lesions by CT scan or assessing infection qualitatively by demonstration of hyphae in infected tissues.

There are numerous applications of animal models of mycoses in the field of antifungal therapy, including evaluation of new antifungal drugs, improvement of *in vitro* tests, confirmation of *in vitro* resistance, evaluation of therapy in rare mycoses and evaluation of antifungal combinations.

Animal experiments are particularly useful in the preclinical development stage of new antifungal drugs, for evaluation of their *in vivo* efficacy as well as their short- and long-term toxicity. In some instances, development of an antifungal drug could be stopped if a lack of *in vivo* efficacy is demonstrated in animal models, despite a good *in vitro* activity (Polak, 1999b). Several studies have demonstrated, for example, the good efficacy of voriconazole in animal models of invasive pulmonary aspergillosis (Murphy *et al.*, 1997) and of disseminated aspergillosis (Kirkpatrick *et al.*, 2002). Because the half-life of voriconazole is short in mice, most of these studies have been performed in guinea-pigs. Similarly, efficacy of posaconazole, a new azole in the late stage of development, has been demonstrated in animal models of several mycoses such as aspergillosis (Petraitiene *et al.*, 2001), cryptococcosis (Perfect *et al.*, 1996), fusariosis (Lozano-Chiu *et al.*, 1999) and zygomycosis (Dannaoui *et al.*, 2003). Animal models of candidosis and aspergillosis have also been used to demonstrate the *in vivo* efficacy of caspofungin (Abruzzo *et al.*, 1997).

Animal experiments are also useful for the technical improvement of *in vitro* antifungal susceptibility methods. Although reference techniques for *in vitro* antifungal susceptibility testing of yeasts and filamentous fungi are now available, some technical problems remain. For example, a trailing is generally observed when testing azoles against *Candida* spp. This trailing could be so important that some isolates are categorized *in vitro* as susceptible after 24 h of incubation and resistant after 48 h. By testing critical isolates in animal models, it has been demonstrated that these isolates were susceptible *in vivo* (Rex *et al.*, 1998).

One important goal of *in vitro* antifungal susceptibility testing is the detection of resistance. Nevertheless, the *in vitro–in vivo* correlation has not yet been established for all drug–organism combinations and this is particularly true for filamentous fungi. For this reason, the clinical relevance of the resistance detected *in vitro* remains unclear and animal experiments could be a way to confirm the resistance *in vivo*. One example was the demonstration that mice infected with isolates of *Aspergillus fumigatus* found resistant *in vitro* to itraconazole did not respond to itraconazole therapy in a murine model of disseminated aspergillosis (Denning *et al.*, 1997; Dannaoui *et al.*, 2001).

Infections due to filamentous fungi are emerging, and a large panel of different species are responsible. However, it is very difficult to evaluate the antifungal efficacy of a new drug in clinical studies with a significant number of patients. For these rare mycoses, animal experiments are then of immense importance. A large number of experimental studies with these rare pathogens have been performed to assess the efficacy of standard and new antifungal drugs, particularly for zygomycosis (Dannaoui *et al.*, 2002, 2003; Sun *et al.*, 2002), fusariosis (Odds *et al.*, 1998; Lozano-Chiu *et al.*, 1999) and scedosporiosis (Odds *et al.*, 1998).

Combinations of two or more antimicrobial drugs are commonly used for the treatment of bacterial and viral diseases. The advantages of combination therapy in comparison with monotherapy include the possibility of decreased emergence of resistant strains, decreased dose-related toxicity as a result of reduced dosage, increased antimicrobial spectrum and, most importantly, antimicrobial synergism. In medical mycology, the only current antifungal combination with documented improvement in clinical outcome over monotherapy is amphotericin B plus 5-fluorocytosine for the treatment of cryptococcal meningitis. Although the benefit over monotherapy has not been demonstrated, this combination is also recommended for the treatment of specific candidal infections. A combination of fluconazole with amphotericin B as a treatment of candidaemia in non-neutropoenic patients has also been demonstrated to be effective. For *in vitro* combination studies, there are no standardized techniques and the definition of synergistic interaction is varied. Thus, an interpretation of *in vitro* results is very difficult (Polak, 1999b). These drawbacks emphasize the usefulness of animal models to evaluate antifungal combinations (Graybill, 2000).

In this chapter, different animal models of disseminated zygomycosis and aspergillosis used for the evaluation of the activity of antifungal drugs either alone or in combination are detailed. With minor modifications, these models can be used for other filamentous fungi. Specific models of candidosis and invasive pulmonary aspergillosis have been detailed recently elsewhere (Andes, 2005; Lewis and Wiederhold, 2005).

6.2 Disseminated zygomycosis in non-immunosuppressed mice

Numerous experimental models of zygomycosis have been developed (Kamei, 2001). These differ by the animal used (guinea-pigs, rats, mice, rabbits or calves), the route of infection (intravenous, intranasal, intrasinusal) and the host status (immunocompetent, neutropoenic, cortisone treated, diabetic). These models have been used for various purposes ranging from physiopathology studies to the development of diagnostic tools. For an evaluation of antifungal activity, most of the studies have been performed on guinea-pigs and mice in either normal or immunocompromised animals (Odds *et al.* 1998; Mosquera *et al.* 2001; Dannaoui *et al.* 2002, 2003; Sun *et al.* 2002; Ibrahim *et al.* 2003; Spellberg *et al.* 2005). These models have proven useful to assess the *in vivo* efficacy of both classic (Odds *et al.* 1998; Mosquera *et al.* 2001; Dannaoui *et al.* 2002) and newly developed eloped antifungal drugs (Sun *et al.* 2002; Dannaoui *et al.* 2003; Ibrahim *et al.* 2003, 2005). In this protocol a disseminated model of zygomycosis in non-immunosuppressed mice is detailed. This model is reproducible and can be used for evaluation of most of the currently available antifungal drugs against the different species of zygomycetes.

Protocol 6.1
Disseminated zygomycosis in non-immunosuppressed mice

Equipment, materials and reagents

Inoculum preparation

Fungal strains to be tested

PD agar slants

Incubator

Sterile cotton swabs

Sterile saline (NaCl 0.9% w/v)

Tween 80

Haemocytometer

1 ml syringe

Nylon filter (pore size of 11 μm) and filter holder

Sabouraud agar plates

Infection

Specific-pathogen-free out-bred female CD1 mice 5 to 7 weeks old, weighing 20–22 gram

Animal facility

Safety cabinet for class 2 pathogens

Heat lamp

Rodent-restraining device

1 ml syringes

6G needles

Drug administration and mortality evaluation

Antifungal drugs: commercialized preparation should be preferred; if not available (e.g. antifungal in development), powder of known potency could be used

Sterile glucose 5% w/v for IV use

Hydroxypropyl beta-cyclodextrin or other solvent for dilution of drugs (Table 6.1)

Feeding tube 20 × 1.5 in.

Syringes, needles

Table 6.1 Most common solvent and route of administration used for the main antifungal drugs used in animal models of filamentous fungi infections

Antifungal drug	Commercial name[a]	Solvent	Route of administration	References
Amphotericin B deoxycholate	Fungizone®	5% glucose	IP, IV	(Dannaoui *et al.* 2001; Ibrahim *et al.* 2003)
Liposomal amphotericin B	Ambisome®	5% glucose	IV	(Ibrahim *et al.* 2003)
Amphotericin B Lipid Complex	Abelcet®	5% glucose	IV	(Spellberg *et al.* 2005)
Itraconazole	Sporanox® oral solution	Water, HPCD[b]	PO, IP	(Denning *et al.* 1997; Odds *et al.* 1998; Dannaoui *et al.* 2001)
Voriconazole	Vfend®	PEG 200	PO	(Kirkpatrick *et al.* 2002)
Posaconazole	—	Methyl-cellulose[c]	PO	(Lozano-Chiu *et al.* 1999; Cacciapuoti *et al.* 2000; Dannaoui *et al.* 2003)
Caspofungin	Cancidas®	Water	IP, IV	(Abruzzo *et al.* 1997; Bowman *et al.* 2001; Ibrahim *et al.* 2005)

[a]Commercial trade name could vary depending on the country

[b]In the commercial oral form, itraconazole is already in solution in hydroxypropyl beta-cyclodextrin (HPCD) and can be diluted in water to a certain extent. Nevertheless, when low concentrations are needed, it is necessary to use beta-hydroxypropyl cyclodextrin for further dilution to avoid precipitation of the drug

[c]Posaconazole powder is prepared as a suspension in 0.4% methylcellulose containing 0.5% polysorbate 80 (Tween 80) and 0.9% NaCl

IP: intraperitoneally; IV: intravenously; PO: per os (by gavage); PEG: polyethylene glycol

Laboratory balance for weighting the animals

Statistical software

Evaluation of organ infection

Carbon dioxide (CO_2) chamber

Forceps, dissecting scissors

10 ml Potter tissue grinder

Blankophor P 0.25 mg/ml in 20% w/v KOH

Fluorescence microscope with appropriate filter (excitation wavelength below 400 nm and barrier filter at 420 nm)

Method

Inoculum preparation

1. Strains are stored as concentrated spore suspensions at −80 °C in 10% (v/v) glycerol. Subculture strains twice on PDA to ensure viability.

2. Culture the strain to be used on PDA slants 5 to 7 days at 35 °C to allow good sporulation. For some zygomycetes species sporulation is hardly obtained and requires special culture media (Padhye and Ajello, 1988).

3. Spore suspension used for infection is preferably prepared the day of infection as some zygomycetes (e.g. *Absidia corymbifera*) lose viability when stored at + 4 °C.

4. Prepare a stock spore suspension in 2 ml of sterile NaCl 0.9% containing 0.05% Tween 80 by swabbing the surface of the culture with a sterile cotton swab.

5. Filter spore suspension with a syringe through a nylon filter (pore size of 11 μm) to get rid of hyphal material. After appropriate dilution (e.g. 1/100), the spore suspension is counted in a haemocytometer. Adjust spore suspension in sterile saline at the required concentration needed for infection.

6. **Viability check:** Prepare a series of 10-fold dilutions of the spore suspension in saline with 0.05% (v/v) Tween 80. Inoculate Sabouraud agar plate in duplicate with 100 μl of the diluted suspension in order to have about 50–100 colonies per plate. Incubate plates at 35 °C and enumerate CFU after 16–24 h of incubation.

Note: Longer incubation time precludes correct CFU counting as zygomycetes grow rapidly.

Infection

7. All experiments should be conducted in accordance with standard animal care regulations. For advice and specific training it is essential to consult the animal facility veterinarian.

8. Before planning experiments, the study design and the protocols have to be approved by the local animal ethical committee.

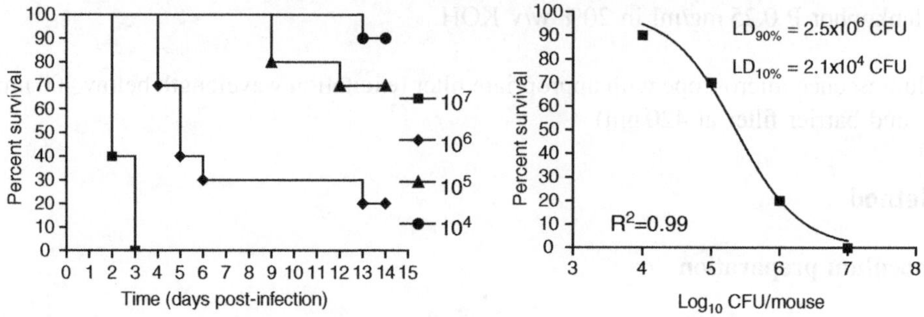

Figure 6.1 Inoculum finding experiment for a filamentous fungi. Mice were infected with four different inoculum sizes and mortality recorded for 14 days. $LD_{90\%}$ and $LD_{10\%}$ is then calculated

9. Mice are housed 5 to 7 per cage. There are minimum space requirements for housing mice in cages. Mice are given food and water *ad libitum.*

10. Animals are kept in the facility for a week in order to acclimatize before the initiation of the experiment.

11. Generally, for the evaluation of antifungal therapy, a lethal infection in control untreated animals is needed. For this reason, preliminary experiments have to be carried out with different inocula to define the lethal dose that produce 90% mortality ($LD_{90\%}$) after 14 days. Three to four inoculum sizes should be tested with a limited number of animals. For each inoculum, the mortality rate is recorded over the timespan of the planned experiment and the $DL_{90\%}$ is calculated (Figure 6.1).

12. All procedures with infected animals have to be carried out in a safety cabinet for class 2 pathogens.

13. Place the cage under a heat lamp. This will induce vasodilation and facilitate intravenous inoculation.

14. Place the mice in a restraint device. Mice are infected by injection of 0.1 ml of the conidial suspension into a lateral tail vein with a syringe equipped with a 26G needle. The inoculum size used is the predetermined $LD_{90\%}$.

15. After inoculation, cages are randomized in the different treatment groups.

16. Identify each mice individually. Although there are several ways of permanent identification (e.g. ear punch or tag, skin tattoo) a temporary identification system is enough for a short-term experiment of 2–3 weeks. Colour marks can be easily made on the tail skin of albino mice with colour markers.

Drug administration and mortality evaluation

17. Depending on the experiment planned, preliminary studies could be required to broadly assess the effective dosages of an antifungal drug against a given isolate.

18. Start drug therapy 2 h after infection. Continue treatment for 7 days and check mice for 1 week after the end of therapy.

19. In most of the studies, a minimum of 10 mice are used for each treatment group for mortality evaluation. Nevertheless, the number of animals per group can be calculated based on of the expected results. It could be useful to get advice from a statistician.

20. It is always necessary to include control mice receiving no active treatment. These mice are given the solvent used for drug preparation under the same volume and by the same way.

21. Antifungal drugs are purchased as commercial preparations and are preferably reconstituted and/or diluted according to the manufacturer's instructions. Solvent used for dilutions depends on the drugs (Table 6.1).

22. Each day, prepare working drug dilutions at the appropriate concentration. Dilutions have to be calculated based on the dose to be used (in mg/kg) and on the maximum volume that can be given to the mice (depending on the way of administration). See an example in Table 6.2.

    Note: Hydrophilic drugs (e.g. fluconazole) and/or drugs with short half-life in mice (e.g. flucytosine) can be diluted in the sole source of drinking water. However, owing to their spectrum of activity, these drugs are more commonly used in animal models of yeast infections.

23. Weigh each mouse individually on a laboratory balance and calculate the volume to be given.

24. Depending on the drug, doses are given by intraperitoneal or intravenous (via lateral tail vein) injection with a 1 ml syringe with 26G needle or orally by gavage with a feeding tube attached to a 1 ml syringe.

25. Check animals for mortality and abnormal clinical signs twice a day. Any mice that are unable to drink or get food must be euthanized. For these mice, death is recorded as occurring on the next day.

26. By the end of the experiment all surviving mice are euthanized by CO_2 asphyxiation.

Table 6.2 Example of drug preparation for amphotericin B and itraconazole treatment

Treatment Group	Antifungal drug	Stock solution	Working solution	Volume and way of administration	Dose delivered
Amphotericin B	Amphotericin B deoxycholate (Fungizone®)	5 mg/ml	0.1 mg/ml (in 5% glucose)	10 µl/g of BW by IP injection OD	1 mg/kg/d
Itraconazole	Itraconazole in HPCD (Sporanox® oral solution)	10 mg/ml	5 mg/ml (in 40% HPCD)	5 µl/g of BW PO by gavage BID	50 mg/kg/d
Control 1	None	—	5% glucose	10 µl/g of BW by IP injection OD	0
Control 2	None	—	40% HPCD	5 µl/g PO by gavage BID	0

HPCD: hydroxypropyl beta-cyclodextrin; BW: body weight; PO: per os; IP: intraperitoneally; OD: once per day; BID: twice per day

27. Enter the survival data for each group of animals in a statistical software program to create survival curves by the method of Kaplan and Meier (Figure 6.2). Comparison of survival curves is best performed by using the log-rank test.

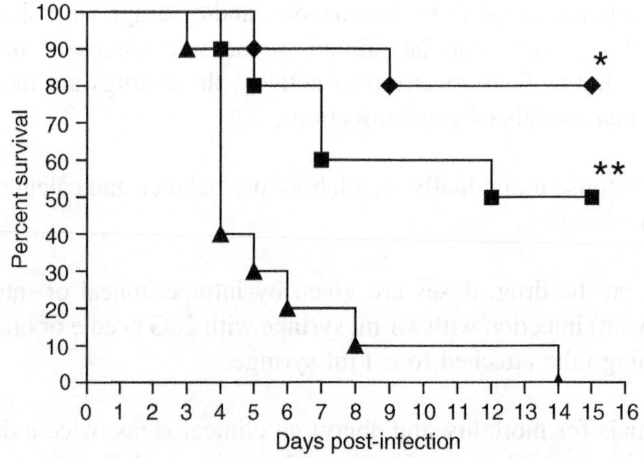

Figure 6.2 Example of mortality experiment in an animal model of disseminated filamentous fungi infection. Mice were treated with an antifungal at dose A (Diamond) and dose B (Square) and efficacy compared to control mice receiving no active treatment (Triangle). Survival curves have been generated by Kaplan–Meier method, and comparison of mortality rates was performed by the log-rank test. * $P < 0.001$, ** $P = 0.0031$

Evaluation of organ infection

Note: There is no satisfactory method for evaluating organ fungal burden in animal models of zygomycosis. Culture of infected tissues is often negative because zygomycetes are non-septated fungi and then extremely sensitive to standard homogenization procedures (Dannaoui *et al.*, 2002). Moreover, in disseminated models of zygomycosis non-germinated spores used for infection are retained in organs for a long period of time and could give falsely positive results when these organs are cultured. Quantitative PCR methods have been developed but PCR amplification could be obtained even if the fungi are not viable. PCR techniques could also be limited by false-positive results related to the presence of non-germinated spores. In this protocol, a qualitative assessment of organ infection by direct examination is detailed.

28. Assessment of organ infection is performed on mice that died during the experiment and on surviving mice, which are sacrificed at the end of the experiment.

29. Aseptically remove kidneys and brain and transfer the organs in sterile tubes.

    Note: Kidney and brain are the two main target organs in this model of disseminated zygomycosis. Additional organs can also be assessed.

30. Add 1 ml of NaCl 0.9% (w/v) to each tube and homogenize in a tissue grinder.

31. Place 50 μl of the homogenate on a microscope slide and allow the preparation to dry on ambient air at room temperature.

32. Add 20 μl of Blankophor P and place a coverslip on top of it.

33. Examine the slide on a fluorescence microscope and record the absence/presence of hyphae and spores. An active infection is defined by the presence of hyphae (Figure 6.3).

6.3 Animal model of disseminated aspergillosis

Animal models of aspergillosis have been widely used for decades to study the pathogenesis of aspergillosis, virulence factors of *Aspergillus* spp., immune response of the host or the development and evaluation of new diagnostic tools (Clemons and Stevens, 2005). In the field of antifungal therapy, different models are currently used such as models of disseminated aspergillosis (either in immunocompetent or immunocompromised animals), models of invasive pulmonary aspergillosis or models of extrapulmonary infection. Mice have been the animals of choice, particularly because of practical and economical reasons, as large numbers of animals are generally needed for any comparison of efficacy in different treatment groups. A wide range of studies

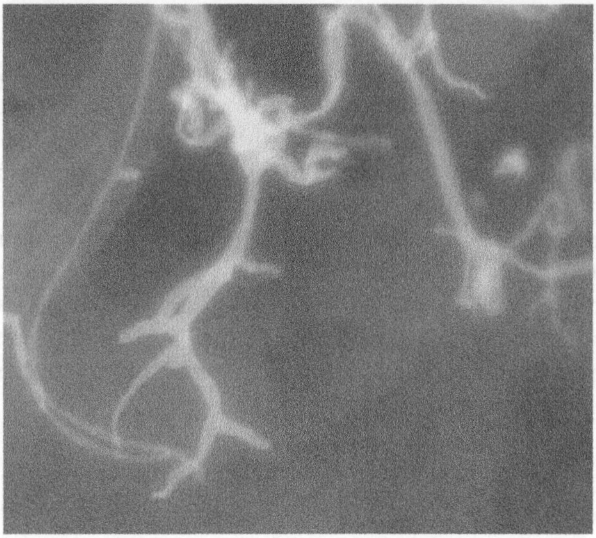

Figure 6.3 Disseminated zygomycosis in non-immunocompromised mice. Direct examination with a fluorescence microscope of a kidney tissue homogenate after staining with Blankophor P

have been performed including the evaluation of new antifungal drugs (Abruzzo *et al.*, 1997; Cacciapuoti *et al.*, 2000), detection and confirmation of antifungal resistance (Denning *et al.*, 1997; Dannaoui *et al.*, 1999, 2001), evaluation of correlations between *in vitro* antifungal susceptibility testing results and *in vivo* efficacy (Odds *et al.*, 1998; Johnson *et al.*, 2000) and the study of various antifungal combinations (Polak 1999B; Steinbach *et al.*, 2003). In this protocol, a murine model of disseminated aspergillosis in neutropoenic mice is detailed. The two major parameters of antifungal efficacy are the mortality rate and the fungal burden in organs.

Protocol 6.2
Disseminated aspergillosis in neutropoenic mice

Equipment, materials and reagents

Inoculum preparation

Same as for Protocol 6.1.

Immunosuppression and infection

Specific-pathogen-free out-bred female CD1 mice 5 to 7 weeks old, weighing 20–22 gram

Cyclophosphamide

Sterile food and water

Filter top for cages

Tetracycline hydrochloride

Calibrated microcapillary pipettes for blood collection

Türk solution (0.01% (w/v) crystal violet in 3% (v/v) acetic acid in water)

Giemsa stain

1 ml syringe with 26G needles

For infection, same materials and reagents as used in Protocol 6.1 are needed.

Drug administration and mortality evaluation

Same as for Protocol 6.1.

Evaluation of organ infection

Carbon dioxide (CO_2) chamber

Forceps, dissecting scissors

10 ml Potter tissue grinder with motor-driven Teflon pestle

Sterile NaCl 0.9% (w/v)

Sabouraud agar plates supplemented with 0.05% chloramphenicol

Method

Inoculum preparation

1. Prepare inoculum in a similar way than for Protocol 6.1.

Immunosuppression

2. Prepare a stock solution of cyclophosphamide at 20 mg/ml in water.

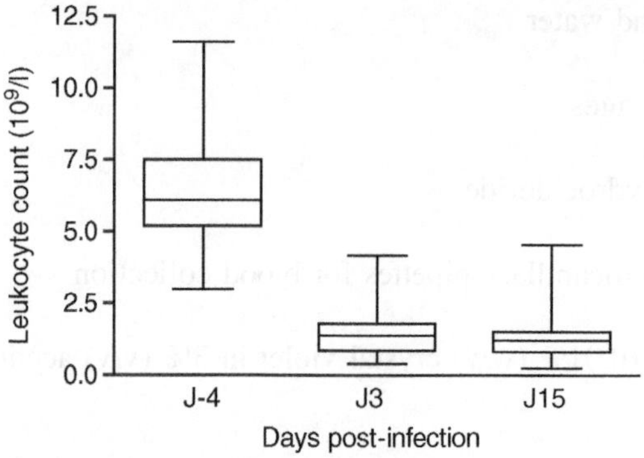

Figure 6.4 Total leukocyte count in mice treated with cyclophosphamide 200 mg/kg at D-4 and repeated injection of 100 mg/kg at days 0, 4, 8 and 12 post-infection. Data pooled from independent experiments with 40 mice at each time-point. A sustained leukopoenia is obtained until day 15. Absolute values for PMNs can be obtained from white blood cells' differential counts

3. To induce neutropoenia, inject mice intraperitoneally with 200 mg/kg of cyclophosphamide (10 μl per gram of body weight of the stock solution) 4 days before infection, and repeat injections of 100 mg/kg the day of infection and 4 days after infection.

   Note: Additional cyclosphosphamide injections can be performed to produce a sustained neutropoenia (Figure 6.4).

4. Mice are housed in cages with filter tops and animals are given sterile food. Sterile drinking water is supplemented with tetracycline at 1 mg/litre to prevent bacterial infections.

5. Determination of PMNs count is performed at different time points to ensure that animals are neutropoenic.

6. After a short-inhalation anaesthesia collect blood from the orbital sinus with a 10 μl calibrated microcapillary pipette. Fill the capillary with more than 10 μl.

7. Place 1 drop on a slide to make a blood smear.

8. Remove excess blood in the capillary to adjust the volume exactly to 10 μl.

9. Make a 1/20 dilution of blood in 190 μl of Türk's solution for erythrocyte lysis.

10. Count leukocytes in a haemocytometer.

11. Stain the blood smear with Giemsa staining.

12. Calculate the absolute number of PMNs from the total leukocyte count and the white blood cell differential count.

Infection

13. Determine the $LD_{90\%}$ (for mortality studies) and $LD_{10\%}$ (for CFU studies) in preliminary experiments for the strain(s) to be used.

14. Infect mice by intravenous injection of the calibrated spore suspension as described in Protocol 6.1.

Mortality studies

15. For mortality studies mice are challenged with $LD_{90\%}$. Antifungal treatment is started 24 h after infection and continued for 7 days. Mortality is recorded for another 1 week and survivors are euthanized on day 14 post-infection.

16. Treatments are given according to Protocol 6.1.

Organ fungal burden studies

Note: Different methods for measuring fungal burden in organs can be used. Recently, quantitative PCR has been developed in animal models of aspergillosis (Bowman *et al.*, 2001) and has been recently detailed (Lewis and Wiederhold, 2005). The most commonly used technique until now is the CFU determination by culturing organs after homogenization.

17. Fungal burden studies have to be preferably planned as separate experiments.

18. Fungal burden determination is performed in the kidneys and brain, which are the target organs in this model. Dissemination to other organs (liver, spleen) can also be assessed.

19. Mice are infected with $LD_{10\%}$ by intravenous injection.

20. Start drug therapy 24 h after infection. Continue treatment for 5 days.

21. Stop treatment 24 h after the last treatment dose to allow for a wash-out period. The wash-out period permits to minimize drug carry-over.

22. Euthanize the mice by CO_2 asphyxiation.

23. Aseptically remove the kidneys and brain and transfer the organs in pre-weighed tubes.

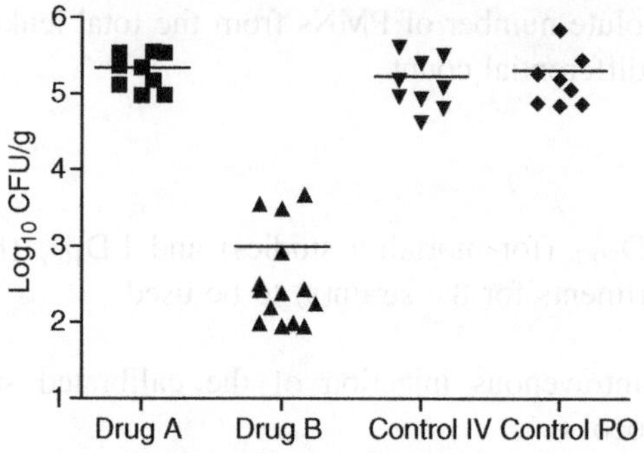

Figure 6.5 Example of CFU determination in brain in an animal model of disseminated filamentous fungi infection. Drug A was ineffective and drug B was active with a significant decrease of CFU/g compared to controls ($P < 0.01$)

24. Weigh the tubes containing the organs and calculate the organs' weight.

25. Homogenize organs in 2 ml of sterile NaCl 0.9% (w/v) at 1000 rpm in a tissue grinder with a motor-driven pestle until all tissues are broken up.

26. Make a 1/10 and 1/100 dilution of the homogenate in sterile NaCl 0.9% (w/v).

27. Plate 100 µl of each dilution (including the neat dilution) on Sabouraud agar plates supplemented with chloramphenicol.

    Note: It is possible to increase the cultured volume to decrease the limit of detection.

28. Incubate plates at 35 °C for 24 h and count CFU.

29. Calculate the CFU/g based on the organ weight, the dilutions used and the CFU count.

30. Transform CFU/g data in $\log_{10}$ values for plotting and comparison (Figure 6.5).

6.4 Study design for evaluation of antifungal combinations therapy in animal models

Although there are reproducible techniques for *in vitro* testing of antifungal combinations (Vitale *et al.*, 2005), these techniques are not fully standardized and it is still necessary to test the clinical relevance of *in vitro* results in animal models. *In vitro* techniques allow for the evaluation of a large number of combinations against a large number of isolates for a given species (Dannaoui *et al.*, 2004). In contrast, because of the important number of animals needed, it is only possible to test a few isolates for selected combinations in animal models. Overall, the techniques (inoculum preparation, infection, endpoints for evaluation) used to study antifungal combinations in animals are similar to those used for the evaluation of monotherapies. The main problem is to properly design the study in order to be able to detect drug interactions. In particular, the efficacy of monotherapies should not be optimal to allow detection of a synergistic interaction nor too low to allow the detection of an antagonism. In this protocol, a typical study design is detailed for the evaluation of potential synergistic combinations between two antifungal drugs in a model of disseminated aspergillosis in mice.

Protocol 6.3
Study design for the evaluation of combination therapy in animal models

Equipment, materials and reagents

Same as for Protocol 6.2. Mice could be either non-immunocompromised or neutropoenic.

Method

1. Determine the $LD_{90\%}$ (for mortality studies) and $LD_{10\%}$ (for CFU studies) in preliminary experiments for the strain(s) to be used.

Mortality studies

2. Test 3 to 4 different dosages of drug A and drug B used as monotherapies in groups of 10 animals. Treat animals for 7 days starting 24 h post infection. Record mortality until day 14 post infection.

3. Plot dose-response curves for monotherapies and determine for both antifungals the 2 drug regimens that give about 50% of survival (high dose) and 10–20% of survival (low dose). These doses will be subsequently used in the combination experiment.

Note: It has to be noted that antifungal dosages are then defined by the model used and could be much lower than dosages used in patients. If higher dosages (more clinically relevant) have to be tested, it is necessary to change the model either by delaying the initiation of therapy or by increasing the inoculum used for infection in order to have a lower efficacy of monotherapies.

D -7	**Strain culture** (PDA –35 °C – 5 to 7days)
D 0	**Inoculum preparation** (viable spores numeration) **infection** (intravenous injection of spores)
D 0 to D7	**Treatment (7 days) and follow-up** Group 1: Drug A (PO) low dose Group 2: Drug A (PO) high dose Group 3: Drug B (IP) low dose Group 4: Drug B (IP) high dose Group 5: Drug A (PO) low dose + Drug B (IP) low dose Group 6: Drug A (PO) low dose + Drug B (IP) high dose Group 7: Drug A (PO) high dose + Drug B (IP) low dose Group 8: Drug A (PO) high dose + Drug B (IP) high dose Group 9: Control (PO) Group 10: Control (IP)
D 14	**Cumulative mortality rate**

Figure 6.6 Design of a typical combination experiment for the detection of a synergistic interaction between two antifungal drugs (a and b). Preliminary experiments are performed to define the doses of antifungal A and B that give about 50% of survival (high dose) and about 10–20% survival when used as a single agent

4. In order to reach statistically significant differences between groups, the number of animals in each treatment group in the combination experiments have to be ideally defined based on the expected results. The best approach is to discuss expected results and hypotheses with a statistician.

5. Include 10 groups of animals for the combination experiment (Figure 6.6): 4 groups for monotherapies, 4 groups for combination therapies and 2 groups of controls.

6. Record mortality and plot survival curves for each treatment groups.

7. Perform statistical comparison (by log-rank test) of survival rates for groups treated by the combinations and by monotherapies.

8. Synergistic interaction is achieved when the survival rate of a combination is significantly higher than survival obtained by both antifungals used alone at the same dosages than in the combination. An antagonistic interaction is achieved when the survival rate of a combination is significantly lower than survival

obtained by any of the two antifungals used alone at the same dosages than in the combination.

CFU studies

9. A similar protocol (steps 2–8), adapted for CFU studies, is used for testing the efficacy of combinations by an evaluation of the fungal burden in target organs.

Note: Using fungal burden in organs could be a better parameter of evaluation (Graybill, 2000) because CFU/g can vary between about 10^6 CFU/g in control animals and about 10 CFU/g in treated animals, leaving a 5 log range for comparison of efficacy. In contrast, mortality studies only allow evaluation over a range of 2 log (0–100%).

6.5 References

Abruzzo, G. K., A. M. Flattery, C. J. Gill, L. Kong *et al.* (1997) Evaluation of the echinocandin antifungal MK-0991 (L-743,872): Efficacies in mouse models of disseminated aspergillosis, candidiasis, and cryptococcosis. *Antimicrob. Agents Chemother.* **41**: 2333–2338.

Andes, D. (2005) Use of an animal model of disseminated candidiasis in the evaluation of antifungal therapy. *Methods Mol Med* **118**: 111–128.

Bowman, J. C., Abruzzo, G. K., Anderson, J. W., Flattery, A. M. *et al.* (2001) Quantitative PCR assay to measure *Aspergillus fumigatus* burden in a murine model of disseminated aspergillosis: demonstration of efficacy of caspofungin acetate. *Antimicrob. Agents Chemother.* **45**: 3474–3481.

Cacciapuoti, A., Loebenberg, D., Corcoran, E., Menzel, F. Jr. *et al.* (2000) *In vitro* and *in vivo* activities of SCH 56592 (posaconazole), a new triazole antifungal agent, against *Aspergillus* and Candida. *Antimicrob. Agents Chemother.* **44**: 2017–2022.

Clemons, K. V. and Stevens, D. A. (2005) The contribution of animal models of aspergillosis to understanding pathogenesis, therapy and virulence. *Med. Mycol.* **43** Suppl 1: S101–110.

CLSI (2002) *Reference method for broth dilution antifungal susceptibility testing of filamentous fungi: approved standard.* Document M-38A, Clinical and Laboratory Standards Institute, Wayne, Pa.

Dannaoui, E., Borel, E., Monier, M. F., Piens, M. A. *et al.*(2001) Acquired itraconazole resistance in *Aspergillus fumigatus. J. Antimicrob. Chemother.* **47**: 333–340.

Dannaoui, E., E. Borel, F. Persat, M. F. Monier and M. A. Piens (1999) In-vivo itraconazole resistance of *Aspergillus fumigatus* in systemic murine aspergillosis. *J. Med. Microbiol.* **48**: 1087–1093.

Dannaoui, E., O. Lortholary and F. Dromer (2004) *In vitro* evaluation of double and triple combinations of antifungal drugs against *Aspergillus fumigatus* and *Aspergillus terreus. Antimicrob. Agents Chemother.* **48**: 970–978.

Dannaoui, E., J. F. Meis, D. Loebenberg and P. E. Verweij (2003) Activity of posaconazole in treatment of experimental disseminated zygomycosis. *Antimicrob. Agents Chemother.* **47**: 3647–36450.

Dannaoui, E., Mouton, J. W., Meis, J. F. and Verweij, P. E. (2002) Efficacy of antifungal therapy in a nonneutropenic murine model of zygomycosis. *Antimicrob. Agents Chemother.* **46**: 1953–1959.

Denning, D. W., Venkateswarlu, K., Oakley, K. L., Anderson, M. J. *et al* (1997) Itraconazole resistance in *Aspergillus fumigatus*. *Antimicrob. Agents Chemother.* **41**: 1364–1368.

Graybill, J. R. (2000) The role of murine models in the development of antifungal therapy for systemic mycoses. *Drug Resist. Update.* **3**: 364–383.

Herbrecht, R., Denning, D. W., Patterson, T. F., Bennett, J. E.R., *et al.* (2002) Voriconazole versus amphotericin B for primary therapy of invasive aspergillosis. *N. Engl. J. Med.* **347**: 408–415.

Husain, S., Alexander, B. D., Munoz, P., Avery, R. K. *et al.* (2003) Opportunistic mycelial fungal infections in organ transplant recipients: emerging importance of non-*Aspergillus* mycelial fungi. *Clin. Infect. Dis.* **37**: 221–229.

Ibrahim, A. S., Avanessian, V., Spellberg B. and Edwards, J. E. Jr. (2003) Liposomal amphotericin B, and not amphotericin B deoxycholate, improves survival of diabetic mice infected with *Rhizopus oryzae*. *Antimicrob. Agents Chemother.* **47**: 3343–3344.

Ibrahim, A. S., Bowman, J. C., Avanessian, V., Brown, K. *et al.* (2005) Caspofungin inhibits *Rhizopus oryzae* 1,3-beta-D-glucan synthase, lowers burden in brain measured by quantitative PCR, and improves survival at a low but not a high dose during murine disseminated zygomycosis. *Antimicrob. Agents Chemother.* **49**: 721–727.

Johnson, E. M., Oakley, K. L., Radford, S. A., Moore, C. B. *et al.* (2000) Lack of correlation of *in vitro* amphotericin B susceptibility testing with outcome in a murine model of *Aspergillus* infection. *J. Antimicrob. Chemother.* **45**: 85–93.

Kamei, K. (2001) Animal models of zygomycosis–*Absidia*, *Rhizopus*, *Rhizomucor*, and *Cunninghamella*. *Mycopathologia.* **152**: 5–13.

Keating, G. M. and Jarvis, B. (2001) Caspofungin. *Drugs* 61: 1121–1129.

Kirkpatrick, W. R., Perea, S., Coco, B. J. and Patterson, T. F. (2002) Efficacy of caspofungin alone and in combination with voriconazole in a Guinea pig model of invasive aspergillosis. *Antimicrob. Agents Chemother.* **46**: 2564–2568.

Kontoyiannis, D. P., Lionakis, M. S., Lewis, R. E., Chamilos, G. *et al.* (2005) Zygomycosis in a tertiary-care cancer center in the era of *Aspergillus*-active antifungal therapy: a case-control observational study of 27 recent cases. *J. Infect. Dis.* **191**: 1350–1360.

Latge, J. P. (1999) *Aspergillus fumigatus* and aspergillosis. *Clin. Microbiol. Rev.* **12**: 310–350.

Lewis, R. E. and Wiederhold, N. P. (2005) Murine model of invasive aspergillosis. *Methods Mol. Med.* **118**: 129–142.

Lozano-Chiu, M., Arikan, S., Paetznick, V. L. Anaissie, E. J. *et al.* (1999) Treatment of murine fusariosis with SCH 56592. *Antimicrob. Agents Chemother.* **43**: 589–591.

Marr, K. A., Carter, R. A., Crippa, F., Wald, A. and Corey, L. (2002) Epidemiology and outcome of mould infections in haematopoietic stem cell transplant recipients. *Clin. Infect. Dis.* **34**: 909–917.

Mosquera, J., Warn, P. A., Rodriguez-Tudela, J. L. and Denning, D. W. (2001) Treatment of *Absidia corymbifera* infection in mice with amphotericin B and itraconazole. *J. Antimicrob. Chemother.* **48**: 583–586.

Murphy, M., Bernard, E. M., Ishimaru, T. and D. Armstrong (1997) Activity of voriconazole (UK-109,496) against clinical isolates of *Aspergillus* species and its effectiveness in an experimental model of invasive pulmonary aspergillosis. *Antimicrob. Agents Chemother.* **41**: 696–698.

Odds, F. C., Van Gerven, F. Espinel-Ingroff, A., Bartlett, M. S. *et al.* (1998) Evaluation of possible correlations between antifungal susceptibilities of filamentous fungi *in vitro* and antifungal treatment outcomes in animal infection models. *Antimicrob. Agents Chemother.* **42**: 282–288.

Padhye, A. A. and Ajello, L. (1988) Simple method of inducing sporulation by *Apophysomyces elegans* and *Saksenaea vasiformis*. *J. Clin. Microbiol.* **26**: 1861–1863.

Perfect, J. R., Cox, G. M., Dodge, R. K. and Schell, W. A. (1996) *In vitro* and *in vivo* efficacies of the azole SCH56592 against *Cryptococcus neoformans*. *Antimicrob. Agents Chemother.* **40**: 1910–1913.

Petraitiene, R., Petraitis, V. Groll, A. H., Sein, T. *et al.* (2001) Antifungal activity and pharmacokinetics of posaconazole (SCH 56592) in treatment and prevention of experimental invasive pulmonary aspergillosis: correlation with galactomannan antigenemia. *Antimicrob. Agents Chemother.* **45**: 857–869.

Polak, A. (1999) The past, present and future of antimycotic combination therapy. *Mycoses.* **42**: 355–370.

Rex, J. H., Nelson, P. W., Paetznick, V. L., Lozano-Chiu, M. *et al.* (1998) Optimizing the correlation between results of testing *in vitro* and therapeutic outcome *in vivo* for fluconazole by testing critical isolates in a murine model of invasive candidiasis. *Antimicrob. Agents Chemother.* **42**: 129–134.

Spellberg, B., Fu, Y., Edwards, J. E. Jr. and Ibrahim, A. S. (2005) Combination therapy with amphotericin B lipid complex and caspofungin acetate of disseminated zygomycosis in diabetic ketoacidotic mice. *Antimicrob. Agents Chemother.* **49**: 830–832.

Steinbach, W. J., Stevens, D. A. and Denning, D. W. (2003) Combination and sequential antifungal therapy for invasive aspergillosis: review of published *in vitro* and *in vivo* interactions and 6281 clinical cases from 1966 to 2001. *Clin. Infect. Dis.* **37** (Suppl 3): S188–224.

Stevens, D. A., Kan, V. L., Judson, M. A., Morrison, V. A. *et al.* (2000) Practice guidelines for diseases caused by *Aspergillus*. Infectious Diseases Society of America. *Clin. Infect. Dis.* **30**: 696–709.

Sun, Q. N., Najvar, L. K., Bocanegra, R., Loebenberg, D. and Graybill, J. R. (2002) *In vivo* activity of posaconazole against *Mucor* spp. in an immunosuppressed-mouse model. *Antimicrob. Agents Chemother.* **46**: 2310–2312.

Vitale, R. G., Afeltra, J. and Dannaoui, E. (2005) Antifungal Combinations. *Methods Mol. Med.* **118**: 143–152.

7

Proteomic analysis of pathogenic fungi

Alan Murphy

7.1 Introduction

The wealth of genomic information now available has led to the rapid development of many fields within the sphere of biological sciences and has revolutionized the study of proteins. It is now possible to study proteins and their functions and interactions within an organism on a large scale. This new and continuously developing field of study is known as 'proteomics'.

Proteomics primarily focuses on the identification of many proteins, from single or multiple extracts, and is useful in describing the effects of an action on hundreds or even thousands of proteins. Proteomics has proven extremely useful to the medical mycologist. The effects of exposure of *Candida albicans* to an antifungal agent have been examined across the whole proteome allowing rapid identification of proteins hitherto unknown to be involved in drug resistance (Hooshdaran *et al.* 2004). This contrasts with traditional studies, which usually rely upon examining the effect of the drug on a small number of predefined proteins (Vandeputte *et al.*, 2005). While both approaches are valid and necessary, the advantages of the whole proteome approach are obvious.

Analysis of the whole proteome has been made possible chiefly through the continued refinement of two-dimensional sodium dodecyl sulphate polyacrylamide gel electrophoresis (2D SDS-PAGE) and the use of mass spectrometry. Traditionally, conventional 2D SDS-PAGE has been employed, but continuing developments and modifications such as 2D difference gel electrophoresis (2D DIGE) (Lilley and Friedman, 2004) may replace 2D SDS-PAGE in the future, although as

Medical Mycology: Cellular and Molecular Techniques. Edited by Kevin Kavanagh

2D DIGE is more expensive it is likely that 2D SDS-PAGE will remain the method of choice for proteomics research in the majority of laboratories for some time. The development of Matrix Assisted Laser Desorption/Ionization-Time of Flight (MALDI-TOF) mass spectrometry and its use in protein identification has heralded a new age in proteomics. These two experimental processes coupled together form the core of proteomics as evidenced by the literature (Shevchenko *et al.*, 1996; Laure *et al.*, 2003; Medina *et al.*, 2004). The proteomic approach to biological systems has examined, for example, drug resistance in yeast (Hooshdaran *et al.* 2004).

The aim of this chapter is to introduce and describe the experimental processes most valuable in proteomic studies. When protein is first extracted from a fungus, the protein content must be quantified. This allows for the equal loading of different samples to 2D SDS-PAGE gels. Following staining and destaining of the gel, spots of interest may be excised and digested using trypsin, in preparation for analysis by mass spectrometry. If a single protein has been purified from an extract and needs to be identified, the sample may be analysed directly by mass spectrometry to accurately determine the exact molecular weight of the protein (providing mass spectrometry grade reagents were used in purification). The protein is then digested in-solution and the peptide sequence analysed by mass spectrometry (Burns *et al.*, 2005; Neville *et al.*, 2005). Should the sample not have been purified using mass spectrometry grade reagents, the sample may be run on a one-dimensional SDS-PAGE gel, digested and subsequently subjected to mass spectrometry (Burns *et al.*, 2005; Neville *et al.*, 2005). A flowchart describing the major experimental processes involved in proteomics is described in Figure 7.1.

2D SDS-PAGE

The 2D SDS-PAGE technique is one of the fundamental procedures used in proteomics research. During 2D SDS-PAGE, the protein sample is first run through a pH gradient in a process known as 'iso-electric focusing' (IEF). During IEF, the proteins migrate along an immobilized pH gradient (IPG) until they reach a point where their net-charge is zero. The gradient is prepared using IPG strips, which are available over both narrow and wide pH ranges depending on the level of separation required by the user. The samples are separated over a number of hours, depending on the properties of the proteins of interest.

During the second stage, the strips are applied to the top of an SDS-PAGE gel and separated on the basis of size. This allows for a greater resolution of protein samples of the same molecular weight since they have already been separated by IEF. While each protein spot on a 2D SDS-PAGE gel is more highly resolved than a protein band on a one-dimensional SDS-PAGE gel, users must be careful not to assume that each spot will represent a single protein, as it is possible that two or more proteins could have the same weight and iso-electric point.

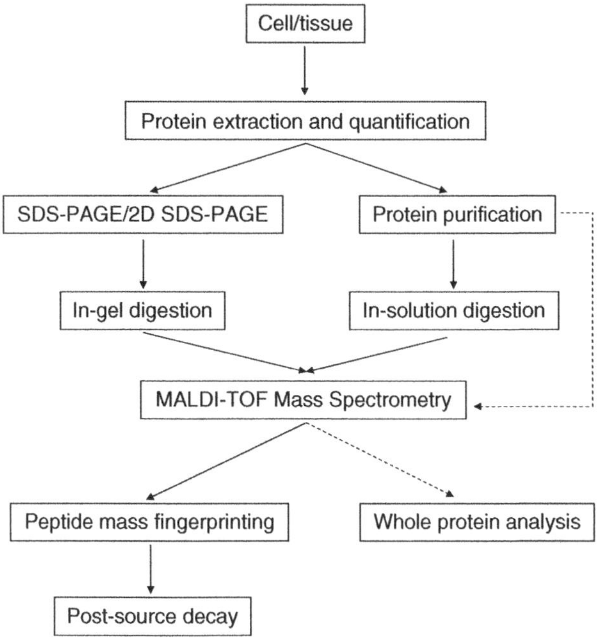

Figure 7.1 Proteomics flowchart. Following protein extraction and quantification, proteins may be further purified and the intact purified protein analysed by MALDI-TOF mass spectrometry. Alternatively, proteins to be analysed by PMF may be digested in-solution or separated by SDS-PAGE/2D SDS-PAGE prior to in-gel digestion

Protocol 7.1
2D SDS-PAGE of protein samples

Equipment, materials and reagents

All reagents can be purchased from the Sigma-Aldrich Chemical Co, St Louis, MO, USA. unless otherwise stated.

IEF buffer (8 M urea, 2 M thiourea, 4% (w/v) CHAPS, 1% (v/v) Triton X-100, 65 mM DTT, 10 mM Trizma® base (Sigma-Aldrich), 0.8% (v/v) ampholyte)

Equilibration buffer (6 M urea, 30% (v/v) glycerol, 2% (w/v) SDS, 50 mM Tris-HCl, pH 6.8)

Method

Note: It may be necessary to load high concentrations of protein extract if the proteins of interest are expressed in very low quantities.

Table 7.1 IEF cycle using the Ettan IPGphor II IEF system (Amersham Biosciences)

	Voltage*	Volt hour*	Conditions
Step 1	50	500	Step and hold
Step 2	250	62.5	Step and hold
Step 3	8000	40 000	Gradient
Step 4	8000	64 000	Step and hold

*The optimal voltage and volt hour must be determined for different sample concentrations.

1. Perform Bradford assay on extracted protein and dilute protein solution to desired concentration.

2. Add protein solution to IEF buffer and solubilize at room temperature over a 30- to 60-min period with occasional vortexing to prevent clumping.

3. The protein concentration should then be quantified using a Bradford assay. As a guide, use 0.1 mg of protein per 100 μl for a 7 cm IEF strip or 0.3 mg of protein per 250 μl for a 13 cm IEF strip.

4. Perform first-dimension separation (IEF) using immobilized pH gradient strips (Amersham Biosciences, GE Healthcare, Little Chalfont, UK). A sample program using an Ettan^{TM} IPGphor II IEF system (Amersham Biosciences) is shown in Table 7.1.

5. After separation store strips at −70 °C until needed.

6. Equilibrate strips for 5 min at −20 °C using 10 ml equilibration buffer containing 2% (w/v) DTT.

7. Equilibrate strips again for 5 min at −20 °C using 10 ml equilibration buffer containing 2.5% (w/v) iodoacetamide.

8. Blot the strips and quickly apply to the top of a previously prepared SDS-PAGE slab gel.

9. Apply cooled, molten sealing solution on top of the IPG strip, add running buffer to the gel chamber and run overnight at room temperature and at 50 V.

10. Gels can be visualized by either Coomassie or silver staining.

7.2 Protein digestion in preparation for mass spectrometry by MALDI-TOF

One of the most common applications of MALDI-TOF is peptide mass fingerprinting (PMF) which has been used to classify bacterial species (Dworzanski and Snyder, 2005), to study multidrug resistance in human cells (Chen *et al.*, 2005), to identify the proteome of plant cell walls (Bayer *et al.*, 2005), to study the structure of viral proteins (Schmitt *et al.*, 2004) and to identify novel diagnostic antigens for systemic fungaemia caused by *Candida albicans* (Pitarch *et al.*, 2004).

When proteins are extracted from an organism and separated by electrophoresis, they are further denatured using a number of methods and then digested, usually with trypsin. This generates a range of peptides for each sample, which provides a 'fingerprint' for identification. The resulting mix of MALDI-TOF monoisotopic m/z peak values are used to search publicly or privately held databases to provide a matching result or a result pointing towards a similar protein, providing the user with more information on an unidentified protein.

Protein digestion in preparation for mass spectrometry is relatively simple but requires extreme care to maintain sample integrity and avoid contamination. When preparing samples, it is essential that all reagents used, including water, are of mass spectrometry grade or equivalent. It is advisable to wear gloves and ensure that long hair is tied back to avoid keratin contamination that may interfere with sample identification following mass spectrometry.

To obtain efficient digestion, protein samples must first be prepared by reduction and alkylation, processes used to denature the protein. Reduction of disulphide bonds is achieved using dithiothreitol (DTT) and results in the partial denaturation of the sample. Cysteine residues are commonly modified by alkylation using iodoacetamide, a process known as carbamidomethylation. Reduction and alkylation are necessary because cysteine residues are highly reactive, and if they are not capped during the alkylation process they are liable to re-form disulphide bonds post-reduction. Reduction and alkylation provide a greater number of available sites for the enzymatic digestion of the sample.

One of the most commonly used enzymes for protein digestion in the laboratory is trypsin, a serine protease. The enzyme cleaves proteins at the carboxyl side of lysine and arginine residues. Using trypsin to digest the protein generates a set of peptides, the masses of which can be determined by mass spectrometry thus building a fingerprint profile that can be used to search a number of databases for information that will aid in protein identification.

Protocol 7.2
Peptide mass fingerprinting (PMF) by MALDI-TOF mass spectrometry

Equipment, materials and reagents

Coomassie-stained SDS-PAGE gel

Ice

Sigmacote® (Sigma-Aldrich, St Louis, MO, USA)

Siliconized 1.5 ml microcentrifuge tubes

Microcentrifuge

Acetonitrile

100 mM NH_4HCO_3

50% (v/v) acetonitrile in 100 mM NH_4HCO_3

10 mM DTT in 100 mM NH_4HCO_3

50 mM iodoacetamide in 100 mM NH_4HCO_3

50% (v/v) acetonitrile in 100 mM NH_4HCO_3

Vacuum centrifuge

5 mM $CaCl_2$

Trypisn digestion buffer (1 μg trypsin/20 μl 50 mM NH_4HCO_3/5 mM $CaCl_2$)

Digestion buffer (20 μl 50 mM NH_4HCO_3/5 mM $CaCl_2$)

Extraction buffer (1% (v/v) TFA, 60% (v/v) acetonitrile)

(All buffers are prepared in 17 MΩ/cm Nanopure® or Milli-Q® (Millipore Corporation, Bedford, MA, USA) water or equivalent).

Protocol 7.2a
In-gel digestion

(This method is based on that described in Bergin *et al.*, 2005)

1. Prepare siliconized 1.5 ml microcentrifuge tubes using Sigmacote or equivalent. Allow to air dry before use.

2. Excise band of interest from the SDS-PAGE gel using a sterile siliconized scalpel blade. Dice the protein band into small pieces of approximately 2 × 2 mm.

3. Place the gel pieces into a 1.5 ml microcentrifuge tube and cover with a minimum amount of 100% (v/v) acetonitrile so as to cover the gel pieces. Incubate at room temperature for 5–10 min. After this stage, the gel pieces should begin to clump and the edges become white in colour.

4. Remove the acetonitrile using siliconized pipette tips. Add a minimum volume of 100 mM NH_4HCO_3 and incubate at room temperature for 5 min.

5. Add an equal volume of 50% (v/v) acetonitrile in 100 mM NH_4HCO_3 and incubate at room temperature for 15 min.

6. Remove all liquid using siliconized pipette tips and dry in a vacuum centrifuge. The gel pieces may appear either white or blue at this stage depending on the size of the gel pieces and the length of incubation during the 100 mM NH_4HCO_3 wash steps.

7. Suspend the gel pieces in 100 µl of 10 mM DTT in 100 mM NH_4HCO_3 and incubate for 1 h at 37 °C.

8. Remove the supernatant and wash the gel pieces *twice* in a minimum volume of 100 mM NH_4HCO_3 for 10 min. Discard the final supernatant.

9. Add 100 µl of 50 mM iodoacetamide in 100 mM NH_4HCO_3 and incubate in the dark for 20 min at 37 °C.

10. Wash once with 100 mM NH_4HCO_3 and once with 50% (v/v) acetonitrile in 100 mM NH_4HCO_3.

11. Dry the gel pieces by vacuum centrifugation for 30 min.

12. Cover the gel pieces in trypsin digestion buffer (1 µg trypsin/20 µl 50 mM NH_4HCO_3/5 mM $CaCl_2$) and incubate at 37 °C for 45 min.

13. Remove the trypsin solution and cover the gel pieces with an equal volume of digestion buffer (20 µl 50 mM NH_4HCO_3/5 mM $CaCl_2$). Incubate overnight at 37 °C.

14. Centrifuge samples for 10 min at 12 000 x *g* and remove supernatant to a fresh (siliconized) 1.5 ml microcentrifuge tube.

15. Extract residual peptides twice from the gel pieces by the addition of 50 µl extraction buffer (1% (v/v) TFA, 60% (v/v) acetonitrile) and incubate for 30 min at 37 °C.

16. Centrifuge for 10 min at 12 000 x *g*; pool supernatants with that from step 14.

17. Concentrate volume to 50 µl by vacuum centrifugation.

Protocol 7.2b
In-solution digestion

Based on method described in Kinter and Sherman, 2000)

If a purified protein sample is readily available for MALDI-TOF, it is not necessary to perform gel electrophoresis. These samples can be digested in-solution, which follows the same procedures as an in-gel digestion (reduction and alkylation).

Equipment, materials and reagents

Trypsin-sequencing grade (Sigma-Aldrich)

Ice

Sigmacote® (Sigma-Aldrich, St Louis, MO, USA)

Siliconized 1.5 ml microcentrifuge tubes

Solubilization buffer (6M urea in 100 mM Tris buffer pH 7.8)

Vortex

Reducing buffer (200 mM DTT in 100 mM Tris, pH 7.8)

Alkylating buffer (200 mM iodoacetamide in 100 mM Tris, pH 7.8)

Digestion buffer (25 μl 0.4 M Tris pH 7.8, 75 μl water, 20 μg trypsin)

Concentrated acetic acid

PH paper

(All buffers are prepared in 17 MΩ/cm Nanopure® or Milli-Q® (Millipore Corporation, Bedford, MA, USA) water or equivalent.)

Method

1. Prepare siliconized 1.5 ml microcentrifuge tubes using Sigmacote or equivalent.
2. Suspend evaporated protein sample to a concentration of 10 mg/ml using 6M urea in 100 mM Tris buffer (pH 7.8).
3. Place 100 μl (1 mg) of the 6M urea in 100 mM Tris, pH 7.8 solution to a 1.5 ml microcentrifuge tube.
4. Add 5 μl of reducing buffer (200 mM DTT in 100 mM Tris, pH 7.8) and mix by gentle vortexing. Incubate for 1 h, in darkness, at room temperature.

5. Add 20 μl alkylation buffer (200 mM iodoacetamide in 100 mM Tris, pH 7.8). Incubate for 1 h, in darkness, at room temperature.

6. Add 20 μl reducing agent and vortex gently. Incubate for 1 h, in darkness, at room temperature.

7. Add 775 μl of water and mix by gentle vortexing, this reduces urea concentration to ~0.6 M allowing trypsin activity.

8. Add ice-cold trypsin digestion buffer (25 μl 0.4 M Tris, pH 7.8, 75 μl water, 20 μg trypsin) to the sample and incubate overnight at 37 °C.

9. Halt digestion by addition of concentrated acetic acid until pH is <6. Spot 1 μl volumes of the sample to pH paper to determine pH.

7.3 MALDI-TOF mass spectrometry

MALDI-TOF mass spectrometry is a powerful tool used for protein identification in proteomics. MALDI has gained in popularity because it is relatively easy to use compared with more traditional mass spectrometry techniques while providing valuable results for protein biologists. This, coupled with the development of online tools for the analysis of MS data, has led to an even greater amount of information becoming available in worldwide protein databases.

MALDI works on the principle of sample ionization by pulsed laser shots. The sample is mixed with a matrix material (which can vary depending upon the type of sample used) and loaded onto a sample plate known as the MALDI target. The most common type of laser used for MALDI is a 337 nm Nitrogen laser. The laser can be set to fire a set number of shots with each shot consisting of a number of pulses. It is normal to employ a few hundred pulses per sample. The short, intense pulses efficiently ionize the matrix/sample material causing them to enter the time of flight (TOF) tube during MALDI-TOF after passing through an electric field. The electric field accelerates the passage of ions through the TOF tube. The mass/charge (m/z) ratio of ions is then determined by the time of flight through the TOF tube.

All MALDI experiments take place within a vacuum. Samples are loaded to the sample chamber, which is then sealed while a vacuum is generated. Most commercial MALDI-TOF machines employ both linear and reflectron detectors for use in MALDI experiments. During linear MALDI-TOF, the ionized particles travel through the TOF tube to a detector at the far end of the tube, which then determines the exact molecular weight of the ions hitting it. This leads to the generation of a broad-based peak, which gives the molecular weight of the ion in question. **Linear MALDI-TOF** (Figure 7.2) is usually used for large molecules, typically intact, native protein.

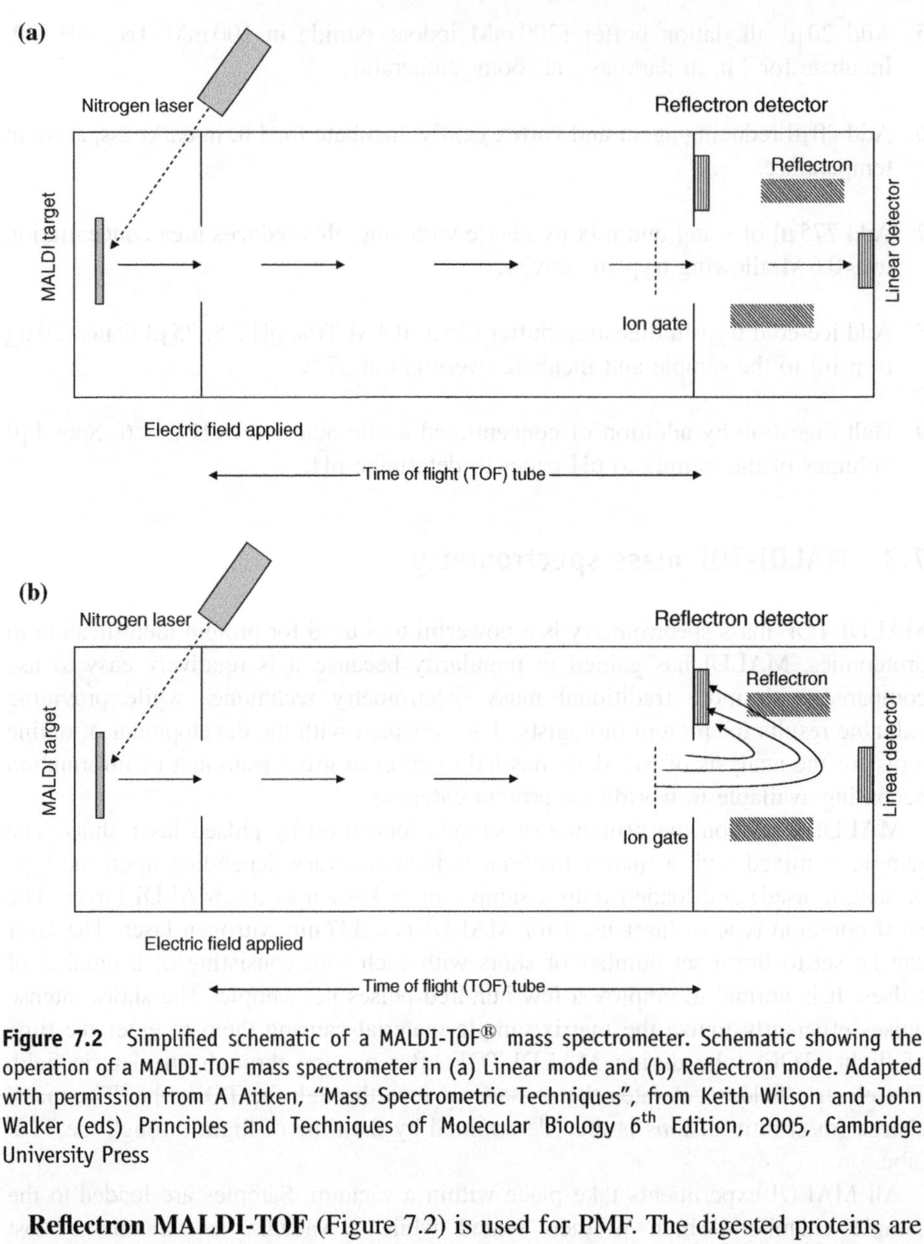

Figure 7.2 Simplified schematic of a MALDI-TOF® mass spectrometer. Schematic showing the operation of a MALDI-TOF mass spectrometer in (a) Linear mode and (b) Reflectron mode. Adapted with permission from A Aitken, "Mass Spectrometric Techniques", from Keith Wilson and John Walker (eds) Principles and Techniques of Molecular Biology 6th Edition, 2005, Cambridge University Press

Reflectron MALDI-TOF (Figure 7.2) is used for PMF. The digested proteins are loaded to the sample chamber and ionized. The sample ions pass through the electric field, are accelerated in the normal manner through the TOF tube but are reflected towards a second detector by the application of an electric charge usually coupled with a procedure called 'delayed extraction' (DE), which is discussed later in this chapter.

These two methods allow MALDI-TOF to be used to analyse the mass of intact proteins, digested proteins and also to sequence peptide fragments.

Table 7.2 Commonly used matrix materials

Anylate	Matrix material
Peptide/Protein Mass < 10 kDa	α-cyano-4-hydroxycinnamic acid
Peptide/Protein Mass > 10 kDa	Sinapic acid 2-(4-hydroxyphenylazo)benzoic acid
Lipids	Dithranol DIT
Oligosaccharide	1-isoquinolinol

MALDI-TOF matrix material

The matrix used in MALDI-TOF is simply a light-absorbing compound. Different matrices allow for different levels of ionization and are used for optimal particle ionization depending on the type of sample to be analysed (Table 7.2). Two of the most commonly used matrices are α-cyano-4-hydroxycinnamic acid (CHCA, MW189.04 Da), generally used with samples of MW < 10 kDa and sinapic acid (MW224.07 Da), generally used with samples of MW > 10 kDa. The matrix must be freshly made, immediately before use, and it is common practice to add standards to be used for internal calibration during the mixing of the matrix material. Samples for MALDI-TOF analysis are usually pipetted onto the MALDI target and matrix applied directly on top of the sample. Excess moisture in the MALDI-TOF sample chamber makes it difficult to achieve a vacuum; so the sample and matrix must be allowed to air dry completely before the target is loaded to the sample chamber. Once the sample and matrix mix have dried, it is possible to further load the target by repeatedly spotting sample followed by matrix and allowing to dry each time. This can be useful in cases where multiple spectra are required from the same sample on the target. If MALDI-TOF spectra are yielding few or poor peaks, it is sometimes advisable to try a different matrix material before beginning the process of protein extraction and digestion of all samples again.

Protocol 7.3
Preparation of matrix for MALDI-TOF

The following short protocols describe the preparation of matrix for MALDI-TOF analysis and the optional addition of standards for use as internal calibrants.

Equipment, materials and reagents

All reagents are from Sigma-Aldrich Chemical Co, St Louis, MO, USA unless otherwise stated.

CHCA (LaserBio Labs, Sophia-Antipolis Technopole, France)

Sinapic acid

Acetonitrile

0.1% (w/v) tri-fluro acetic acid (TFA)

MALDI-TOF internal calibrants

Note: Matrix solution must be made freshly, immediately before use.

Method

Preparation of α-cyano-4-hydroxycinnamic acid (CHCA)

1. Weigh 10 mg CHCA matrix to a sterile microcentrifuge tube.
2. Add 1 ml of 50% (v/v) acetonitrile, 0.1% ([v/v] final concentration) TFA.
3. Vortex for 30 sec, centrifuge for 60 sec at 12 000 x *g*.
4. Transfer supernatant to a new microcentrifuge tube.
5. Optional: Add 1–3 μl internal calibrants, vortex for 30 sec.
6. Use immediately.

Preparation of sinapic acid

1. Weigh 10 mg of sinapic acid to a sterile microcentrifuge tube
2. Add 1 ml of a 30% (v/v) acetonitrile solution, 0.1% ([v/v] final concentration) TFA
3. Vortex for 30 sec, centrifuge for 60 sec at 12 000 x *g*
4. Transfer supernatant to a new microcentrifuge tube
5. Optional: add 1–3 μl internal calibrants, vortex for 30 sec.
6. Use immediately.

Preparation of internal calibrants for MALDI-TOF

1. Prepare 100 pmol stocks of calibrants to be used by dilution in 0.1% (v/v) TFA. If using insulin, dissolve in 1% (v/v) TFA. Bradykinin or insulin oxidized B chain should be dissolved in 0.05% (v/v) TFA, 50% (v/v) acetonitrile.

2. If using a standard for the first time, perform serial dilutions to 10 pmol, 1 pmol and 100 fmol respectively, in the appropriate solution. These can then be used for sensitivity testing to optimize for use in future experiments.

3. When using the standard in experiments, dilute to the desired concentration as determined in step 2 and add 1–3 μl to the matrix material and vortex for 30 sec before use.

7.4 Peptide mass fingerprinting (PMF)

By far the most common application of MALDI is peptide mass fingerprinting (PMF). During PMF, digested proteins, mixed with the matrix material are applied to a MALDI target undergoing ionization and entering the time of flight (TOF) tube after passing through an electric field. The electric field accelerates the ionized particles allowing them to be separated along the TOF tube according to their mass and charge (m/z).

Resolution of MALDI-TOF results is less than optimal if particles of the same m/z arrive at the detector at slightly different times. This happens when freshly ionized particles of the same m/z enter the TOF tube at different velocities. The principle of delayed extraction (DE) is used to maximize the resolution of MALDI-TOF results. When DE is applied to a sample, ions reaching the end of the TOF tube pass through a series of electric fields of varying strength known as an 'ion gate'. Ions travelling at a higher velocity arrive at the ion gate first but travel deeper into the field before being reflected towards the detector. Ions of the same m/z travelling at a lower velocity arrive at the ion gate later but do not penetrate the field as deeply and are therefore reflected earlier. This has the net effect of allowing ions of the same m/z travelling at different initial velocities to arrive at the detector at the same time. This allows for maximal peak resolution.

When performing MALDI-TOF, it is normal practice to obtain at least three spectra from each sample to be analysed. This is due to a number of factors including the nature of matrix crystallization, the ratio of peptides or protein to matrix and improved or reduced ionization as matrix material is removed from the target. All of these factors contribute to the success of sample analysis. It is often the case that the first spectra produced yields of very few or no peaks only for the second spectra to yield a high number of peaks and vice versa. It is almost impossible to predict in advance how these factors will affect the success of an experiment, and therefore it is strongly advised to produce more than one spectrum for each sample.

Post-source decay (PSD) for peptide sequencing

It is possible to sequence peptides using MALDI-TOF post-source decay (PSD). During PSD, a reflectron spectrum is generated as for PMF and the user chooses a peak of interest for further analysis. PSD has the advantage over Edman degradation sequencing of peptides/proteins in that PSD does not suffer the effects of N-terminal blocking. When a suitable peptide peak has been chosen, the user can alter the machine settings so that only ions of the m/z of the selected peak are allowed into the TOF tube. This is achieved by the application of an electric charge known as the ion gate, which will not allow ions other than the selected m/z to pass through. The sample undergoes a second round of analysis during which the power of the laser is increased and the ions are accelerated into the TOF tube. The outcome of this treatment is the breakdown of the peptide into its constituent amino acids.

The PSD process can be further enhanced through the application of chemically assisted fragmentation (CAF) (Flensberg *et al.*, 2005). During CAF, the samples are treated prior to MALDI-TOF analysis. The CAF process introduces a sulfonic acid group to the peptide N-terminus and involves modification of the lysine residues of peptides such that the lysine residues are converted to homoarginine. The net effect is that two protons are introduced to the peptides, one in the C-terminal side and one on the peptide backbone. When the peptides are fragmented, the y-ion retains a positive charge while the b-ions do not and therefore only the y-ions are detected.

Protocol 7.4
Post-source decay (PSD) and chemically assisted fragmentation (CAF)

Post-source decay relies on the fragmentation of peptide chains to their individual amino acid components.

Equipment, materials and reagents

This protocol describes peptide sequencing PSD by use of the Ettan™ CAF MALDI sequencing kit (Amersham Biosciences, GE Healthcare, Little Chalfont, UK).

Ettan CAF MALDI sequencing kit (Amersham Biosciences)

C_{18} ZipTip® (μZT) (Millipore Corporation, Bedford, MA, USA)

0.1% (v/v) trifluoroacetic acid (TFA)

50% (v/v) acetonitrile: 0.5% (v/v) TFA

80% (v/v) acetonitrile: 0.5% (v/v) TFA

Siliconized 1.5 ml microcentrifuge tubes

(All buffers are prepared in 17 MΩ/cm Nanopure® or Milli-Q® (Millipore Corporation, Bedford, MA, USA) water or equivalent.)

Method

Sample loading to µZT

1. Take 90 µl of a protein digest solution and dry by vacuum centrifugation.
2. Resuspend sample in 5 µl 0.1% (v/v) TFA.
3. Prepare µZT for use by aspirating 10 µl, 50% (v/v) acetonitrile, 0.5% (v/v) TFA and dispensing to waste 5 times.
4. Equilibrate the µZT by aspirating, 0.1% (v/v) TFA and dispensing to waste 5 times.
5. Load the sample to the µZT by aspirating and dispensing *slowly* 10 times. Retain the µZT as the sample has now bound to the column membrane.

Lysine modification of sample

1. Add 100 µl of water to the 8.6 mg vial of lysine modifier supplied.
2. Mix 2 µl of the lysine modifier solution with 8 µl of the lysine modifier buffer and load to the µZT by aspirating and dispensing 5 times. Do not dispense all liquid from the µZT on the final cycle as some liquid must remain above the C18 plug within the µZT.
3. Remove the µZT, being careful not to allow air into the sample. Place the µZT in a 1.5 ml microcentrifuge tube closing the lid and leaving at room temperature overnight.
4. Following the overnight incubation, dispense the lysine modification solution from the µZT.

Sulphonation of the N-terminus by CAF reaction

1. Wash the µZT with water by aspirating and dispensing 5 times; this removes traces of lysine modifier.
2. Make fresh CAF reagent by adding 60 µl of CAF buffer to 1 of the CAF reagent vials supplied with the kit. Mix well by aspirating and dispensing. The reagent should be used within 20 min.

3. Dispense 10 µl of the CAF solution into a 1.5 ml microcentrifuge tube and vortex gently to remove air bubbles.

4. Immediately load the CAF solution to the µZT by aspirating and dispensing *slowly* 10 times.

5. Leave the CAF solution in the µZT for 3 min, removing the µZT from the pipette at this time, leaving the CAF solution in the µZT. Continue with the next sample.

6. Dispense the CAF solution from the µZT into the microcentrifuge tube.

Reversal of unwanted side reactions

1. Add 1 µl of the stop solution to the unloaded CAF reagent. Mix and load to the µZT by aspirating and dispensing 10 times.

Eluting the sample

1. Wash the µZT with 0.1% (v/v) TFA aspirating and dispensing *to waste* 5 times.

2. Elute the sample in 5 µl of 80% (v/v) acetonitrile, 0.5% (v/v) TFA, aspirating and dispensing 10 times.

3. Dry the sample by vacuum centrifugation or under a stream of nitrogen gas.

7.5 Interpreting MALDI-TOF result spectra

One of the reasons for the popularity of MALDI-TOF in biological sciences is the relative ease at which result spectra can be analysed. To obtain a meaningful result from a spectrum does require training, but MALDI-TOF spectra can be interpreted without specialist training as a mass spectrometrist.

Examples of MALDI-TOF spectra are shown in Figures 7.3–7.5. A PMF spectrum usually ranges from 500 m/z to approximately 3000 m/z (Figure 7.3), although for practical purposes it is better to search for peaks between 800 m/z and 2500 m/z.

This is because at <800 m/z peptide peaks can be difficult to distinguish from those generated by matrix materials and realistically a peak of this m/z is likely to represent a peptide of at most three amino acids. Above 2500 m/z peaks are likely to be of very low intensity; this is due to the fact that a peak of this value would represent a peptide of at least 20 amino acids and these are relatively few in number due to the specificity of trypsin cleavage. At >2500 m/z peaks are likely to be indistinguishable from baseline noise due to their low numbers unless an enzyme that cuts less frequently

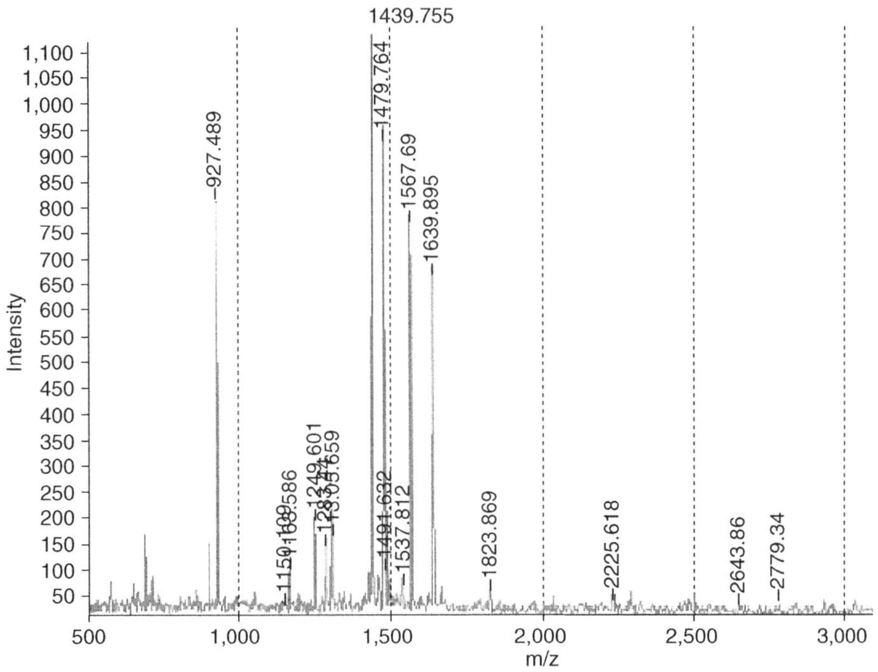

Figure 7.3 MALDI-TOF® PMF of trypsin-digested BSA

than trypsin is used. Trypsin is used most often because it is well characterized and understood and yields a large number of useful peaks for the identification of most proteins.

Each peak spreads across a narrow m/z range and is characteristically split into four or more peaks when analysed in detail (Figure 7.4). The peak furthest on the left is referred to as the monoisotopic peak due to its representation of Carbon-12 (Figure 7.4). As a general, although not exclusive, rule below 2000 m/z the monoisotopic peaks are most intense with the other peaks decreasing in intensity respective to one another in a left-to-right fashion. Above 2000 m/z the monoisotopic peak appears reduced in intensity and the peaks tend to take the shape of a bell curve (Figure 7.5). The m/z values from a PMF spectrum are usually reported using the m/z at the zenith of the monoisotopic peaks. Above approximately 2500 m/z, the m/z is reported as an average as it is increasingly difficult to identify the monoisotopic peak.

A good MALDI-TOF spectrum will yield at least ten unambiguous peaks, although it is possible to identify a protein with fewer peaks. Many peaks can be hidden in the background noise due to their low intensity, and searching the spectrum thoroughly can often reveal these peaks. The use of standards within the matrix adds reassurance of the mass accuracy of the result as these peaks are usually very intense due to the amount loaded when mixing with the matrix material. Overloading with standard can flatten the spectra obtained obscuring the peaks, which is why it is important to zoom in and take a closer look at the spectra rather than choosing the most obvious peaks

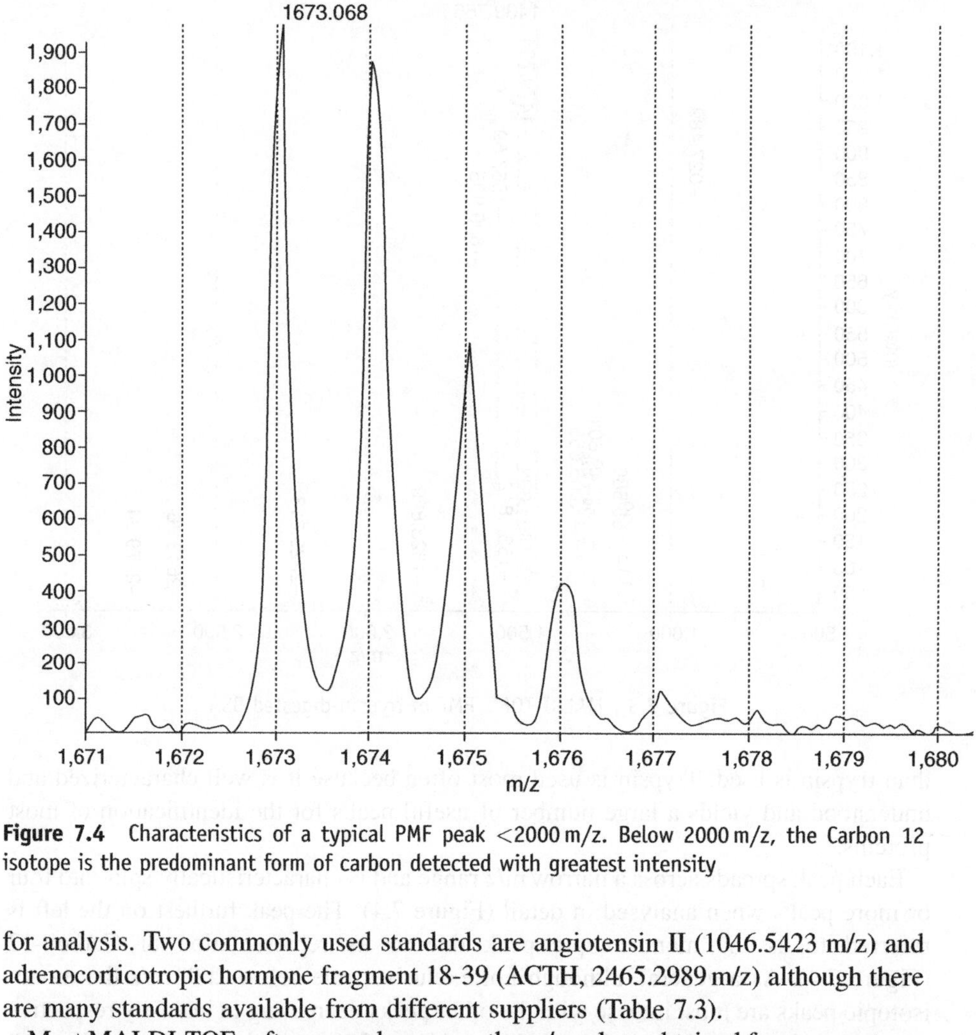

Figure 7.4 Characteristics of a typical PMF peak <2000 m/z. Below 2000 m/z, the Carbon 12 isotope is the predominant form of carbon detected with greatest intensity

for analysis. Two commonly used standards are angiotensin II (1046.5423 m/z) and adrenocorticotropic hormone fragment 18-39 (ACTH, 2465.2989 m/z) although there are many standards available from different suppliers (Table 7.3).

Most MALDI-TOF software packages use the m/z values obtained from a spectrum to interrogate databases chosen by the user, based upon the criteria employed to generate the sample. When using this software, incorrect parameter choices can lead to misleading results, no results or at worst the incorrect identification of one protein as another. There are several tools available online that allow a user to input the m/z values obtained from a spectrum and interrogate publicly available databases; examples include MS-Fit from the Protein Prospector suite of tools (http://prospector.ucsf.edu/) and *Pep*MAPPER (http://wolf.bms.umist.ac.uk/mapper/). These tools require the user to import a list of masses and detail the enzyme used for digestion, the number of missed cleavages allowed and the methods of cysteine modification employed. A list of all possible matches from the database chosen are then displayed with the number of matching peaks.

It is also possible to generate theoretical digests based on a protein sequence using online tools. This can be useful, for example in identifying proteins modified in the

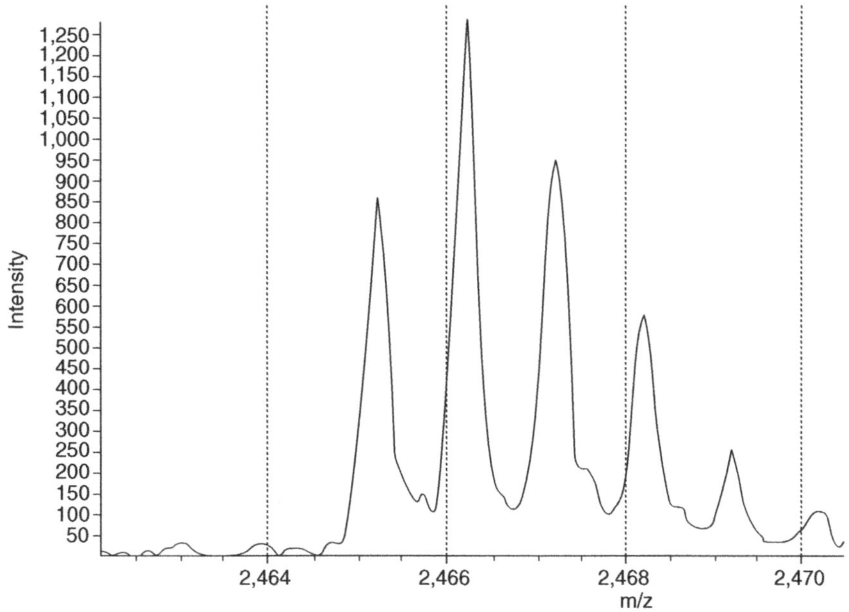

Figure 7.5 Characteristics of a typical PMF peak >2000 m/z. Above 2000 m/z, the Carbon 12 isotope is no longer the predominant form of carbon detected and the peak characteristics change as intensities alter

laboratory to contain certain tags or conserved motifs expected in a number of samples. The MS-DIGEST tool available on the Protein Prospector website (http://prospector.ucsf.edu/) is one such example. This tool requires the user to enter a protein sequence as well as the methods the software is to use for the *in silico* digestion. The user is required to choose the enzyme to be used for digestion and the methods to employ for cysteine modification as well as other modifications to be considered. This generates a list of hypothetical monoisotopic peaks that can be used to search the peaks in the sample spectrum. This is especially useful when searching for peaks obscured by background noise and/or very intense neighbouring peaks.

Table 7.3 Commonly used MALDI-TOF® standards

Standard	m/z	Monoisotopic/Average
Bradykinin Fragment 1–7	757.3997	monoisotopic
Angiotensin II (human)	1046.5423	monoisotopic
ACTH Fragment 18–39 (human)	2465.1989	monoisotopic
Insulin-oxidized B chain (bovine)	3494.6513	monoisotopic
Cytochrome c (equine)	12 361.96	average
Apomyoglobin (equine)	16 952.27	average
Aldolase (rabbit muscle)	39 212.28	average
BSA	66 430.09	average

7.6 Overview

The completion of the *Saccharomyces cerevisiae* genome project in 1996 (*Goffeau et al.*, 1996) heralded the arrival of the genomic era in biological sciences. Since then, numerous viral, bacterial, fungal, plant and animal genomes have been sequenced and the first drafts of the complete human genome have been completed. There are a number of important sequencing projects ongoing that will mean greater availability of data for medical mycologists, these include the *C. albicans* (Braun *et al.*, 2005) and *A. fumigatus* sequencing projects.

The completion of these projects should make it easier for protein biologists to determine the identity of proteins from these species. This should also make it easier to predict the presence of similar proteins in many different medically important fungi.

The development of proteomics has facilitated the examination of the entire proteome of an organism, and this is slowly leading to the building of protein–protein interaction maps (Rual *et al.*, 2005), or 'interactomes' as they are being termed. These maps of protein–protein interaction are allowing scientists to build a picture of how exactly one set of occurrences impact upon another during pathogenesis.

A conventional method of determining antifungal drug susceptibilities is to treat one culture of yeast or fungi with an antifungal agent while maintaining a control, drug-naive culture. Using this approach, specific proteins suspected of being involved in resistance are targeted for analysis, but this type of analysis cannot fully explain what happens during resistance. A proteomic approach involves extracting the whole-cell protein from both cultures, performing 2D SDS-PAGE, looking for differences in the protein spots between the treated and untreated gels and subsequent digestion and MALDI-TOF analysis. This approach enables the researcher to quickly determine what proteins are affected by the addition of a drug and develop an idea of global effects within the fungal cell.

Proteomics has been used to compare strain-specific differences within a single bacterial species (Sullivan and Bennett, 2005). This approach could also be used to determine the differences between virulent and avirulent strains of yeast or fungi and may shed further light on why certain strains of less common yeast and fungi colonize human tissue.

An examination of the protein content of mitochondrial membranes in *Neurospora crassa* has identified proteins whose localization within the fungus had previously been unknown (Schmitt *et al.*, 2005). Pathogenic yeast and fungi can easily be analysed to determine the presence or absence of known virulence factors using the same approach.

High-throughput proteomics is also rapidly changing the way in which researchers work. Automated spot pickers and machines that automatically digest protein samples can be coupled with MALDI-TOF mass spectrometers. This type of analysis can yield results quickly, but as always it is the well-designed experiment that yields the most accurate result.

7.7 References

Bayer, E. M., Bottrill, A. R., Walshaw, J., Vigouroux, M. *et al.* (2005) Arabidopsis cell wall proteome defined using multidimensional protein identification technology. *Proteomics.* DOI:10.1002/pmic.200500046.

Bergin, D., Reeves, E. P., Renwick, J. Wientjes, F. B. and Kavanagh, K. (2005) Superoxide production in haemocytes of *Galleria mellonella*: Identification of proteins homologous to the NADPH oxidase complex of human neutrophils. *Infect. Immun.* **73**: 4161–4173.

Braun, B. R., van Het Hoog, M., d'Enfert, C., Martchenko, M. *et al.* (2005) A human-curated annotation of the *Candida albicans* genome. *PLoS Genet.* **1**(1): 36–57.

Burns, C., Geraghty, R., Neville, C., Murphy, A. *et al.* (2005) Identification, cloning, and functional expression of three glutathione transferase genes from *Aspergillus fumigatus. Fungal Genet. and Biol.* **42**(4): 319–327.

Chen, C. Y., Jia, J. H., Zhang, M. X., Meng, Y. S. *et al.* (2005) Proteomic analysis on multi-drug resistant cells HL-60/DOX of acute myeloblastic leukemia. *Chin. J. Physiol.* **48**(3): 115–120.

Dworzanski, J. P. and Snyder, A. P. (2005) Classification and identification of bacteria using mass spectrometry-based proteomics. *Expert Rev Proteomics.* **2**(6): 863–878.

Flensburg, J., Tangen, A., Prieto, M., Hellman, U., and Wadensten, H. (2005) Chemically-assisted fragmentation combined with multi-dimensional liquid chromatography and matrix-assisted laser desorption/ionization post source decay, matrix-assisted laser desorption/ionization tandem time-of-flight or matrix-assisted laser desorption/ionization tandem mass spectrometry for improved sequencing of tryptic peptides. *Eur. J. Mass. Spectrom.* **11**(2): 169–179.

Goffeau, A., Barrell, B. G., Bussey, H., Davis, R. W. *et al.* (1996) Life with 6000 genes. *Science.* **274**(5287): 546, 563–567.

Hooshdaran, M., Barker, K. S., Hilliard, G. M., Kusch, H. *et al.* (2004) Proteomic analysis of azole resistance in *Candida albicans* clinical isolates. *Antimicrob. Agents. Chemother.* **48**(7): 2733–2735.

Kinter, M. and Sherman, N. E. (2000) *Protein Sequencing and Identification Using Tandem Mass Spectrometry*, John Wiley & Sons, Chichester.

Laure, F., Roberts, M. A., Fay, L. B. (2003) Matrix-assisted laser desorption/ionization time-of-flight mass spectrometry in clinical chemistry. *Clinica. Chimica. Acta.* **337**: 11–21.

Lilley, K. S. and Friedman, D. B. (2004) All about DIGE: Quantification technology for differential-display 2D-gel proteomics. *Expert. Rev. Proteomics.* **1**(4): 401–409.

Massoumeh, Z., Hooshdaran, Barker, K. S., Hilliard, G. M., Kusch, H. *et al.* (2004) Proteomic analysis of azole resistance in *Candida albicans* clinical isolates. *Antimicrob. Agents. Chemother.* **48**(7): 2733–2735.

Medina, M. L., Kiernan, U. A. and Francisco, W. A. (2004) Proteomic analysis of rutin-induced secreted proteins from *Aspergillus flavus. Fungal. Genet. Biol.* **41**(3): 327–335.

Neville, C., Murphy, A., Kavanagh, K., and Doyle, S. (2005) A 4′-Phosphopantetheinyl transferase mediates non-ribosomal peptide synthetase activation in *Aspergillus fumigatus. Chembiochem.* **6**(4): 679–685.

Pitarch, A., Abian, J., Carrascal, M., Sanchez, M. *et al.* (2004) Proteomics-based identification of novel *Candida albicans* antigens for diagnosis of systemic candidiasis in patients with underlying hematological malignancies. *Proteomics.* **4**(10): 3084–3106.

Rual, J. F., Venkatesan, K., Hao, T., Hirozane-Kishikawa, T. *et al.* (2005) Towards a proteome-scale map of the human protein–protein interaction network. *Nature.* **437**(7062): 1173–1178.

Schmitt, S., Glebe, D., Tolle, T. K., Lochnit, G. *et al.* (2004) Structure of pre-S2 N- and O-linked glycans in surface proteins from different genotypes of hepatitis B virus. *J. Gen. Virol.* **85**(7): 2045–2053.

Schmitt, S., Prokisch, H., Schlunck, T. *et al.* (2005) Proteome analysis of mitochondrial outer membrane from *Neurospora crassa*. *Proteomics.* DOI: 10.1002/pmic.200402084.

Shevchenko, A., Jensen, O. N., Podtelejnikov, A. V., Sagliocco, F. *et al.* (1996) Linking genome and proteome by mass spectrometry: Large-scale identification of yeast proteins from two-dimensional gels. *Proc. Natl. Acad. Sci. USA.* **93**(25): 14440–14445.

Sullivan, L. and Bennett, G. N. (2005) Proteome analysis and comparison of *Clostridium acetobutylicum* ATCC 824 and Spo0A strain variants. *J. Ind. Microbiol. Biotechnol.* **25**: DOI: 10.1007/s10295-005-0050-7.

Vandeputte, P., Larcher, G., Berges, T., Renier, G. *et al.* (2005) Mechanisms of azole resistance in a clinical isolate of *Candida tropicalis*. *Antimicrob. Agents Chemother.* **49**(11): 4608–4615.

8

Extraction and detection of DNA and RNA from yeast

Patrick Geraghty and **Kevin Kavanagh**

8.1 Introduction

The isolation, purification and characterization of DNA or RNA is critical for every modern molecular technique. This chapter outlines several methods for isolating DNA and RNA from yeast and several widely used approaches to detecting the presence or expression levels of specific genes. There are many methods now described in the literature for the isolation of RNA and DNA, some of which are also described in other chapters in this book. DNA may be isolated for use in DNA mapping, for making new vector or plasmid constructs or used to amplify a desired sequence and introduce it into competent cells. In addition, DNA may be required for the construction of genomic libraries, cDNA synthesis or PCR.

DNA can be purified from many fungi by physically breaking down the cell wall and precipitating the DNA with an alcohol in the presence of a high salt concentration. A variety of methods can be used to disrupt cells and include French press, vortexing cells in the presence of glass beads, grinding cells under liquid nitrogen or sonication (use of ultrasound pulses). Fungal cell walls can also be degraded by exposure to lytic enzymes such as glucanases and chitinases (Kavanagh and Whittaker, 1996) – many of which are available commercially. In recent years, a number of kits have become available commercially, and these can be very efficient for isolating and purifying DNA. However, it is important to note that the optimum method for isolating DNA from one fungus may not work for another species or strain of the same species so it is advisable to optimize the technique for each fungus under study (Fredricks *et al.*, 2005).

Medical Mycology: Cellular and Molecular Techniques. Edited by Kevin Kavanagh

High-quality RNA is required for many molecular biological techniques; however, the isolation and use of purified RNA in the laboratory is complicated as RNA is significantly more labile than DNA (Chworos *et al.*, 2004). Like DNA, RNA-extraction kits are commercially available from many sources. RNA is extremely susceptible to degradation by endogenous and exogenous ribonucleases (RNases), which are both ubiquitous and difficult to inactivate in the presence of RNA (Alifano *et al.*, 1994). Detailed procedures for working with RNA are described in this chapter.

The choice of using total RNA or mRNA (i.e. poly(A) + RNA) depends largely upon the downstream application(s) for the isolated material. Most molecular-biology applications such as RT-PCR, Northern blotting and RNase protection assays do not ordinarily require mRNA and usually work quite well with high-quality total RNA (as described in this chapter). Improved sensitivity and more-accurate results may be obtained using mRNA in techniques such as cDNA library construction and Northern blot detection of rare messages (Wipf *et al.*, 2003; Neville *et al.*, 2005).

The organic extraction of nucleic acids with acidified phenol:chloroform is the most commonly employed method, but samples obtained may contain contaminants which need to be removed. RNA-isolation techniques generally involve steps to eliminate DNA. Enzymatic steps such as the use of DNase or RNase are outlined here for the removal of nuclease contamination; this is particularly important prior to performing techniques such as RT-PCR (Burns *et al.*, 2005). Also, protein RNase inhibitors are often used in enzyme reactions to sequester RNases that may remain in an RNA sample or that may be introduced with other reagents or while pipetting (Carriero *et al.*, 2003).

Methods for the detection of the presence or levels of expression of a gene are explained in detail in this chapter. Nucleic acid blotting involves using labelled single-stranded DNA or RNA to hybridize and detect complementary single-stranded nucleic acid that has been fixed to a solid membrane surface. Traditionally, the target nucleic acid is presented in a complex mixture of nucleic acid fragments that have been electrophoresed to separate the fragments by relative size prior to their transfer and fixation to nylon or nitrocellulose membrane.

Northern blotting (named by analogy to Southern blotting) is used to characterize one or more specific mRNA transcript in a sample of total RNA. The RNA is denatured, size-separated by electrophoresis and transferred to a nylon membrane, where it is hybridized and detected with labelled, complementary, target-specific DNA or RNA probes. Northern blots may be used to quantify a target, determine mRNA transcript size, detect alternative splice variants of a gene and identify closely related species.

Factors that influence hybridization efficiency and specificity in Northern and Southern analyses include temperature, ionic strength, destabilizing agents, mismatched base pairs, duplex length, viscosity and base composition. High-salt concentrations favour hybridization reactions while decreased salt levels and/or increased detergent concentration or temperature increase hybridization specificity and reduce background. Traditionally, the DNA or RNA probe in Southern and Northern blotting procedures is labelled with radioactive phosphorus (e.g. ^{32}P) or sulphur (e.g. ^{35}S) by incorporation of radiolabelled nucleotides during its synthesis.

In recent years, alternative, non-isotopic methods have largely replaced radioactive systems but may not always result in better (cleaner) results.

8.2 The extraction of yeast DNA with the aid of phenol: chloroform

There are many protocols in use for the extraction of whole-cell DNA from fungi but the conventional and most reliable method involves the use of phenol and chloroform. Many scientific product providers have specific kits and reagents that provide a fast and reliable means of extracting DNA, which is clean and available in high quantities. The technique described in Protocol 8.1 yields large amounts of DNA, which can be used for many molecular techniques and results in good yields of high-purity DNA.

Protocol 8.1 Whole-cell DNA extraction from *C. albicans* using phenol: chloroform

Equipment, materials and reagents

Stationary phase yeast cells (approximately 1×10^9 total)

10 mM EDTA, pH 8.0

Spheroplasting buffer (1 M sorbitol (Sigma-Aldrich), 0.1 M EDTA, 1.5 mg/ml lyticase (Sigma-Aldrich) and 0.05 M dithiothreitol, pH 7.5)

Lysing buffer (50 mM EDTA, 50 mM Tris (pH 8), 0.4 ml of 10% (w/v) SDS and 0.1 ml of a 30 mg/ml proteinase K solution)

2 mm diameter glass beads

Phenol-chloroform-isoamyl alcohol (25:24:1) solution

Ice-cold 95% (v/v) ethanol

1 mg/ml RNase (can be prepared by suspending 1 mg of Ribonuclease A in 50 mM potassium acetate and boiled for 20 min to remove any DNase contamination)

TE buffer, pH 7.4 (10 mM Tris-HCl, pH 7.4 and 1 mM EDTA, pH, 8.0)

50 x TAE (24.2% (w/v) Tris-HCl, 5.71% (v/v) glacial acetic acid, 0.05 M EDTA, pH 8.0)

Method

This method is adapted from that described by Wach *et al.* (1994).

1. Harvest stationary phase cells of *C. albicans* by centrifugation, wash in 10 ml of 10 mM EDTA, (pH 8.0), resuspend in 4 ml of spheroplasting buffer and incubate at 37 °C for 1–2 h. Alternatively, once cells are suspended in spheroplasting buffer without a digestive enzyme, they can be disrupted using a vortex mixer (5 min) or bead beater (30 sec) in the presence of 5 gram of glass beads. It is advisable to vortex cells for 30 sec and then place on ice for 30 sec prior to the next phase of agitation since the heat generated by this process can degrade DNA and RNA.

2. Add 4 ml of lysing buffer to cell suspension and maintain at 65 °C for 30 min.

3. Add 1 volume of phenol-chloroform-isoamyl alcohol (25:24:1) to the cell suspension, which is mixed for 5 min, followed by centrifugation at 17 400 x *g* for 15 min.

4. Collect the upper aqueous layer and wash with 1 volume of chloroform-isoamyl alcohol (24:1). This is repeated until no white flocculent interphase remains. At this stage, the DNA extraction is complete but the solution will contain many impurities and must undergo several **purification** steps.

5. Precipitate the DNA with two volumes of ice-cold 95% (v/v) ethanol and incubate at −20 °C overnight or −70 °C for 1 h. Centrifuge at 17 400 x *g* at 4 °C for 15 min to pellet the DNA.

6. Washing the DNA with 5 ml of 70% (v/v) ethanol followed by centrifugation at 17 400 x *g* at 4 °C for 15 min will further remove any remaining phenol in the DNA sample. Following ethanol precipitation, allow the DNA pellet to air dry.

7. Resuspend the DNA in 4 ml of sterile water with 150 μl of a 1 mg/ml solution of RNase and incubate at 37 °C for 30 min.

8. Re-precipitate the DNA by once again adding two volumes of 95% (v/v) ethanol, 1/10 final volume of a 3 M sodium acetate solution (pH 5.2) and incubate at −20 °C overnight or −70 °C for 1 h.

9. Centrifugation at 17 400 x *g* for 15 min will pellet the DNA, followed by air drying.

10. Dissolve the pellet in either TE buffer (for long-term storage) or sterile water and store at −20 °C. It may be difficult to dissolve the pellet completely, but incubation at 50 °C for 30 min will greatly aid this.

11. The **quality** and **quantity** of the DNA should be examined before proceeding with further experiments. DNA concentration can be determined by measuring the absorbance at 260 nm in a spectrophotometer using a quartz cuvette. The DNA isolated using the extraction method described above will need to be diluted 1:100 in sterile water or TE before reading. An absorbance of 1 unit at 260 nm corresponds to 50 µg genomic DNA per ml at pH 7.0 (remember to adjust concentration by dilution factor). A pure sample of DNA has an OD260/OD280 value between 1.9–2 units; anything less indicates protein contamination.

12. DNA can also be examined by running a few µl of the sample on a 0.8% (w/v) agarose gel to visualize the amount of DNA present. This is useful for visualizing any RNA contamination. An agarose gel can be prepared by adding 0.8% (w/v) agarose to 1 x TAE, boiling and allowing to set. This DNA can be cut with a variety of restriction enzymes (Figure 8.1).

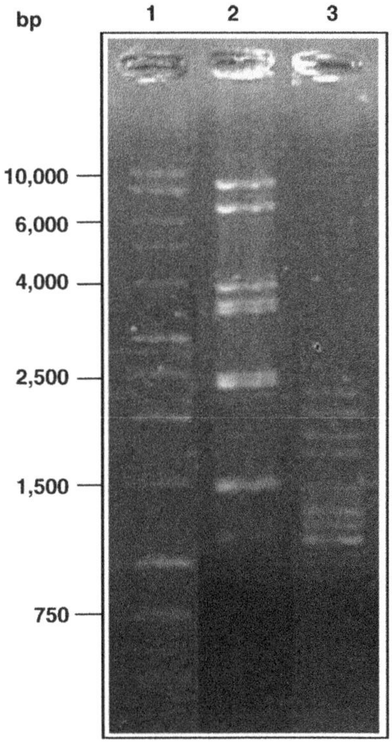

Figure 8.1 Whole-cell DNA from *C. albicans* that underwent RFLP analysis. Whole-cell DNA was extracted as described in Protocol 8.1 and was digested for 3 h at 37 °C with *Ecor*V (lane 2) and *Hinf*I (lane 3). Lane 1 molecular weight marker

Protocol 8.2
Rapid extraction of DNA from *C. albicans* colonies for PCR

Equipment, materials and reagents

C. albicans cells

YEPD agar plates (same as YEPD broth but containing 2% (w/v) agar)

Heating block

Method

This method is adapted from that described by Deak *et al.* (2000).

1. Plate-out approximately 250 cells on a YEPD plate and culture overnight at 37 °C.
2. Collect colonies (4 or 5 will usually suffice) of cells and wash with sterile water.
3. Resuspend cells in 100 μl of sterile water.
4. Heat cells to 95 °C for 10 min and then centrifuge at 16 000 x *g* for 5 min. The supernatant is used as the DNA template for PCR analysis. 2 μl of supernatant should be sufficient in a 50 μl PCR reaction. (Figure 8.2)

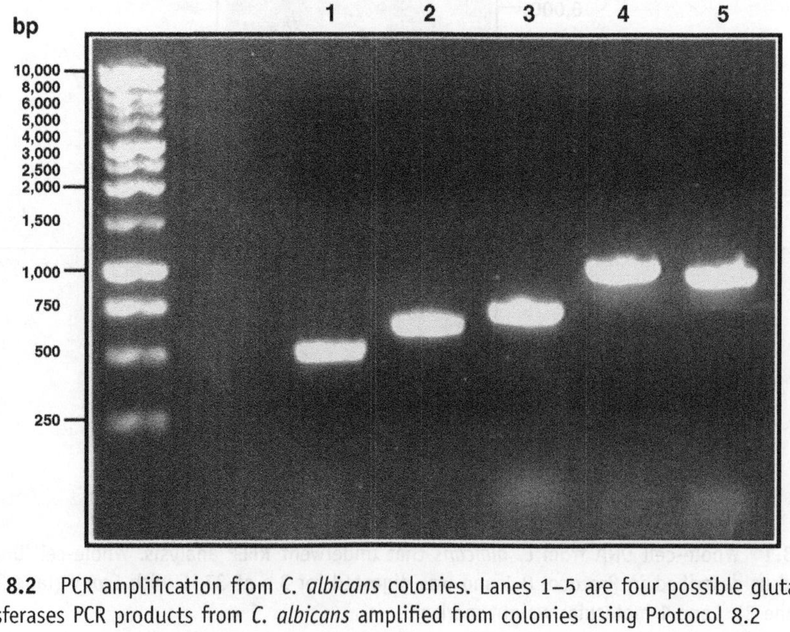

Figure 8.2 PCR amplification from *C. albicans* colonies. Lanes 1–5 are four possible glutathione S-transferases PCR products from *C. albicans* amplified from colonies using Protocol 8.2

8.3 Detection of yeast DNA using radio-labelled probes

Southern blotting is a widely used technique that allows analysis of specific DNA sequences. In the 1970s, E. M. Southern developed this method for locating a particular sequence of DNA within a complex mixture. DNA is normally cut into fragments by restriction enzyme digestion and electrophoretically separated on an agarose gel as described in Protocol 8.1 and Figure 8.1. Southern blotting involves the transfer of the fragmented DNA to a membrane (usually nylon or nitrocellulose) by means of capillary-transfer. DNA of interest is detected by hybridizing with a probe labelled with a radioactive isotope or one that is labelled chemiluminescently. This can be visualized by means of staining or by autoradiography. Southern blots are used in gene discovery and mapping studies to monitor gene incorporation and to assess evolutionary relationships among organisms.

A standard protocol for Southern blotting is described here. This method uses radiolabelled cDNA as a probe as it yields consistent, clean results.

Protocol 8.3
DNA detection by Southern blotting

Equipment, materials and reagents

Agarose gel containing DNA that previously underwent restriction enzyme digestion and electrophoretic separation

Denaturation buffer (0.5 M NaOH and 1.5 M NaCl)

Neutralization buffer (1.5 M NaCl, 0.5 M Tris-HCl, pH 7.4)

Nitrocellulose paper (NCP) (many companies supply this, including Sigma-Aldrich)

20 x SSC (0.3 M sodium citrate, 3 M NaCl, pH 7 with 1 M NaOH)

3 MM Whatmann paper (many companies supply this, including Sigma-Aldrich)

Paper towels

1 kg weight

Glass plate and glass basin (both treated to remove RNases)

UV transilluminator

Purified cDNA PCR products

^{32}P labelled dCTP or ^{33}P labelled dCTP (^{33}P gives cleaner results and is safer to use)

Non-specific nucleic acid/radio-labelled isotope labelling kit (Promega Prime-A-Gene® (Promega Corporation, Madison, WI, USA) kit described here) (containing nuclease-free water, reaction buffer, BSA, 10 mM dATP, 10 mM dGTP, 10 mM dTTP and Klenow)

Pre-hybridization solution (4 x SSC, 50% (v/v) formamide, 1% (v/v) SDS, 10% (w/v) dextran sulphate, 200 µg/ml denatured sheared salmon sperm DNA (suspended in water, passed through a syringe and boiled for 15 min))

Hybridization oven and appropriate hybridization bottles or bags

Wash buffer (2 x SSC and 0.5% (w/v) SDS)

Photographic cassette and Kodak Biomax™ (Anachem Biosciences, Luton UK) Ms film

Method

Caution: This method requires extreme care when handling **radio-labelled** products. The appropriate Government licence and Institutional training is required before the handling of such products is attempted.

1. An agarose gel containing DNA that underwent restriction digestion and electrophoresis is suspended in denaturing buffer for 30 min with gentle shaking, to denature double-stranded DNA.

2. Submerge the gel in neutralization buffer for a further 30 min with gentle shaking. Capillary-transfer is performed overnight using 20 x SSC and the DNA is fixed to the NCP as illustrated in Figure 8.3.

3. Soak three 3MM Whatmann paper sheets in 20 x SSC and leave the edges submerged over a glass container filled with 20 x SSC. Extreme care is required here to remove all air bubbles between the gel and the NCP paper. This can easily be performed by rolling a cylindrical object over the top layer of Whatmann paper several times.

4. Place the gel face-down on top of the filter paper sheets with the area surrounding the gel covered with plastic film. The pre-soaked (in 20 x SSC) NCP is placed onto the gel, followed by three sheets of soaked 3MM Whatmann paper. Ten to

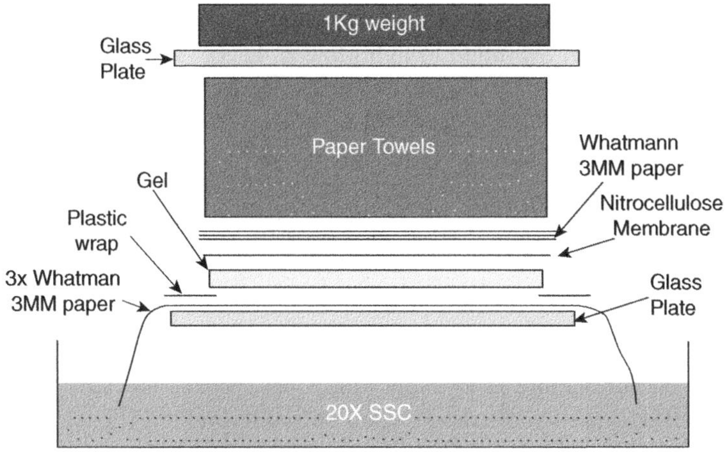

Figure 8.3 Capillary transfer

20 layers of dry paper are placed on top of the gel/NCP/filter paper followed by a 1 kg weight.

5. DNA transfer can take place, by capillary-transfer, for 24 h at room temperature. The orientation of the wells can be marked out on the NCP and the NCP is rinsed in 10 x SSC for 1 min before exposing both sides of the NCP on a UV transilluminator for 1 min (exposure on an UV light-box for the same period will also fix the DNA to the NCP if a transilluminator is not accessible). The NCP can be stored at −70 °C in several layers of filter paper in a heat-sealed bag, until required.

6. The DNA used for a probe will depend on the particular gene of interest. It is recommended to use a PCR product that has been amplified from cDNA (see Protocol 8.6 for cDNA synthesis) and has been purified. There are many companies that produce PCR product purification kits (e.g. Promega's Wizard™ (Promega Corporation, Madison, WI, USA) PCR product purification kit).

7. Add the PCR product (5 μl) to 25 μl of nuclease-free water and denature (boiled for 2 min), cool for 2 min on ice and mix briefly. Add reaction mixture (10 μl 5 x buffer, 2 μl BSA, 0.66 μl 10 mM dATP, 0.66 μl 10 mM dGTP and 0.66 μl 10 mM dTTP) to the denatured PCR product and place on ice. The PCR reaction mixture is placed in a 1 cm thick Perspex box, behind a 1 cm Perspex shield before adding 1 μl Klenow and 5 μl ^{32}P dCTP. The reaction tube must be kept in a Perspex box overnight at room temperature, as recommended by Promega for the labelling reaction to take place.

8. Place the NCP in a hybridization bottle or bag with 10 ml of pre-hybridization solution. The bottle is placed in a hybridization oven for 3 h at 42 °C. The

^{32}P-labelled probe is heated to 100 °C for 3 min and added to the pre-hybridization mixture and left overnight to hybridize at 42 °C.

9. Wash the NCP 5 times with wash buffer (in the order that the wash buffers are listed in above), twice at room temperature and 3 times at 55 °C. Lay the NCP between 2 sheets of transparency film and place into a photographic cassette. Kodak Biomax Ms film is placed onto the NCP and exposed at −20 °C for 24 h. The film can be developed. Alternatively, a phosphoimager can be used instead of film.

10. Blots can be stripped of probe by washing with 6 x SSC, 50% (v/v) formamide at 65 °C for 30 min, before re-probing with a new probe. The stripped blots will need to undergo pre-hybridization again before re-probing (Figure 8.4).

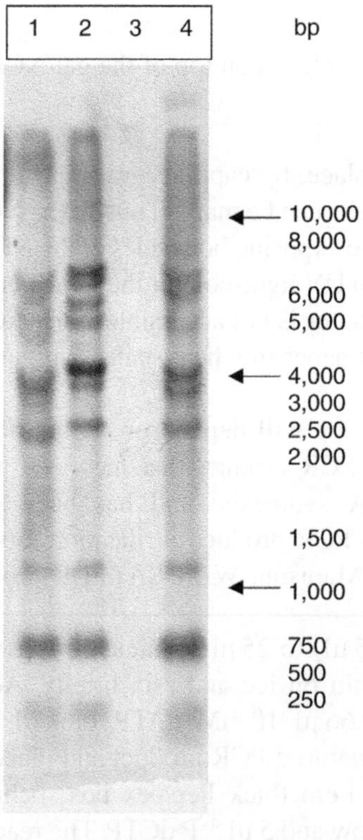

Figure 8.4 Southern blot of *Ca3* repetitive region in *Candida* species. DNA was extracted from various yeast strains and exposed to the restriction enzyme EcoR1. Samples were run on an agarose gel overnight and transferred to nitrocellulose paper. The blot was exposed to the radioactive *Ca3* probe. The *Ca3* probe identifies specific sequences in the *C. albicans* genome (and no other *Candida* species) and as such is a useful tool in strain differentiation. Lane 1: *C. albicans* 10231; Lane 2: *C. albicans* MEN; Lane 3: *C. glabrata*; Lane 4: *C. albicans* ATCC 44990

8.4 Extraction of whole-cell RNA using two different protocols

RNA is a biological macromolecule that has several critical functions. Messenger RNA (mRNA) is transcribed from DNA and serves as the template for protein synthesis. Protein synthesis is carried out by ribosomes, which consist of ribosomal RNA (rRNA) and proteins. Amino acids are delivered to the ribosome on transfer RNA (tRNA) molecules. RNA is also involved in its own processing. Messenger RNA accounts for about 1–5% of the total RNA content of the cell.

Compared to DNA, RNA is very unstable, which is due to the presence of ribonucleases, that actively breakdown RNA at a faster rate than DNase act upon DNA. There are many different methods used to isolate RNA from fungal cells. Most methods in use at the moment are fast and give clean RNA. Kits can be purchased from commercial sources that can be used to isolate mRNA or total RNA in less than an hour. The two methods that are described here are very different in approach but both result in the isolation of total RNA from yeast. The first method (the phenol:chloroform extraction procedure) is one of the best-established techniques for isolating total RNA from yeast. It can be time-consuming and labour-intensive but results in very high yields of relatively clean RNA. The second method is a rapid method that results in the isolation of total RNA in a very short time but with a lower yield. It is important that the RNA extracted is free from contamination, as purified full-length RNA is required for many molecular techniques, including cDNA cloning, analysing gene expression and for microarrays utilizing cDNA.

Good housekeeping rules on handling RNA

Ribonucleases (RNases) are stable and active enzymes that are difficult to inactive. Small amounts of RNases are effective in destroying relatively large amounts of RNA. Specific procedures must be implemented to eliminate all possible RNase contamination from plastics, glassware and buffers. Several steps can be implemented to minimize the introduction of RNases into RNA samples during or following isolation. **Aseptic techniques** should always be used when working with RNA. The human body is an abundant source of RNases; therefore wearing laboratory gloves and coat throughout the experimental procedure will limit contamination. Frequent changing of gloves, keeping samples contained within tubes whenever possible and on ice can minimize RNA breakdown. The use of sterile, disposable plastic tubes is recommended as most commercially available sterile tubes will be RNase-free. Other plastic ware can be treated with 0.1 M NaOH, 1 mM EDTA followed by RNase-free water to inactivate RNases. Chloroform may be used to treat plastics, if they are chloroform-resistant. Glassware must be treated to inactive RNases. This can be achieved by rinsing glassware with a detergent, followed by baking at 240 °C for 4 h. Glassware can, alternatively, be treated with 0.1% diethylpyrocarbonate (DEPC), which is an inhibitor of RNases. Electrophoresis apparatus should be cleaned with detergent, rinsed with RNase-free water and with ethanol. Water can be treated with DEPC overnight at 37 °C and autoclaved (this is explained in greater detail in Protocol 8.4). All solutions can be prepared with RNase-free water.

Protocol 8.4
The extraction of whole-cell RNA from yeast using phenol: chloroform

Equipment, materials and reagents

Cells to be analysed (50 ml of exponential phase cells, initial cell density approximately 5×10^6/ml)

Extraction buffer (0.1 M Tris-HCl, 0.1 M LiCl, 1% (w/v) SDS and 0.01 M d-dithioreitol, pH 7.5)

Diethylpyrocarbonate (DEPC)

YEPD medium (2% (w/v) glucose, 2% (w/v) bactopeptone, 1% (w/v) yeast extract)

Acid-washed glass beads (wash with concentrated HCl followed by several rinses with water and dry in an oven)

Phenol:chloroform (1:1)

Bench-top centrifuge or microfuge and the appropriate RNase-free tubes

6 M LiCl

10 x formaldehyde agarose (FA) gel buffer (20 mM MOPS, 1 mM EDTA, 5 mM sodium acetate, pH 7)

FA gel running buffer (1 x FA gel buffer and 2% (v/v) 12.3 M formaldehyde)

Agarose electrophoresis apparatus

5 x RNA sample buffer (0.25% bromophenol blue, 4 mM EDTA, 0.9 M formaldehyde, 20% (w/v) glycerol, 30.1% (v/v) formamide and 4 x FA gel buffer)

Heating block or water bath

Method

Prior to beginning any experiments with RNA, all equipment and materials must first be treated to remove all RNases as stated above. The best approach is to bake all glassware at 240 °C for 4 h before use. Water can be treated with DEPC by adding 0.1% (v/v) DEPC and mixing overnight, followed by incubation at 37 °C for 4 h and autoclaving. All buffers should be prepared in heat-treated bottles (or RNase-free plastic ware) with DEPC treated water and with chemicals that were weighed without the use of a spatula. Gloves are used at all times. All tips must not have been touched by hand (or RNase-free purchased), and it is recommended to autoclave before use. Take as much precaution as possible to minimize degradation of the RNA samples.

1. Inoculate 100 ml of YEPD with a 1 ml aliquot of an overnight culture of *C. albicans* and incubate at 30 °C for 3–4 h, or until cells have grown to an OD_{600} of approximately 0.6 units (early to mid-exponential phase).

2. Harvest the cells by centrifugation at 2056 x *g* for 5 min and resuspended in 3 ml of extraction buffer, and mix by vortexing. Mix the suspension vigorously for 5 min upon the addition of 6 gram of acid-washed glass beads and 5 ml phenol:chloroform (1:1). To minimize thermal breakdown of nucleic acids, the sample should be held on ice for 30 sec, vortexed for another 30 sec and then returned to the ice.

3. The cell mixture is centrifuged for 15 min at 5000 x *g* and the upper aqueous phase is transferred to a fresh tube. A further 2 phenol:chloroform washes are performed until no white interphase remains.

4. The upper layer is mixed with 2 volumes of absolute ethanol and stored at −20 °C overnight to aid precipitation of the RNA.

At this stage, the RNA is extracted but purification is required to remove contamination (e.g. DNA). Any of the −20 °C incubations can be replaced by incubation at −70 °C for a considerably shorter time (an hour would suffice at −70 °C).

5. Collect the RNA by centrifugation at 11 600 x *g* for 10 min and resuspend in 100 μl DEPC-treated water. The RNA is re-precipitated with the addition of 2 volumes of 6 M LiCl and by 2-hour storage at −20 °C.

6. The RNA is collected by centrifugation (as above), followed by a wash with 6 M LiCl and resuspending in 100 μl DEPC-treated water.

7. Re-precipitated the RNA with 10 μl 3 M sodium acetate, 2 volumes of absolute ethanol and store at −20 °C for 2 h. The RNA is pelleted by centrifugation, resuspended in 100 μl DEPC-treated water and stored at −70 °C until required.

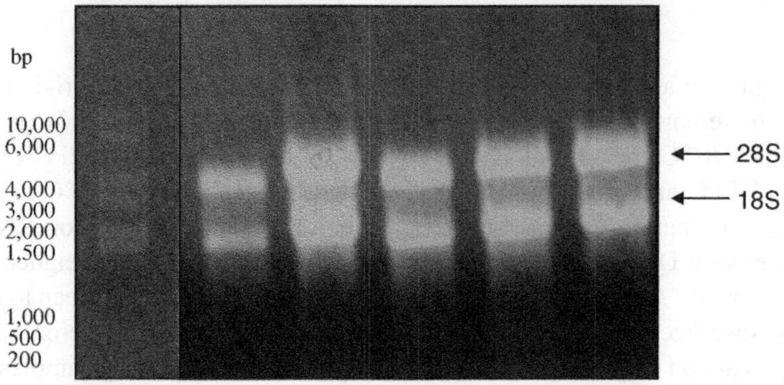

Figure 8.5 Typical profile of yeast total RNA using the phenol/chloroform extraction method. The 18S and 28S rRNA bands are the most striking bands observed in yeast RNA

8. The RNA concentration is calculated as described in Protocol 8.1 for DNA but with 1 unit at 260 nm being equivalent to 42 μg of RNA. The quality of RNA can also be calculated by reading absorbance at 260/280 nm.

9. The quantity of RNA can also be visualized on a formaldehyde gel. A 1% (w/v) agarose gel, consisting of 5% (v/v) FA and DEPC-treated water, is melted and allowed to cool to approximately 55 °C. Once the gel solution is cool, 1.8% (v/v) formaldehyde and a small amount of ethidium bromide (2 μl of a 10 μg/ml stock solution) are added and the molten gel can be poured into the gel apparatus. Upon solidification, the gel is placed in FA gel running buffer within the gel rig for 30 min before any samples are loaded onto it.

10. The samples are aliquoted into 60 μg total RNA and 4 μl of 5 x RNA sample buffer is added to each sample. The samples are heated at 65 °C for 5 min and cooled on ice for 2 min before loading onto the gel. The gel is run at 5–7 V/cm until the bromophenol blue front is about 65% through the gel. The gel can then be examined under UV (Figure 8.5).

Protocol 8.5
Rapid extraction of whole-yeast-cell RNA

Equipment, materials and reagents

Cells to be analysed (exponential phase)

Vortex mixer

DEPC-treated water and RNase-free equipment

YEPD medium (2% (w/v) glucose, 2% (w/v) bactopeptone, 1% (w/v) yeast extract)

Spectrophotometer and a quartz cuvette

Microfuge and the appropriate RNase-free tubes

DEPC-treated water

0.1% *N*-lauroyl sarcosine sodium salt

Acetate/SDS solution (1% SDS in 10 mM EDTA, 50 mM sodium acetate, pH 5.1 adjusted with acetic acid)

Agarose electrophoresis apparatus and other materials for running RNA on an agarose gel, mentioned in Protocol 8.4.

Heating block or water bath

Ice

UV light source (transilluminator)

Method

This method is adapted from that described by Rivas *et al.* (2001).

Note: The same precautions should be taken when undertaking this protocol as described in Protocol 8.4 to minimize RNA degradation.

1. Take 1 ml of cells (approximately 5×10^6 cells/ml) and harvest by centrifugation at 6000 x *g* for 5 min at room temperature.

2. Resuspend the cells in 200 μl of 0.1% *N*-lauroyl sarcosine sodium salt and collect by centrifugation at 10 000 x *g* for 5 min at room temperature.

3. Resuspend the pellet in 100 μl acetate/SDS (pH 5.1) solution and place in a heating block at 100 °C for 5 min (pH of 5.1 is required to obtain DNA-free RNA).

4. Dilute the suspension with 100 μl DEPC-treated water and centrifuge at 7000 x *g* for 5 min at room temperature. Collect the supernatant, as it contains the RNA.

5. RNA samples can be quantified by reading at 260/280 nm on a spectrophotometer as mentioned in Protocol 8.4 and by visualizing the 18S and 28S rRNA bands on

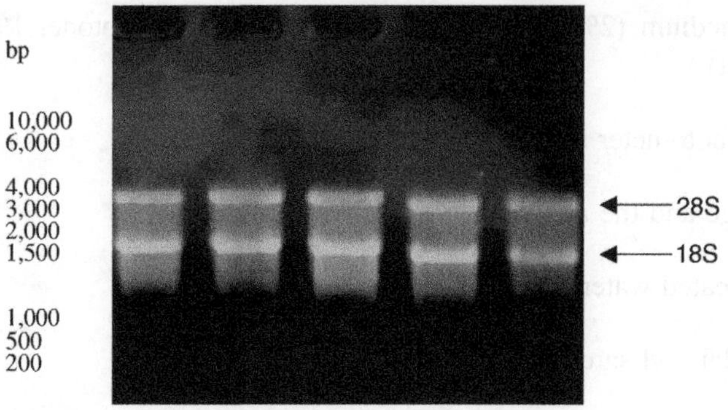

Figure 8.6 RNA extraction from *C. albicans* cells using the rapid whole-cell RNA extraction method. Lane 1: Control; Lane 2: Doxorubicin (20 μg/ml) for 4 h); Lane 3: Doxorubicin (20 μg/ml) (24 h); Lane 4: Amphotericin B (0.2 μg/ml)(4 h); Lane 5: Amphotericin B (0.2 μg/ml) (24 h)

an agarose gel. The RNA samples are stored at −70 °C to avoid RNA degradation by RNAses (Figure 8.6).

8.5 Detection and expression levels of specific genes by the examination of mRNA in yeast

The analysis of RNA by hybridization technology (Northern blotting and gene microarray analysis) or RT-PCR can provide a good examination of gene-expression profiles. The two methods (Northern blotting and RT-PCR) described give a semi-quantitative evaluation of gene expression. The first method (Northern blotting) is still regarded as the best approach for investigating alterations in mRNA levels. This method is long, labour-intensive and potentially hazardous since it may require the use of radioisotopes. The second method (RT-PCR) yields faster results.

Northern blot analysis involves size separation of the RNA species by electrophoresis and transferring these RNA bands onto a membrane, followed by identification with labelled gene-specific oligonucleotides or DNA probes. The size and abundance of the desired RNA is determined. The predominant rRNA in a total RNA preparation can also be detected by UV absorbance as a measure of intactness as described in the previous section.

The usual PCR technique copies a piece of DNA and greatly amplifies the copy number for analysis or storage. RT-PCR starts with mRNA or total RNA, makes a cDNA complementary strand using reverse transcriptase and then amplifies the product. The product is used to determine the presence of the mRNA and for size determination, sequencing or for quantitation by quantitative-PCR.

Specific RNA species in an un-fractionated preparation can be measured by immobilizing a sample in a spot (dot blot) or in a manifold slot (slot blot) with detection by a labelled DNA probe that hybridizes to the immobilized RNA. Another important RNA detection method is the use of *in situ* hybridization, which is used to localize the expression of a particular mRNA among different cells in a tissue or cell preparation. A cell preparation is fixed on a slide and the RNA content probed with radiolabelled or fluorescent probe. Neither dot blot, slot blot nor *in situ* hybridization are examined in this chapter, but all three can be utilized by employing similar hybridization methods to those described for Northern blot analysis.

Protocol 8.6
Examining mRNA content as a means of investigating gene-expression profile by northern blot analysis

Equipment, materials and reagents

Nitrocellulose paper (NCP)

Formaldehyde gel containing RNA samples of interest

0.5 M NaOH

20 x SSC (0.3 M sodium citrate, 3 M NaCl, pH 7 with 1 M NaOH)

3 MM Whatmann paper

Paper towels

1 kg weight

Glass plate and glass basin (both treated to remove RNases)

UV transilluminator

Purified PCR products

^{32}P-labelled dCTP

Non-specific nucleic acid/radiolabelled isotope labelling kit (containing nuclease-free water, 5 x buffer, BSA, 10 mM dATP, 10 mM dGTP, 10 mM dTTP and Klenow)

Church's pre-hybridization solution (0.5 M NaH_2PO_4, pH 7.2, 7% (w/v) SDS, 1 mM EDTA, pH 8, 200 µg/ml denatured sheared salmon sperm DNA) (Church and Gilbert, 1984)

Wash buffers (First – 2 x SSC and 0.1% (w/v) SDS, Second – 1 x SSC and 0.1% (w/v) SDS, Third – 0.5 x SSC and 0.1% (w/v) SDS and finally – 0.1 x SSC and 0.1% (w/v) SDS)

Densitometry apparatus and software

Method

1. The blotting of the RNA, which has undergone electrophoresis, to the NCP is undertaken by capillary-transfer. Rinse the formaldehyde gel with DEPC-treated water for 10 min, 0.5 M NaOH for 15 min and 10 x SSC for a further 10 min.

2. Blotting the separated RNA to the NCP is performed by capillary-transfer as described in the Southern blotting protocol (Protocol 8.3)

3. Probe construction is also the same as described in Protocol 8.3.

4. The NCP is placed in a hybridization bottle with 20 ml of Church's pre-hybridization solution and is placed in a hybridization oven for 3 h at 65 °C. The reacted radiolabelled probe is boiled for 3 min, cooled, added to the pre-hybridization mixture and left overnight to hybridize.

5. The hybridization buffer is removed from the bottle and stored at −20 °C for up to a month (if re-probing is required). The NCP undergoes 4 washes with 4 different wash buffers (listed above) at 65 °C.

6. Place the NCP between 2 sheets of transparency film and into a photographic cassette. Kodak Biomax Ms film is placed onto the NCP and exposed at −70 °C for at least 24 h. Exposure for up to a week gives best results when using photographic film, which is usually equivalent to a 24 h exposure on a phosphoimager (Figure 8.7).

7. The specific intensity of each sample band can be semi-quantified by densitometry. A housekeeping gene should be examined on your blot and the other samples can be standardized by comparing densitometry values. This will enable standardization if equal mRNA was loaded.

8. Blots can be stripped of the probe by washing with 6 x SSC, 50% (v/v) formamide at 65 °C for 30 min, before re-probing with a new probe. It is important not to let the blot completely dry before stripping.

Protocol 8.7
Examining mRNA content as a means of investigating gene-expression profile by RT-PCR analysis

Equipment, materials and reagents

Total RNA (recovered from one of the RNA extraction methods above)

Deoxynucleotide mix (e.g. Promega)

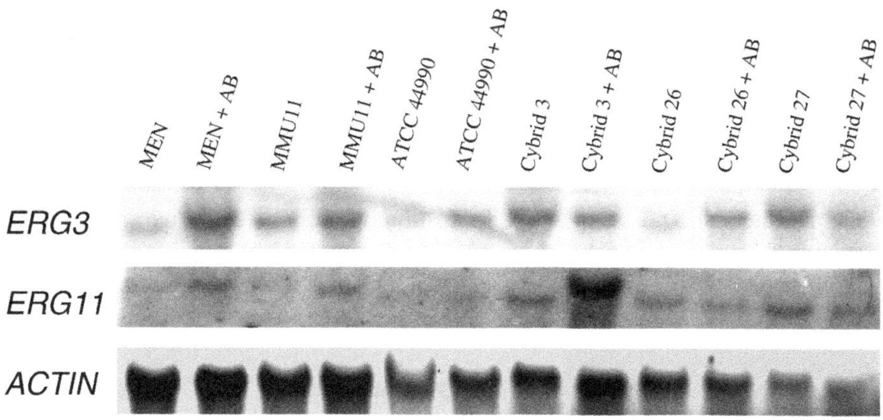

Figure 8.7 Northern blot analysis of the ergosterol biosynthesis genes (*ERG3* and *ERG11*) and the *ACTIN* gene from untreated and amphotericin B treated *C. albicans* cells. Northern blot analysis was performed as described in Protocol 8.6. (*C. albicans* MEN: untreated cells, MEN + AB: cells exposed to 0.25 μg/ml amphotericin B for 4 h)

Random primers (Roche-Oligo-p[dT]15 primer)

Reverse transcriptase (Promega-M-MLV™, Promega Corporation, Madison, WI, USA)

Reverse transcriptase reaction buffer

DNase, its appropriate reaction buffer and stop solution (both the reaction buffer and stop solution usually accompany DNase, e.g. Promega)

RNasin ribonuclease inhibitor (e.g. Promega)

Method

1. Treat RNA (2 μg) with DNase and its appropriate buffer at 37 °C for 15–30 min. Addition of stop solution (1 μl) and heating at 70 °C for 10 min inactivates the DNase.

2. Incubate the RNA with the random primers for 10 min at 70 °C and allow cooling on ice for 5 min.

3. The RNA is reverse-transcribed at 37 °C for 2–3 h with 1 mM deoxynucleotide mix, RNasin ribonuclease inhibitor (protects RNA from degradation), reverse transcriptase reaction buffer and 1 μl M-MLV reverse transcriptase. Store at −70 °C until required.

4. cDNA (2 μl) is amplified by PCR in a 50 μl volume. A typical PCR sequence would be 5 min at 95 °C followed by 30 cycles at 95 °C for 1 min during denaturation, primer-annealing at 55 °C for 1 min and extension for 1 min at 72 °C. This can be followed by further extension at 72 °C for 10 min and a holding step at 4 °C. These conditions will amplify most gene fragments under 1 kb.

5. PCR products should be quantified densitometrically at cycle numbers between 10 and 40 to determine the appropriate cycle number at which the exponential amplification of products occurs, and to identify the cycle number at which

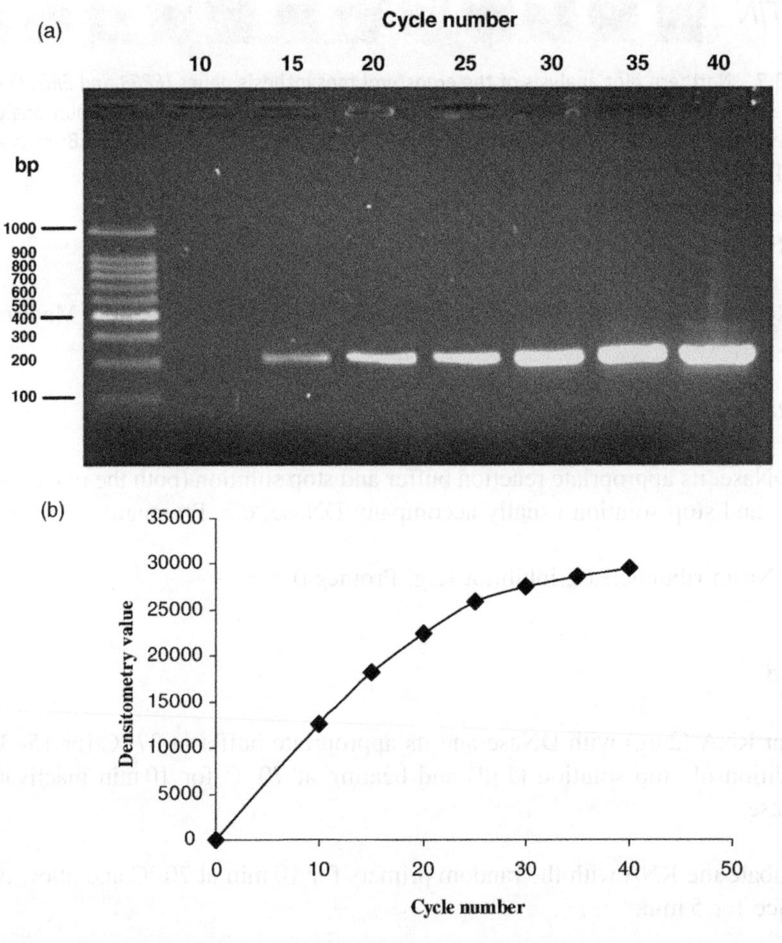

Figure 8.8 *β-Actin* curve (a) and densitometry (b) to determine the appropriate cycle number at which exponential amplification of products occurs. 2 μl of cDNA was amplified using primers specific for a 220 bp fragment of β–Actin at serial different PCR cycle numbers (10, 15, 20, 25, 30, 35 and 40). 15–20 cycles were determined to be the appropriate cycle number at which exponential amplification of products occurs

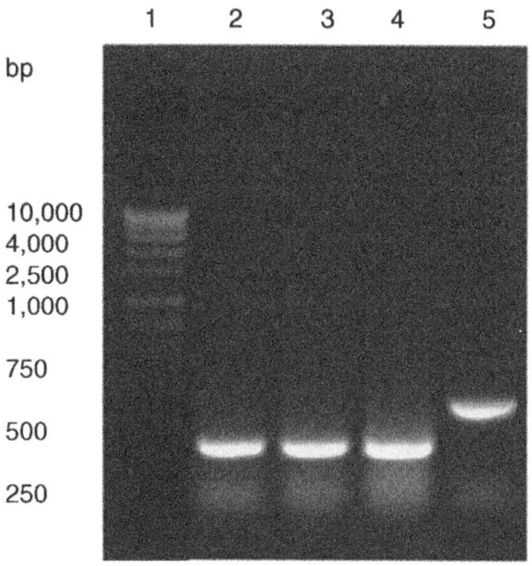

Figure 8.9 RT-PCR of *CALMODULIN* gene expression in *A. fumigatus*. Lane 1: molecular marker; Lane 2: cDNA expression of *CALMODULIN* at 24 h; Lane 3: 48 hrs; Lane 4: 72 h; and Lane 5: gDNA control. cDNA product size is 348 bp where as gDNA is 617 bp

sufficient discrimination is possible to accurately quantify alterations in gene expression. This step is extremely important, as a continuous amplification of a product will give an overexaggerated result (e.g. a gene may be sufficiently amplified at 20 cycles for densitometry purposes but if it undergoes another 10 cycles a quantified discrimination of results may not be possible) (see Figure 8.8).

6. PCR products should be sequenced to verify gene identity when using newly constructed primers. This service is available through many biotechnology companies (e.g. MWG, Milton Keynes, UK, GENOSYS, Dorset, UK, LARK, Cambridge, UK).

7. PCR products are run on a 1% (w/v) agarose gel containing 0.5 μg/ml ethidium bromide. The specific intensity of each sample band can be semi-quantified by densitometry.

8. A housekeeping gene (e.g. actin) should be examined by RT-PCR and the other samples can be standardized by comparing densitometric values (see Figure 8.9).

8.6 References

Alifano, P., Bruni, C. B. and Carlomagno, M. S. (1994) Control of mRNA processing and decay in prokaryotes. *Genetica* **94**(2–3): 157–172.

Burns, C., Geraghty, R., Neville, C., Murphy, A. *et al.* (2005) Identification, cloning, and functional expression of three glutathione transferase genes from *Aspergillus fumigatus. Fungal Genetics and Biology* **42**(4): 319–327.

Carriero, S. and Damha, M. J. (2003) Inhibition of pre-mRNA splicing by synthetic branched nucleic acids. *Nucleic Acids Research* **31**(21): 6157–6167.

Church, G. M. and Gilbert, W. (1984) Genomic sequencing. *Proceedings of the National Academy of Sciences of the United States of America* **81**(7): 1991–5.

Chworos, A., Severcan, I., Koyfman, A. Y., Weinkam, P. *et al.* (2004) Building Programmable Jigsaw Puzzles with RNA. *Science* **306**(5704): 2068–2072.

Deak, T., Chen, J. and Beuchat, L. R. (2000) Molecular characterization of *Yarrowia lipolytica* and *Candida zeylanoides* isolated from Poultry. *Applied and Environmental Microbiology* **66**(10): 4340–4344.

Fredricks, D. N., Smith, C. and Meier, A. (2005) Comparison of six DNA extraction methods for recovery of fungal DNA as assessed by quantitative PCR. *Journal of Clinical Microbiology* **43**(10): 5122–5128.

Kavanagh, K. and Whittaker, P. A. (1996) Application of protoplast fusion to the non-conventional yeasts. *Enzyme and Microbial Technology* **18**: 45–51.

Neville, C., Murphy, A., Kavanagh, K. and Doyle S. (2005) A 4′-phosphopantetheinyl transferase mediates non-ribosomal peptide synthetase activation in *Aspergillus fumigatus. Chembiochem: A European Journal of chemical biology* **6**(4), 679–85.

Rivas, R., Vizcaàno, N., Buey, R. M., Mateos, P. F. *et al.* (2001) An effective, rapid and simple method for total RNA extraction from bacteria and yeast. *Journal of Microbiological Methods* **47**(1): 59–63.

Wach, A., Brachat, A., Pohlmann, R. and Philippsen, P. (1994) New heterologous modules for classical or PCR-based gene disruptions in *Saccharomyces cerevisiae*. *Yeast* **10**(13): 1793–1808.

Wipf, D., Benjdia, M., Rikirsch, E., Zimmermann, S. *et al.* (2003) An expression cDNA library for suppression cloning in yeast mutants, complementation of a yeast his4 mutant, and EST analysis from the symbiotic basidiomycete *Hebeloma Cylindrosporum Genome/Génome* **46**(2): 177–181.

9

Microarrays for studying pathogenicity in *Candida albicans*

André Nantel, Tracey Rigby, Hervé Hogues and **Malcolm Whiteway**

9.1 Introduction

Candida albicans is an important human pathogen that can cause a range of chronic or acute infections of mucosal membranes. Vaginitis and thrush are two common manifestations of such infections. In addition, immunocompromised patients, such as those undergoing chemotherapy, suffering from AIDS or undergoing immunosuppressive therapies following organ transplantation, can develop systemic *Candida* infections with high mortality rates (Pappas *et al.*, 2003). Drugs to treat *Candida* infections are currently aimed at a relatively small number of molecular targets. Azole based compounds are directed at the ergosterol biosynthetic pathway and disrupt membrane function, amphotericin B also targets the membrane, and the recently developed echinocandins act by disrupting the synthesis of the cell wall. While amphotericin B can cause severe side effects, azoles and echinocandins target functions that are found in fungi and not in humans, and thus are less problematic for the host. However, the fact that azoles are fungistatic rather than fungicidal, together with the growing development of resistant populations of pathogens, has led to a desire to identify new molecular targets for antifungal therapies (Sanglard and Odds, 2002).

Molecular genetics has been a powerful technology to dissect cellular function in a variety of organisms. This tool has been spectacularly successful in determining the molecular bases of many cellular activities within the model yeast *Saccharomyces cerevisiae*. The application of molecular genetics to a fungal pathogen such as *C. albicans* could bring the same level of sophistication to the dissection of antifungal

Medical Mycology: Cellular and Molecular Techniques. Edited by Kevin Kavanagh

targets. Several of the properties of *S. cerevisiae* that have made yeast such a useful model organism are also found in *C. albicans*. In particular, *C. albicans* can be relatively efficiently transformed with exogenous DNA, and this permits the genetic engineering of strains through the production of clonal transformants with modified genetic properties (Berman and Sudbery, 2002). However, the diploid nature of *C. albicans*, and the absence of an efficient meiotic process, has greatly hampered the molecular genetic analysis of *C. albicans*. Although *C. albicans* genes can be disrupted by genetic engineering, each of the two alleles must be eliminated to generate a null mutation, making each manipulation time-consuming. Extensive genetic screens for recessive mutations are not yet possible, independent mutations cannot be easily combined through crosses and suppression studies are essentially limited to the identification of dominant mutations, which are subsequently often difficult to study.

Although significant progress is being made on improving the genetic manipulation of *C. albicans*, genomic-based tools using the recently determined genome sequence provide an attractive approach to address questions about gene function that are not dependent on any direct genetic analysis of the cells. Transcriptional profiling is an important example of such technology. This tool can provide a comprehensive picture of the changes in gene expression exhibited by *C. albicans* cells subjected to an environmental challenge, a genetic modification or some other controlled manipulation. Genes whose expression are found to vary in response to the controlled manipulation can be linked functionally to the process under study, and thus networks of related genes can be connected without the need for genetic analysis. Because one of the goals of antifungal drug development is to identify fungal-specific functions, the ability to place novel genetic elements into functional networks is a powerful attribute of the array technology.

9.2 DNA microarrays

DNA microarrays consist of multiple DNA species distributed in an orderly pattern on a solid surface. These arrays represent a popular technique to investigate questions such as gene expression on a genomic level. However, while the profiling of gene expression patterns is one of the primary uses of microarrays, they can be used in other ways, for example in comparative genomics and in genome assembly.

Three types of whole-genome arrays have been produced for *C. albicans*. The most common type is based on the two-colour technology developed at Stanford University (DeRisi *et al.*, 1996). In general, these arrays are constructed by physically spotting PCR amplicons on a coated glass microscope slide. In use since 2001, amplicon arrays have been produced by Incyte (De Backer *et al.*, 2001; Rogers and Barker, 2002), our group at the BRI-NRC (Enjalbert and Whiteway, 2005; Bachewich *et al.*, 2005; Nicholls *et al.*, 2004; Lee *et al.*, 2004; Harcus *et al.*, 2004; Enjalbert *et al.*, 2003; Nantel *et al.*, 2002; Cowen *et al.*, 2002; Andes *et al.*,

2005), Eurogentec and the Galar Fungail Consortium (Barker *et al.*, 2004; Coste *et al.*, 2004; Garcia-Sanchez *et al.*, 2004; Copping *et al.*, 2005; Fradin *et al.*, 2005; Liu *et al.*, 2005), the Johnson and Fink labs at UCSF and MIT (Bennett *et al.*, 2003; Chauhan *et al.*, 2003; Lorenz *et al.*, 2004; Hromatka *et al.*, 2005; Kadosh and Johnson, 2005), the Berman and Hoyer labs at UMN and UIL (Bensen *et al.*, 2004; Selmecki *et al.*, 2005; Zhao *et al.*, 2005b) and the Jiang lab in China (Cao *et al.*, 2005; Xu *et al.*, 2005). A variant of the amplicon arrays are produced instead by spotting long (70 mer) oligonucleotides such as those that are distributed commercially by Operon (Zhao *et al.*, 2005a). Finally, the Agabian lab has been using custom oligonucleotide GeneChips based on Affymetric technology (Lan *et al.*, 2002; Lan *et al.*, 2004; Murillo *et al.*, 2005). All of these microarrays used preliminary genomic information from the version 4 and 6 *C. albicans* genome assemblies.

9.3 Building a second-generation 2-colour long oligonucleotide microarray for *C. albicans*

With the release of the version 19 genome assembly (Jones *et al.*, 2004), the annotation of the *C. albicans* genome (Braun *et al.*, 2005) and the impending release of the final assembly (see http://candida.bri.nrc.ca), we felt it necessary to produce a new generation of *C. albicans* microarray that is based on updated genomic information.

About 40% of *C. albicans* genes contain frequent short low complexity elements, also called STR (short tandem repeats), the periodicity of which is always a small multiple of 3. These repeated sequences, even when they are short (<20 bp), can produce detectable non-specific hybridization signals due to the number of genes that have an equivalent STR but also because of the degenerate binding modes that these sequences can form. The challenge of producing a good set of probes for *C. albicans* is to select a sequence region for each gene that will avoid any STR and offer the best specificity within gene families. Each probe sequence should also have minimal tendencies for self-annealing and should contain a similar G/C ratio which will ensure a homogeneous signal intensity over the entire array. Finally, the selected region should also be located within a few hundred base pairs from the 3′ end of the ORF in order to maximize overlap with cDNA fragments that are produced from oligo(dT) primers.

The design strategy differs depending on the length of the probes. If amplicons are used (>250 bp), the self-annealing and GC content can rarely be optimized. In this case, the specificity of the amplicon sequence and the quality of the PCR primers are the main criteria. If long oligonucleotide probes (60–100 bp) are used, self-annealing and GC content becomes more critical, while for shorter oligos (<40 bp) the precise melting temperature must be considered in order to maintain the homogeneity of the signal across the array. Our approach for probe design was to use a sliding window of

```
ANN:                                     48   48   58   72   62   48   48   48   40   40   40   25   25   20
ALN:  1111111111111111111111111111111111111111111111111111111111111111111111111111116689299999999999999999
SEQ:  TTGTGATGATTTAGAAATTAAAGTTAGTGATCCAATGGTGAAATTTTCTGAAACTTGTATTGAAAATGGTTACATTAGAACATCAACAACGACAACAACA
NGC:                                     19   18   18   19   21   22   22   22   20   23   24   24   23   23

ANN:     36   66   66   66   66   92   72   72   66   66   66   76   86   98   86   80   78   58   64   74
ALN:  9999999999432221111213555555555541111111111111111111111111111111111111111111111111111111111111111111
SEQ:  ACAACAACAACCAACGAGGATAAAGATAAAGATAAAGATTCTTTATTATCGATGACTATAATTGTTGAACCTATTATAGATTATAAATTCAGTCATGATA
NGC:     22   22   22   22   21   20   20   19   19   16   14   15   16   16   17   17   16   15   14   14

ANN:     74   74   74   74   74   82   82   82   82   82   66   56   56   56   42   42   42   56   56   56
ALN:  1112222222222222221222222222222111111111111111111111111111111111111111111111111111111111111111111111
SEQ:  TTGAAATTGGGAAATTAAAATTTGATAATATTGATATTGATTCTAAACAATTGATCAAAATATTGAAAACAGAATATGGTTGGGATTCTTTAGCTGCTAG
NGC:     13   13   13   14   14   13   13   13   13   15   16   17   20   21   22   23   25   25   25   25

ANN:     56   56   64   76   84   98   98   98   98   98   84   76   64   62   62   62   58   58   68   78
ALN:  1111111111111111111111111111111111111111111111111111111111111111111111111111111111111111111112222222
SEQ:  ATCTCTTTGGGCAATTGGACCAATTAATGATTTACAAAATCCTAGTATTTTATTAAATGATACATTAAATCAACACCATCAACAAGACAACAACAACATC
NGC:     24   26   24   21   21   20   17   17   18   17   16   17   18   18   19   16   19   20   20   21

ANN:     78   68   60   66   66   66   66   66   66   60   50   50   50   50   50   50   56   56   50   50
ALN:  2222221111111111111111111111111111111111111111111111111111111111111111111111111111111111111111111111
SEQ:  ATTGAATCAATTAAATCGTCGATAATTTCTGGATTTAAATGGAGTATTAATGAAGGACCATTATGTGAGGATCAATTTCGAAATGTTCAATTTACAATCA
NGC:     21   22   19   20   21   20   21   21   22   23   21   21   21   21   23   24   25   23   20   22

ANN:     50   42   52   64   64   64   82  102  106  106  106  106  106  102  102  102   96   56   66   66
ALN:  1111111111111111111111111111111111111111112245999999999999999999865544211111111111111111111111111111
SEQ:  TCGATATTCCCGCAGACAATAATAATAAAACTCCACCGCTGGATAATAATAATAATAATAATAAATTATTATTATCTCCAGCACAAATAATCCCATTAAT
NGC:     23   24   22   21   20   19   18   16   17   16   15   18   18   17   16   16   17   19   20   23

ANN:     66   66   64   64   60   38   38   52   52   60   60   68   80   80   74   74   74   88   88   88
ALN:  1111111111111111111111111111111111111111111111111111111111111111111111111111111111111111111111111111
SEQ:  GAGAAGAGCATGTCATAATGCAATCACTAATGCCATACCAAAATTAATGGAACCTATTTATCAATTAAATGTAATTTGTTCATATAAAGCCATCAATGTT
NGC:     25   25   24   23   23   21   21   20   18   17   18   18   18   15   14   14   12   10   14   16

ANN:     88   78   74   74   62   58   58   58   36   38   38   38   38   38   38   38   38   38   38   38
ALN:  1111111111111111111111111111112323999999999999999999999999999999999999999999998111111111111111111111
SEQ:  ATAAAACATTTATTATTAAATAAAAACCCGCAGCAACAGCAACAACAACATCAGCAGCAGCAACAACAACAACAACAACGACGTGGTGAAATTGATACTG
NGC:     19   19   20   22   23   23   24   25   27   30   32   33   31   30   29   31   31   30   27   26

ANN:     56   56   56   64   64   64   64   64   64   64   74   76   76   76   76   76   52   62   62   62
ALN:  1111111111111111111111111111111111111111111111111111111111111111111111111111111111111111111111111111
SEQ:  TAACCCCAATACCAGGAACACCTTTATTTTCTATTAAAGGTTATTTACCAGTAATTGATTCAATTGGGATATTAACTGATATTAAATTAAATACTCAAGG
NGC:     26   25   24   22   21   21   22   20   20   17   15   14   17   18   17   20   18   18   18   20

ANN:     62   62   62   62   62   62   62   62   62   62   56   50
ALN:  1111111111111111111111111111111111111111111111111111111111111111111111111111111111111111111111111111
SEQ:  TCAAGCTATAGGATCATTAAGATTTAATCATTGGGAAATTGTTCCTGATGAATTAAGTGAAGAATTTATAATTAAAACCAGGAAAAGAAAAGGTATATAA
NGC:     17   20   20   20   22   22   19   17   18   17   18   19
```

ANN: self-annealing score, ALN: specificity mask, SEQ: orf sequence, NGC: GC count.
In bold are sequences identified as Short Tandem Repeats while the selected 70 me probe sequence is in bold and underlined.

Figure 9.1 Sliding window profile for the design of 70 bp microarray probe for gene orf 19.144

the desired probe length (70 bp), which is moved from the 3′ toward the 5′ end of the target sequence. At each position a self-annealing term is computed as well as the GC count. A pre-computed cumulative mask of all the significant alignments found in the genome was made to favour regions that are unique for each gene family member (Figure 9.1).

This mask highlights the sequence specificity as well as the low complexity regions. At each position a global score is assigned. If the offset from the 3′ end of the ORF is too large, an increasing penalty is added to the score. As the window slides upward, the optimal position is retained and the resulting oligonucleotide sequence is extracted. In the end, we were able to produce specific probes for 6263 *C. albicans* genes. Owing to their high sequence similarity to other gene family members, it was impossible to design specific probes for the remaining 91 genes. These long oligonucleotide *C. albicans* microarrays are available for distribution to academic researchers and more information is available on our Web page (http://www.bri.nrc.gc.ca/services/microarray/index_e.html).

Protocol 9.1
Isolation of *C. albicans* RNA

Cell preparation

Reproducibility is essential in microarray experiments, and yeast cells are extremely responsive to even slight differences in their environment. Preparing the cells should be the most rigorous part of the following protocol. Try to keep track of everything. Documentation could help you out later if your microarray experiments produce unexpected results.

Glassware and plasticware used to prepare medium and grow *C. albicans* should always be rinsed 3 times with deionized water (i.e. cylinders, bottles, flasks, beakers etc.). Check the pH of the culture media. For YPD, it should be between 6.0 and 6.5. Buffer with NaOH or HCl if needed.

Method

1. *C. albicans* should be grown on a plate and streaked for single colonies. Do not use colonies older than 2 weeks. Pick a single colony and grow overnight in fresh YPD at the lowest temperature of the experiment (i.e. if one compares a 25 °C culture with 30 °C culture, grow the cells at 25 °C).

2. When studying morphological changes, take a photograph of the starter culture and dilute it in the appropriate medium to an OD_{600nm} of 0.05 to 0.1. Use a flask that is at least 2 times larger than the final culture volume.

3. Collect *Candida* cells when OD_{600nm} reaches 0.6 to 0.8 (not more!) (2–3 generations). Past an OD_{600nm} of 0.8, cells start to change their transcription profiles in response to limiting nutrients. We use OD_{600nm} as a good representative of the total metabolizing volume of yeast present in the culture, e.g. morphological and size differences are better taken into account than with direct cell counting. Never add fresh media to your cultures before harvesting in order to bring down the OD_{600nm}; you want your cultures to have both the same nutrient status and OD_{600nm} when you harvest them.

4. Take another photograph at the end of the incubation time for documentation.

5. Harvest cells by centrifugation or filtration (0.45 μm pore size filter).

6. For centrifugation, spin down for 2 min (usually in 50 ml Sarstedt tubes or 250 ml conical tubes) at 2500–3000 x *g* at room temperature.

7. Pour off supernatant and plunge tubes into dry ice plus ethanol or liquid nitrogen for 30 sec. Harvest no more than 2 samples at a time, and plunge both into the

freezing liquid simultaneously. Cooling and starving the cells must be kept at a minimum as yeast quickly change their transcriptional profiles in response to either stress.

8. Documentation: the final OD_{600nm} of the culture, growing temperature and carbon source must be recorded for the microarray database. Size profiles, cell counts and photographs of the cells are also a good idea, especially when studying arresting Centre for Disease Control Strains. With CADS strains or others, which tend to revert, check to ensure your cultures haven't been taken over by revertants by plating 100 ml of your final culture, before pelleting, on selective media (i.e. grow CDC mutants overnight at 37 °C; no colonies should appear).

Protocol 9.2
Isolation of total RNA using the hot phenol method

This method is very similar to the one described by Kohrer and Domdey (1991).

Method

For a 250 ml of culture at 0.6–0.8 OD_{600nm}.

1. Thaw the cell pellet in 20 ml SAB buffer (50 mM sodium acetate, 10 mM EDTA, adjust pH to 5.0 with glacial acetic acid) and transfer the suspension into 2 new sterile 50 ml Sarstedt tube (10 ml per tube). Place the tubes in a 65 °C water bath, add 1/10 volume 10% w/v SDS (1 ml per tube) and 10 ml/tube (1 volume) of warm SAB-equilibrated phenol (65 °C).

2. Vortex the pellet every 2 min for 10 min in 10 sec bursts. Let the cells/lysates sit into the 65 °C water bath between the bursts.

3. Cool the samples down to room temperature by incubating on ice. Beware that SDS can precipitate.

4. Centrifuge at room temperature for 10 min at 2200 x g.

5. Remove carefully the lower organic phase by pipetting and leave the pellet of broken cells.

6. Place the tubes into the aqueous bath and add 10 ml of the 65 °C warm phenol and repeat step 2.

7. Use ice to cool the samples down to room temperature.

8. Centrifuge for 10 min at 2500 x *g* at room temperature to separate both phases.

9. Repeat step 5–8.

10. Remove the upper aqueous phase and transfer it to new sterile 50 ml Sarstedt tubes. Extract the supernatant with 1 volume (10 ml) chloropane by vortexing 2 min in 30 sec bursts at room temperature.

11. Centrifuge for 4 min at 2500 x *g* at room temperature.

12. Transfer the upper water phase in new sterile 50 ml Sarstedt tubes and add 10 ml chloroform:isoamylalcohol (1 volume). Vortex and separate the phases by centrifugation for 4 min at 2500 x *g* at room temperature.

13. Remove the upper phase and transfer it into a new sterile 50 ml Sarstedt tube. Add 1/10 volume 3M sodium acetate (1 ml) and 3 volume 95% (v/v) EtOH (30 ml). Mix well and precipitate RNA at least for 1 h or overnight at −20 °C.

14. Centrifuge at 4 °C for 20 min at 7500 x *g*. Wash the pellet with cold 70% EtOH. Remove as much 70% EtOH as possible, spin down remaining 70% EtOH and remove by pipetting. Do not dry the pellet. Resuspend the RNA pellet in 200 µl DEPC-treated water. RNA can be stored at −80 °C at this point.

15. Quantitate RNA: analyse 5–10 µl RNA (in 1 ml TE-SDS 0.1%) on spectrophotometer using WL scan. Calculation 1 $OD_{260nm} = 40$ µg RNA. Calculate OD 260/280 (should be 1.8 or higher).

16. Qualitate RNA: run an 0.8% RNase-free agarose gel and load 2 µg RNA per lane. As running buffer use 1 x MOPS buffer (dilute the 10 x MOPS buffer with DEPC-treated water). Before using the gel system, rinse it with 1% HCl for 1 h to degrade RNase and wash 3 times with DEPC dH_2O.

Protocol 9.3
Isolation of Poly-A+ mRNA

For isolation of mRNA, we use Oligotex mRNA Mini Kit from Qiagen Incorporated (Valencia, CA, USA) (cat. no. 70022), or Micro-FastTrack mRNA isolation kit by Invitrogen (Invitrogen Corporation, Carlsbad, CA, USA) (cat. no. 45-0036). Typically, 500 µg of total RNA gives a yield of 2–10 µg of mRNA. Quantitate RNA with RiboGreen or an Agilent Bioanalyser.

Direct-cDNA labelling synthesis protocol

Equipment, materials and reagents

5 x First Strand Reaction Buffer (SuperScript™ II Invitrogen cat. no. 180604-014)

0.1 M DTT

SuperScript™ II (200 U/μl) (SuperScript™ II, Invitrogen cat. no. 18064-014)

20 mM dNTP mix (6.67 mM each dATP, dGTP, dTTP) (Invitrogen cat. no. 10297-018)

2 mM dCTP (Invitrogen cat. no. 10297-018)

1.5 μl AncT mRNA primer (p54, p55 and p56 (1:1:1) 100 pmol

1–5 ng Control RNA (we use an artificial *Arabidopsis* transcript)

RNase A (0.05 mg/ml) (RNase A; USB cat. no. 70194Y)

RNase h (0.05 units/μl) (RNase H; Life Technologies cat. no. 18021-014)

Dig Ease Hybridization buffer (Roche™, Roche Group, Basel, Switzerland)

Cyanine 3-dCTP (1.0 mM) (Perkin-Elmer™/NEN cat. no. NEL 576, PerkinElmer LAS (UK) Ltd Beaconsfield, UK)

Cyanine 5-dCTP (1.0 mM) (Perkin-Elmer™/NEN cat. no. NEL 577)

DNase/RNase-free distilled water pH 7–8.5 (e.g. Sigma-Aldrich cat. no. W-4502)

100% isopropanol (Fisher Scientific™ (Fisher Scientific UK, Loughborough, UK) cat. no. A451-1)

NaOAc pH 5.2, 3 M

QIAquick PCR Purification Kit™ (Qiagen Incorporated, Valencia, CA, USA cat. no. 28104)

Cover slips, No. 1½ × 24 × 60 mm (Fisher Scientific™ cat. no. 12-548-5P, Grace Biolabs cat. no. HS6024)

10% SDS

20 x SSC

1% BSA

Bakers tRNA (10 mg/ml)

Salmon sperm DNA (10 mg/ml)

Hybridization chamber (Corning cat. no. 2551)

50 ml Coplin jar (e.g. Sigma-Aldrich cat. no. S5516)

Sterile, nuclease-free 1.5 ml microcentrifuge tubes

Sterile, nuclease-free aerosol barrier pipette tips

Heating block or waterbath temperature set to 70 °C and 95 °C

Microcentrifuge

Circulating waterbath, temperature set to 42 °C or humid chamber at 42 °C

Vortex mixer

SpeedVac™ (Thermo Electron Corporation, Waltham, MA, USA) rotory dessicator

UV spectrophotometer and 0.1 ml volume quartz cuvettes (1 cm path length) *optional*

Note: We use home-made AncT primers for our labelling reactions. Random or oligo(dT) primers can also be used in the labelling reaction with mRNA. Random primers are recommended if some degradation of the RNA sample is suspected. If total RNA is used as a template for cDNA synthesis, oligo(dT) primers must be used to avoid labelling the rRNA and the tRNA present in the total RNA preparation. The yield of cDNA using mRNA and random primers is generally greater than with total RNA with oligo(dT) primers.

Possible sources: Random primers (9 mer; Stratagene cat. no. 300309) for mRNA use 2.5 μg. oligo(dT) primer (oligo(dT)12-18 Primer, Life Technologies cat. no. 18418-012); for total RNA use 2 μg.

Corning Hyb chambers are usable in both the humidity chamber and water bath. Be sure to follow manufacturers' recommendations for use. There are many other commercially available chambers; home-made versions are also acceptable.

Critical guidelines for synthesis of fluorescent-labelled cDNA probes

- The critical step in making fluorescent-labelled cDNA is the quality of the starting material. The RNA must be clean. If phenol was used in the RNA isolation, column purification must follow. All traces of phenol must be removed from the RNA prior to probe synthesis.

- We recommend a final column purification and DNase treatment of RNA. For total RNA use Qiagen RNeasy™ (Qiagen Incorporated, Valencia, CA, USA) (cat. no. 74104) or Stratagene Absolutely RNA RT-PCR Miniprep kit™ (Startagene, Amsterdam, The Netherlands) (cat. no. 400800).

- The RNA concentration and purity should be determined by the absorbance at A_{260} and $A_{280.}$ Analyse 5–10 µl RNA (in 1 ml TE-SDS 0.1%) on spectrophotometer. 1 OD260 nm = 40 µg RNA. Calculate OD 260/280 (should be 1.8 or higher).

- The purity and integrity of the RNA should be confirmed by gel electrophoresis or on an Agilent BioAnalyser. For total RNA, specific bands of highly abundant rRNA should be visible at approximately 1.9 and 5 kb. The bands should be sharp and clear; smearing of the rRNA bands on the gel indicates that the RNA has degraded. In addition, high molecular weight bands (>9000 kb) indicate DNA contamination of the RNA sample.

- For Poly-A+ mRNA, a faint smear in the range of 0.5 to 2 kb should be detectable.

- If the buffer used to bind the DNA to the Qiagen PCR Purification column is not slightly acidic (less than pH 7), the cDNA will bind poorly, resulting in low yields.

- Exposure to light will cause photobleaching of the cyanine dyes. Cy3 (Amersham Biosciences, GE Healthcare, Little Chalfont, UK) is particularly sensitive; therefore, wherever possible, protect the cyanine-labelled nucleotides, labelled probe and hybridization probe from light. Once the cyanine-labelled cDNA is synthesized it is stable for several weeks at 4 °C if protected from light.

- Labelling Poly-A+ mRNA requires 2–4 µg, total RNA requires 20 µg.

- The RNA should be stored at −80 °C until used.

Method

Synthesis of Cyanine 3- and Cyanine 5-labelled cDNA

1. Combine on ice in 1.5 ml Eppendorf tubes:

   a. 1–2 µg mRNA or 10–20 µg total RNA

   b. 1–5 ng Control RNA (we are using artificial *Arabidopsis* transcripts)

   c. 1.5 μl AncT mRNA primer (p55, p56 and p57 (1:1:1) 100 pmol)

   d. DEPC dH_2O to final volume of 18.5 μl (Cy3) or 19.5 μl (Cy5).

2. Incubate at 70 °C for 10 min.

3. Add on ice:

   a. 8 μl 5 x first strand buffer

   b. 4 μl 0.1 M DTT

   c. 3 μl 20 mM dNTP-dCTP (6.67 mM each dATP, dGTP, dTTP)

   d. 1 μl 2 mM dCTP

   e. 2 μl 1 mM Cy3 or 1 μl 1 mM Cy5 + 1 ml DEPC H_2O

   f. 2 μl fresh SuperScript II reverse transcriptase.

Important!

For efficiency and to minimize pipetting errors when setting up the cDNA synthesis reactions, prepare one master mix of the components above for all reactions. Keep the master mix on ice until ready to aliquot into each reaction tube. Both total RNA and mRNA can be used as a template for cDNA probe synthesis. Total RNA is easier and less expensive to prepare than mRNA. Label the two different samples to be analysed (i.e. control and experimental) on an array with different fluors (i.e. Cyanine 3 or Cyanine 5). It is also suggested that you perform reciprocal labelling reaction. One labelling reaction generates enough probe for one yeast slide.

4. Mix gently, give a quick spin and incubate in a pre-warmed water bath at 42 °C.

5. After 2 h, give the reaction a quick spin to collect evaporated solution and add 2 μl of SuperScript™ II reverse transcriptase. Incubate for an additional hour.

6. Give solutions a quick spin and add 1 μl RNAse A (0.05 mg/ml) and 1 μl RNAse H (0.05 units/μl). Incubate at 37 °C for 15–30 min.

7. Purify the probes using Qiagen columns:

   a. Top up the reaction to 50 µl with DEPC H_2O

   b. Add 2.7 µl of 3 M NaOAc pH 5.2

   c. Add 200 µl PB buffer

   d. Apply solutions to columns

   e. Centrifuge for 15 sec at 13 000 x *g*

   f. Wash 4 times with 600 µl of PE buffer (centrifuge for 15 sec each time)

   g. Centrifuge tubes for 1 min to eliminate traces of buffer

   h. Transfer the column to a new microcentrifuge tube

   i. To elute the sample, add 30 µl of H_2O (pH 7.0–8.5) to the centre of the column, leave at RT for 1 min or longer, and then centrifuge for 1 min at 13 000 x *g*

   j. Repeat with an additional 30 µl and elute in the same tube second elution the same way.

   k. Final eluted volume should be approximately 60 µl.

8. At this point, you may want to determine the efficiency of incorporation of the Cy dyes into the probe in order to determine whether you wish to proceed with the experiment. You may choose to skip this option and proceed with the concentration and hybridization of the probe. Eluted probes are usually slightly blue (Cy5) or pink (Cy3) if the labelling was successful. For a more accurate determination, the following steps should be performed.

Important!

Elution efficiency is dependent on pH. The maximum elution efficiency is achieved between 7.0 and 8.5. Make sure that the pH value of the water is within this range.

Protocol 9.4
Determination of the efficiency of incorporation of the probe

1. The optical density of the purified probes should be measured on a spectrophotometer at 260 nm, 550 nm (for Cy3) and 650 nm (for Cy5). Measurements

should be done using undiluted probe directly in an ultra microcuvette. The probe should not be diluted for spectrophotometry, because of its low initial concentration. Diluting the probe prior to assay may give inaccurate readings because of the low absorbance. The amount of probe required for spectrophotometry will depend on the cuvette used. The probe used for spectrophotometry should be recovered from the cuvette and used in the hybridization reaction. Clean the cuvette thoroughly between samples with 0.1 N HCl to prevent cross contamination.

2. Calculate the amount of cDNA produced, pmole of dye incorporated and the efficiency of incorporation. If measuring a probe that was previously stored (i.e. contains 1 mM DTT), the blank used for spectrophotometry should also contain 1 mM DTT. Frequency of incorporation in this application is defined as the number of labelled nucleotides incorporated per 1000 nucleotides.

Amount of cDNA probe:

$$\text{OD260} \times 37 \times \text{total volume of probe}(\mu\text{l}) = \text{ng of probe}$$

pmole of dye incorporated:

$$\text{Cy3}: \ \text{OD550 nm} \times (\text{total volume of probe})/0.15 = \text{pmol of Cy3 dye incorporated}$$

$$\text{Cy5}: \ \text{OD650 nm} \times (\text{total volume of probe})/0.25 = \text{pmol of Cy5 dye incorporated}$$

Efficiency of incorporation: (# labelled nucleotides per 1000 nucleotides)

$$\text{f.o.i} = \text{pmole of dye incorporated} \times 324.5/\text{amount of cDNA probe (ng)}$$

Important!

Do not use the probe if there is less than 15 pmol of dye incorporated or if the efficiency of incorporation of the probe is less than 10. Optimal amounts are 20 pmol of dye incorporated with an efficiency of incorporation of between 20–50 labelled nucleotides per 1000 nucleotides.

If the measured concentration of probe is unusually high, there may still be RNA present in the probe. If this is a chronic problem, increase the length of RNase digestion in step 5. Probes can also be checked by gel electrophoresis for quality and size.

Concentration of the probe

1. Following purification with the QIAquick™ columns, the purified, labelled cDNA samples may be concentrated as follows:

2. Dry the solutions under vacuum in rotary dessicator until volume is reduced to about 5 µl (approximately 60 min).

3. Do not use heat during drying, to prevent degradation of the cyanine dyes; do not overdry the probes.

4. Either proceed directly to hybridization or freeze on dry ice and store at −80 °C.

Protocol 9.5
Hybridization to DNA microarrays

Pre-hybridization

1. Prepare a pre-hybridization buffer and sterilize using a syringe filter (0.22 µm). Preheat buffer to 42 °C before use.

2. Pre-hybridization buffer:

    a. 5 x SSC

    b. 0.1% SDS

    c. 1% BSA

3. Pipette 15 µl aliquots of the pre-hyb solution onto 3 separate areas of the printed slide surface (left centre and right side). Try to cover as much of the printed array area that will be used for the hybridization experiment. Use non-latex gloves for all manipulations of the slides and coverslips.

4. Hold a 22 mm × 60 mm coverslip over the array, gradually lower one end of the coverslip and allow the solution to wick across the entire length of the slide. After the pre-hybridization buffer has covered the entire array, adjust the coverslip's position and the edge of the slide accordingly. Try to remove any air bubbles that may be trapped under the coverslip by tapping gently; smaller bubbles will disappear during the pre-hybridization process.

5. Place the array into a 42 °C prewarmed humid chamber or water bath for at least 1 h.

Hybridization

1. Prepare the hybridization buffer; filter using a syringe filter (0.22 µm).

2. Hybridization buffer:

   a. 400 µl DIG Easy buffer

   b. 20 µl baker's yeast tRNA (10 mg/ml)

   c. 20 µl sonicated salmon sperm DNA (10 mg/ml)

3. Combine the appropriate Cy3-cDNA and Cy5-cDNA targets intended for hybridization to a single array in a 1.5 ml microcentrifuge tube.

4. Add 30–60 µl of hybridization buffer, mix well and incubate at 95 °C for 5 min, then on ice for 1 min.

5. Do a quick spin to recover solution, incubate at 42 °C until ready to use.

6. Remove the coverslip from the prehybridization by dipping the array in filtered water. Shake the slide gently to loosen the coverslip. With time the coverslip will slide free of the slide's surface. Wash twice with water. Spin dry in a centrifuge at 1000 x *g* for 5 min, or dry under a clean air stream.

7. Add the probe onto the coverslip in 3 aliquots onto separate areas of the printed slide surface (left, centre and right side); try to cover as much of the printed array area that will be used for the hybridization experiment.

8. Hold a 22 mm × 60 mm coverslip over the array; gradually lower one end of the coverslip and allow the solution to wick across the entire length of the slide. After the hybridization buffer has covered the entire array, adjust the coverslip's position and the edge of the slide accordingly. Try to remove the air bubbles that may be trapped under the coverslip by gently tapping onto the coverslip; smaller bubbles will disappear during the hybridization process.

9. Incubate immediately at 42 °C overnight, or for 12–16 h.

Washing

1. To remove coverslip submerge slide in a Coplin jar, or a slide holder, containing 1 x SSC, 0.2% SDS wash buffer preheated to 42 °C. Shake the slide gently to loosen the coverslip. With time the coverslip will slide free of the slide's surface.

2. After the coverslip is removed, place slide in 1 x SSC, 0.2% SDS preheated to 42 °C and agitate for 10 min.

3. Wash the slide once in 0.1 x SSC, 0.2% SDS at RT and agitate for 10 min.

4. Wash the slide twice in 0.1 x SSC, agitating for 5 min at room temperature.

5. Finally, dip the slide several times into a Coplin jar filled with water. Spin dry the slides in a centrifuge at 1000 x *g* for 5 min, or dry under a clean air stream.

6. Scan arrays as soon as possible. Stores slides in the dark; they may be stable for 2 or more weeks.

9.4 Experiment design

Many reviews are available on the design, manipulation, interpretation and validation of microarray experiments (Conway and Schoolnik, 2003). For example, the October 2002 issue of *Nature Genetics* did an excellent job of covering these issues (Chuaqui *et al.*, 2002; Quackenbush, 2002; Churchill, 2002; Slonim, 2002).

The most important factor in planning for microarray experiments is providing for sufficient funding and biological material to ensure that each comparison includes enough replicates to yield reliable data and that attention is given to prevent statistical confounding. Microarray results suffer from the curse of dimensionality; so many data points are measured in each individual assay that even a small error rate, or an imbalance in experimental design, can produce a significant number of false-positive results. It is also important to consider that not all replicates are created equal. For example, probes are often spotted in more than one copy on the same microarray.

Most often, these are present as duplicate spots and produce highly similar data, especially when these duplicate spots are situated in the same area of the microarray as they are equally affected by localization-dependent artefacts. In our experience, duplicate spots are useful in flagging problem areas, but they should not be considered as replicates. The average or median ratio of duplicated spots should be used as a single individual measurement in subsequent analysis. Technical replicates, in which the same RNA preparations are hybridized to more than one microarray, are commonly encountered. Most often, these include dye swaps in which, for example, one array would be hybridized with Cy5-labelled RNA from a mutated strain and Cy3-labelled RNA from the wild-type.

The technical replicate would then be hybridized with Cy3-labelled RNA from the mutated strain and Cy5-labelled RNA from the wild-type. Researchers will sometimes attempt to reduce variability by pooling multiple RNA samples and hybridizing the same pair of RNA pools to multiple microarrays. The main flaw in this strategy is the inability to determine whether some genes are inherently more variable than others. In cases where RNA pooling cannot be avoided, more reliable data can be obtained by measuring several independently produced pools and reducing the number of technical replicates.

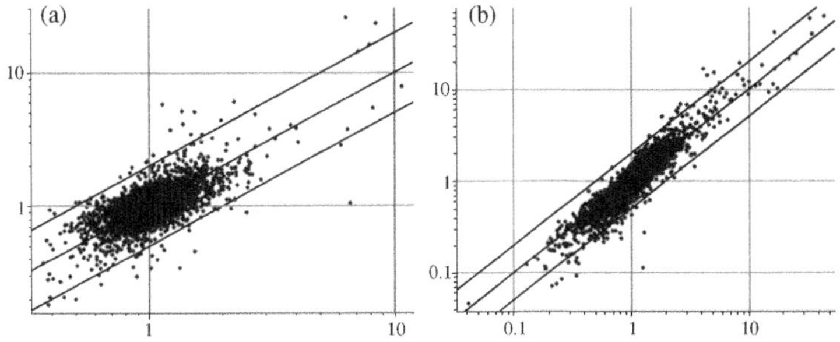

Scatter plots showing the correlation between the fluorescence ratios obtained from biological replicates of a study of the yeast-to-hyphae transition on *C. albicans* long oligonucleotide microarrays. Cy5/Cy3 experiments are in the x axis while the Cy3/Cy5 experiments are on the y axis. Data from panel (a) was obtained from only 2 hybridization (one per axis) while panel (b) shows the improved correlation that can be obtained from 6 hybridizations (3 per axis).

Figure 9.2 Improved data quality from an increase in replicate hybridizations

The most statistically significant type of replicate (and the one with the most variability) is called the 'biological replicate'. In these, each microarray is hybridized with independently produced mRNA preparations from independently produced cultures. We should also emphasize that, if possible, independent mRNA preparations should be obtained for both the experimental and the control samples. Because of the greater variability encountered in biological replicates, and to prevent statistical confounding due to dye-bias, we believe that, as a strict minimum, microarray experiments shouldn't even be considered unless they contain at least four biological replicates composed of two Cy3/Cy5 and two Cy5/Cy3 comparisons. The exact number of replicates that should be performed is often more than four and is dependent on numerous factors including the scale of transcriptional changes between the two conditions under study, the sensitivity of the strain to its environment, the talent of the experimenter and the quality of the microarrays. A scatter plot of the fluorescence ratios that result from two (or more) experiments (Figure 9.2) will usually show a cloud of experimental noise with reproducible changes in transcript abundance arranged in a rough diagonal.

As more and more replicates are added, the size of the non-specific cloud becomes smaller allowing more and more of the significantly modulated transcripts to rise up from the noise. Transcripts that are shown to be significantly modulated from a series of biological replicates are much more likely to be successfully validated.

The objective of most transcriptional profiling experiments is to measure the effects of different growth conditions, the consequences of a genetic change (a mutation or a change in the expression level of one or more genes) or the response to an external stimuli (a drug, for example). Of course, there are many situations where comparing a mutant to a wild-type strain, under normal growth conditions may not provide informative results. Consequently, many experimental set-ups will require a combination of two or more parameters. In such a situation, a closed-loop comparison strategy is the most effective in providing sufficient data to permit effective

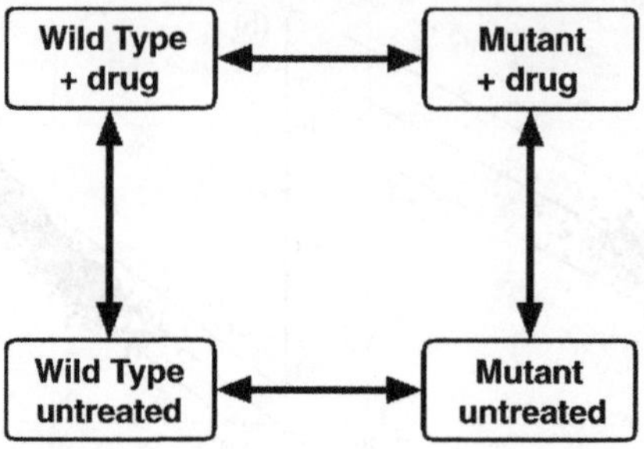

Figure 9.3 Close loop strategy for comparing the response of a mutant to an external stimuli

interpretation. For example, the strategy illustrated in Figure 9.3, where the goal is to study the response of a mutant to an external stimuli, allows the segregation of basal changes in transcript abundance from those that result from an inability of the mutant to respond to the environmental change (see Lee *et al.*, 2004 for example).

Dealing with experimental outliers

The successful completion of a microarray experiment requires the execution of several challenging manipulations over the course of two or more days. Like any other experiment, failures do occur, and the inclusion of low-quality data may severely impair statistical analysis (garbage in, garbage out). Upon the completion of several hybridization experiments, researchers should use global clustering algorithms, such as Principal Components Analysis or hierarchical clustering (see below), to separate the data sets produced by individual microarray hybridizations. Outliers, that is microarrays that are not grouped with other replicates of the same experimental condition, should be identified and steps have to be taken to ensure that these unusual results are not caused by an obvious experimental artefact. Thus the importance of keeping adequate notes during the production of the biological samples. Furthermore, inefficient dye incorporation and low RNA quality as well as cases of extreme localization-dependent bias can severely impair the quality of the data.

Selection of statistically significant transcripts

The next step often involves a reduction of the data set to a more manageable size, usually by masking data from genes that don't show statistically significant changes in transcript abundance. The use of a fold-change cutoff is by far the worst method by which to reduce data. Since this simple method does not take reproducibility into consideration, one gene could be modulated 10-fold in one experiment, not change in three replicates and still pass the magical two-fold cutoff that is, unfortunately, still seen today. Although it is becoming obvious that microarrays have a limited dynamic range and tend to underestimate the real change in transcript abundance, we have often observed cases of small (even 1.3-fold) but highly reproducible changes in fluorescence ratios that were validated in subsequent experiments. The most appropriate techniques for the identification of significantly modulated transcripts use the data's own variability to identify the genes with the most consistent changes. These include the Welsh t-test with a Benjamini and Hochberg correction (Hochberg and Benjamini, 1990) or the permutation methods that are available in the SAM package (Tusher *et al.*, 2001). Unfortunately, many statistical methods are overly strict when applied to microarray data since they are based on the assumption that the expression level of one gene is independent of the expression levels of the others. One needs only to look at the remarkable co-regulation of *C. albicans* genes whose products encode ribosomal proteins to realize that this assumption is clearly false. Nevertheless, these tools are useful in ranking modulated transcripts by order of significance.

Identification of interesting patterns and correlations

Although an exhaustive analysis of all of the statistical methods that can be applied to transcriptional profiling data could easily fill up the rest of this book, some methods have been, in our hands, more informative than others. These include hierarchical clustering in which individual genes and experiments are organized on a two-dimensional grid in such a way that similar data sets find themselves in close proximity (Eisen *et al.*, 1998). Lists of modulated transcripts are routinely compared to Gene Ontology classifications and the components of metabolic and signalling pathways. Furthermore, correlations between gene lists of significantly modulated transcripts with other lists from different profiling experiments have been very productive in our hands. For example, profile correlations have allowed us to identify a link between cAMP signalling and the response to osmotic stress (Enjalbert *et al.*, 2003; Harcus *et al.*, 2004). We have also determined that cells that lack the Efg1p and Cph1p transcription factors will inappropriately activate their heat-shock response pathway under conditions that should normally induce hyphal growth (Nantel *et al.*, 2002). Principal Components Analysis (Raychaudhuri *et al.*, 2000) is another useful method for identifying interesting correlations between gene lists and transcriptional profiles (Lee *et al.*, 2004; Bachewich *et al.*, 2005).

Data validation

The best control for a series of microarray experiments is to use the information from the resulting profiles to develop additional biochemical or physiological investigations. The modulation of a group of transcripts whose products are implicated in the same pathway or process can usually be confirmed independently. By themselves, northern blotting or reverse-transcriptase PCR of randomly selected genes have very little value except in those rare cases where the phenotype change can be explained by the change in transcript abundance of a limited number of genes or when a microarray probe cannot distinguish between gene family members. As for validating the global gene identification and annotation of the microarray, one needs only to look at genes encoding ribosomal proteins as these are highly co-regulated in *C. albicans*. Finally, it is highly recommended that changes in transcript abundance be correlated with the position of each gene on individual chromosomes. Selmecki *et al.* (2005) have recently used comparative genome hybridization to highlight significant levels of genomic instability in *C. albicans* laboratory strains that have been exposed to the stress of transformation and counter-selection on 5-fluoroorotic acid. The recent release of a completed *C. albicans* genome assembly (http://candida.bri.nrc.ca) should greatly facilitate these verifications.

9.5 Microarray-based studies in *C. albicans*

There are currently many examples where transcription profiling technology has been applied to virulence studies in *C. albicans*. Because of the difficulty in applying classical genetics to the analysis of *C. albicans* virulence, this approach has proven to be an attractive strategy to investigate fungal function and virulence. Early studies of expression profiles were focused on providing a compendium of transcripts modulated by the process under study. These gene lists were used to create hypotheses, often quite speculative, about the underlying logic of the observed modulations. Subsequent work has then built on these speculations to identify some of the regulatory circuits that control the observed patterns of expression. This process is iterative: patterns of expression are observed, candidate regulatory elements defined and then patterns of expression tested after modulation of the proposed regulators. Ultimately, this approach will provide a relatively sophisticated picture of the physiology of cellular behaviour, with the obvious caveat that transcriptional effects are only a component of the overall response of a cell to its environment and developmental state.

Antifungal drugs

The initial use of microarrays in the analysis of *C. albicans* was a study directed at establishing the expression profiles of the fungal cells responding to the antifungal azole itraconazole (De Backer *et al.*, 2001). In this study, glass-slide spotted arrays were used to compare transcript profiles of cells in the presence or absence of the antifungal drug. Among the more than 150 genes with a greater than 2.5-fold increase in gene expression after a 24-hour drug treatment were a substantial number of genes involved in the ergosterol biosynthesis pathway, an observation that was consistent with the known target of the azole compounds.

This approach of examining the consequences of drug treatments with microarrays has recently been expanded to establish, on a standard *C. albicans* strain, the global effects of compounds representative of each of the major antifungal drugs (Liu *et al.*, 2005). In this study, the azole used was ketoconazole, and the treatment was for three hours, so the results were not directly comparable with the data from the previous itraconazole study. However, among the 60 genes that were detected as showing enhanced expression in response to ketoconazole, 11 genes, including a number of the ergosterol biosynthesis genes, were also found to be up-regulated by itraconazole treatment. This suggests a common theme between the studies. However, whether the substantial number of genes that are differentially affected in the two reports represent real differences due to the timing of the treatment and/or unique characteristics of each drug or simply reflect noise in the system will require more detailed investigation. In addition to the ketoconazole treatment, Liu *et al.* (2005) examined amphotericin B, the echinocandin caspofungin and flucytosine, a toxic pyrimidine analogue. In each case, a substantial effect on gene expression was observed in response to the chemical treatment. 5-FC was similar to ketoconazole in primarily inducing gene expression, as 383 genes were up-regulated and only 56 showed repression. Many of the up-regulated genes were involved in DNA repair and nucleotide metabolism, consistent with an impact of a toxic nucleotide analogue. In the case of amphotericin B and caspofungin treatment, the predominant effect was to shut off gene expression; amphotericin B resulted in down-regulation of 169 genes and up-regulation of 87, while caspofungin shut down almost 400 genes and activated 81. Because amphotericin B and the azoles act on fungal cells with a somewhat similar molecular mechanism, it is perhaps surprising that the profiles of these two compounds do not overlap more. Although both compounds cause changes in sterol biosynthesis gene expression, amphotericin B tends to cause down-regulation, while ketoconazole treatment causes up-regulation.

These drug treatment studies have developed large data sets of differentially expressed genes, and some general patterns have been discerned. However, by themselves they are not sufficient to confirm the underlying regulatory mechanisms triggered by exposure to the antifungals. However, in addition to examining the direct consequences of drug treatment, microarrays have been used to examine the transcriptional behaviour of strains that have been selected to be resistant to specific antifungal compounds. If the observed increases and decreases in gene expression observed in the transient treatment studies represent the adaptation of the cells to the presence of the drug, it is reasonable to anticipate that selected resistant cells may have fixed the expression levels of key genes at the preferred activated or inactivated states. A population genetics study examined the consequences of long-term adaptation to the antifungal drug fluconazole, and observed that in a series of selected populations there were distinct routes to adapt to the presence of the drug (Cowen *et al.*, 2002). Drug resistance was accompanied by changes in gene expression that persisted in the absence of the drug, and cluster analysis identified three distinct patterns of gene expression underlying the adaptation. Each pattern exhibited an enhanced expression of a drug efflux pump such as *CDR2* or *MDR1*, as well as a set of coordinated gene expression changes. A second study examined the pattern of gene expression in clinical isolates that acquired resistance to azoles (Rogers and Barker, 2003), and also noted the up-regulation of the efflux pumps in concert with patterns of enhancement or the repression of other genes. Genes involved in oxidative stress response were found to be up-regulated in many of the isolates examined. A further study on an amphotericin B resistant isolate of Sc5314 that was identified as exhibiting cross-resistance to fluconazole (Barker *et al.*, 2004) found up-regulation of ergosterol biosynthesis genes and certain stress response genes. The membrane composition of the resistant isolate was modified in its sterols, with higher levels of eburicol and lanosterol in place of ergosterol.

Overall, several genes that were up-regulated in response to acute drug treatments, in particular efflux pumps and sterol biosynthesis genes, were found to be constitutively up-regulated in some drug-resistant isolates. This observation provides evidence that the transcriptional profiling is providing a biologically relevant picture of the physiology of drug-treated cells; more recently, attempts have been made to define the controlling elements underlying these expression profiles. The key role played by the efflux pumps in regulating response to antifungal drugs has directed research at identifying the transcription regulators of the pumps. A trans-acting sequence termed a 'DRE' (drug-response element) was identified upstream of some of the responding genes. A zinc-cluster transcription factor termed Tac1p (transcription activator of *CDR* genes) was identified that could bind the DRE sequence (Coste *et al.*, 2004); disruption of *TAC1* blocked the drug-inducibility of a number of genes identified as being transcriptionally activated in response to antifungal drugs, and an allele of *TAC1* was identified that led to constitutive up-regulation of the efflux pumps in a drug-resistant clinical isolate. Thus a combination of transcriptional profiling using microarrays and classic molecular biology is starting to uncover the regulatory mechanisms that underlie the development of drug-resistant *C. albicans*, a topic with clear relevance to fungal pathogenicity.

Morphogenesis

Drug resistance is not the only *C. albicans* trait that has been investigated by microarrays. The morphological plasticity of *C. albicans* cells has been one of the defining features of this fungus, and is identified as a key component of its opportunistic pathogenicity because blocks to the morphological transitions have proven to reduce virulence (Lo *et al.*, 1997; Rocha *et al.*, 2001). The first whole-genome analysis of the yeast-to-hyphal transition used spotted glass slide amplicon arrays and the two-colour labelling technique (Nantel *et al.*, 2002). This approach confirmed the inducibility of a variety of genes that had been identified as being hyphal-specific through directed studies (Birse *et al.*, 1993; ; Bailey *et al.*, 1996; Staab and Sundstrom, 1998), and showed a variety of new genes of predicted or unknown function to be part of the hyphal-inducing regulon. The transcription factors Efg1p and Cph1p, previously identified as being required for the morphological transition (Lo *et al.*, 1997), were found to be essential for the proper transcriptional response to hyphal-inducing signals. This is not true of all mutants that block the morphological transition, for example defects in the *MYO5* gene block the morphological transition to hyphae in response to inducing signals without dramatically modulating the overall transcriptional response to these signals (Oberholzer, personal communication).

Several earlier studies of gene expression patterns in *C. albicans* cells had used spotted filters to investigate this morphological transition. Analysis of the roles of candidate transcription repressors such as Nrg1p, Tup1p and Mig1p made use of partial genomic arrays containing approximately 2000 genes spotted on nitrocellulose (Murad *et al.*, 2001a, 2001b). In these studies deletions of the transcription regulators were found, not surprisingly, to have large-scale consequences on the patterns of gene expression. Tup1p, Nrg1p and Mig1p all served primarily as repressors, but a simple model that Tup1p served as a general repressor that was targeted to specific subsets of genes by Mig1p and Nrg1p did not fit the expression profiles observed. For example, there were genes regulated by Tup1p that were not influenced by deletion of either Mig1p or Nrg1p, while at the same time there were genes whose expression was de-repressed in all three mutant backgrounds (Murad *et al.*, 2001a). Thus an apparently highly complex regulatory circuit must function to control the negative regulation of hyphal development.

The consequences of mutations in the transcription factors Efg1p, Cph1p and Cph2p were also assessed using filter arrays, but these arrays contained only a representative 10% of the genome (Lane *et al.*, 2001). In this study Cph1p and Efg1p were found to be required for a set of hyphal-induced genes, and that Cph2p and Efg1p were involved in the regulation of yet another transcription factor, Tec1p, also involved in hyphal formation.

More recently, the role of the negative regulators in hyphal development has been expanded through the use of whole-genome spotted-glass arrays (Kadosh and Johnson, 2005). A set of 61 genes was identified that showed filamentation-induced expression – this list confirmed a majority of the genes identified in the previous study (Nantel *et al.*, 2002), and expanded the list by improving the relative purity of the yeast and hyphal cell populations. The transcriptional repressors Tup1p, Nrg1p and Rfg1p were critical to the proper regulation of the majority of these hyphal-regulated genes. As was observed in the previous partial genome studies, the patterns of expression were consistent with Nrg1p and Rfg1p working together with Tup1p to coordinate the expression of specific gene sets, but with each factor influencing the expression of unique gene sets as well.

The Ssn6p regulator, implicated in the Tup1p regulatory circuit in *S. cerevisiae*, has also been examined by whole-genome profiling in *C. albicans* (Garcia-Sanchez *et al.*, 2005). This study showed that the transcriptional consequences of loss of Ssn6p function and loss of Tup1p function were not highly correlated, and that this distinction was also found in the phenotypic consequences of the deletions. *Tup1* mutants, as previously noted (Braun and Johnson, 1997) were hyper-filamentous, while the *ssn6* mutants exhibited characteristics of cells programmed for the epigenetic switching of states. Therefore, the transcriptional profiles of strains mutant in a series of integrated negative regulators have been established through a number of studies using different technologies in different laboratories. These transcriptional regulators are implicated in pathogenesis, as many of the tested mutants show reduced virulence. Because these studies have used different array platforms, and in particular different genome annotations, it is not trivial to compare the data sets to determine the quality of the data overlap. The recent coordinated annotation effort (Braun *et al.*, 2005), coupled with the establishment of the *Candida* Genome Database as the central depository of genomic information for *C. albicans*, should improve the ability to assess data sets from different studies.

Biofilms

There are other pathogenesis-related behaviours of *C. albicans* cells that have been subjected to extensive microarray analysis. Biofilms of *C. albicans* represent a complex cellular assembly with significant implications in pathogenicity (Douglas, 2002; Kumamoto, 2002). Cells in biofilms exhibit an increased resistance to antifungal drugs (Ramage *et al.*, 2002), biofilms on dentures are an important medical problem (Nikawa *et al.*, 2000) and biofilms forming on catheters provide a common route of re-infection for hospital patients (Andes *et al.*, 2005). Because a biofilm consists of cells in a variety of morphological and physiological states, during microarray analysis care must be taken to identify the biofilm-specific components of the expression profile. As well, although there are general characteristics of a *C. albicans* biofilm, there are many distinct conditions that create physically similar structures, so there is no *a priori* reason to predict that there should be a single unique biofilm-specific profile.

An initial investigation into the transcription profiles of biofilms grown on plastic surfaces suggested there were significant differences in gene expression when the biofilm-producing cells were compared with planktonic cells (Garcia-Sanchez *et al.*, 2004). Biofilms that were formed under a variety of growth conditions were compared with matched planktonic cultures, and the biofilm profiles were closely correlated, suggesting there was a core biofilm-specific gene expression pattern. A group of over 300 genes were identified that characterized the biofilm. Many of these genes were implicated in amino acid biosynthesis, and the key biosynthesis regulator Gcn4p was identified as being critical for biofilm production. Because biofilms contain a significant proportion of hyphal cells, the patterns of biofilm development, and the associated gene expression, was determined in both wild-type and hyphal-development-compromised cells that lacked the transcription factors Efg1p and Cph1p. This analysis suggested that, although there was a significant component of the biofilm transcriptome which was dependent on the hyphal cells, a large number of the biofilm-specific transcripts were independent of the hyphal-regulating transcription factors.

A more recent investigation of the transcriptional consequences of biofilm formation focused on the early stages of the process (Murillo *et al.*, 2005). This work established a connection with sulfur metabolism, and found that ribosomal subunit gene expression as part of the Ribi regulon modulated in a biofilm-specific manner. The influence of farnesol on biofilm-specific transcription patterns has also been examined (Cao *et al.*, 2005); this latter study found an effect on hyphal-specific genes, as well as heat shock and cell-wall-maintenance genes.

Although direct genomic profiling can provide a list of differences between cells in biofilms and planktonic *C. albicans*, understanding the process fully requires an identification of the unique regulators of the process. A recent mutant search for genes implicated uniquely in biofilm formation and not in hyphal development has uncovered a zinc-cluster transcription factor termed Bcr1p (Nobile and Mitchell, 2005). Disruption of the *BCR1* prevents the formation of biofilms, and an analysis of the transcription profile of the mutant cells shows that the expression of a variety of genes involved in hyphal development and adhesion is blocked. *BCR1* expression is itself under the control of Tec1p, a transcription factor initially characterized for its influence on hyphal development that is also required for biofilm development. Intriguingly, loss of *BCR1* does not block hyphal development, only biofilm development. This establishes that while hyphal formation is required for biofilm development, it is not sufficient, and directs the search for key effectors of biofilm formation within the group of genes dependent on Bcr1p for proper expression.

9.6 Conclusion

Microarray analysis has been initiated for other processes implicated in aspects of fungal pathogenicity. These include the response to stresses such as oxidative, osmotic and heat shock (Enjalbert *et al.*, 2003), the response to macrophage engulfment (Lorenz *et al.*, 2004), to pH changes (Bensen *et al.*, 2004) and the response to nitrosative stress (Hromatka *et al.*, 2005). As these studies develop from the identification and classification of the gene sets modulated by the external challenges to an investigation of the underlying regulatory circuits controlling the expression of the gene sets, our knowledge of *C. albicans* pathogenesis will advance greatly. Ultimately, an understanding of these transcriptionally controlled regulatory circuits should help uncover exploitable weaknesses in the cells response to environmental stimuli, so the transcription regulators now being identified through array studies may lead to new antifungal drug targets in the future.

9.7 References

Andes, D., Lepak, A., Pitula, A., Marchillo, K. and Clark, J. (2005) A simple approach for estimating gene expression in *Candida albicans* directly from a systemic infection site. *J. Infect. Dis.* **192**: 893–900.

Bachewich, C., Nantel, A. and Whiteway, M. (2005) Cell cycle arrest during S or M phase generates polarized growth via distinct signals in *Candida albicans*. *Mol. Microbiol.* **57**: 942–59.

Bailey, D.A., Feldmann, P. J., Bovey, M., Gow, N. A. and Brown, A. J. (1996) The *Candida albicans* HYR1 gene, which is activated in response to hyphal development, belongs to a gene family encoding yeast cell wall proteins. *J. Bacteriol.* **178**: 5353–60.

Barker, K.S., Crisp, S., Wiederhold, N., Lewis, R. E .*et al.* (2004) Genome-wide expression profiling reveals genes associated with amphotericin B and fluconazole resistance in experimentally induced antifungal resistant isolates of *Candida albicans*. *J. Antimicrob. Chemother.* **54**: 376–85.

Bennett, R. J., Uhl, M. A., Miller, M. G. and Johnson, A. D. (2003) Identification and characterization of a *Candida albicans* mating pheromone. *Mol. Cell. Biol.* **23**: 8189–201.

Bensen, E. S., Martin, S. J., Li, M., Berman, J. and Davis, D. A. (2004) Transcriptional profiling in *Candida albicans* reveals new adaptive responses to extracellular pH and functions for Rim101p. *Mol. Microbiol.* **54**: 1335–51.

Berman, J., and Sudbery, P. E. (2002) *Candida albicans*: A molecular revolution built on lessons from budding yeast. *Nat. Rev. Genet.* **3**: 918–30.

Birse, C. E., Irwin, M. Y., Fonzi, W. A. and Sypherd, P. S. (1993) Cloning and characterization of ECE1, a gene expressed in association with cell elongation of the dimorphic pathogen *Candida albicans*. *Infect. Immun.* **61**: 3648–55.

Braun, B. R. and Johnson, A. D. (1997) Control of filament formation in *Candida albicans* by the transcriptional repressor TUP1. *Science* **277**: 105–9.

Braun, B. R., van Het Hoog, M., d'Enfert, C., Martchenko, M. *et al.* (2005) A human-curated annotation of the *Candida albicans* genome. *PLoS. Genet.* **1**: 36–57.

Cao, Y. Y., Cao, Y. B., Xu, Z., Ying, K. *et al.* (2005) cDNA microarray analysis of differential gene expression in *Candida albicans* biofilm exposed to farnesol. *Antimicrob. Agents. Chemother.* **49**: 584–9.

Chauhan, N., Inglis, D., Roman, E. Pla, J. *et al.* (2003) *Candida albicans* response regulator gene SSK1 regulates a subset of genes whose functions are associated with cell wall biosynthesis and adaptation to oxidative stress. *Eukaryot. Cell.* **2**: 1018–24.

Chuaqui, R. F., Bonner, R. F., Best, C. J., Gillespie, J. W. *et al.* (2002) Post-analysis follow-up and validation of microarray experiments. *Nat. Genet.* **32** Suppl: 509–14.

Churchill, G. A. (2002) Fundamentals of experimental design for cDNA microarrays. *Nat. Genet.* **32** (Suppl): 490–5.

Conway, T. and Schoolnik, G. K. (2003) Microarray expression profiling: Capturing a genome-wide portrait of the transcriptome. *Mol. Microbiol.* **47**: 879–89.

Copping, V. M., Barelle, C. J., Hube, B., Gow, N. A. *et al.* (2005) Exposure of *Candida albicans* to antifungal agents affects expression of SAP2 and SAP9 secreted proteinase genes. *J. Antimicrob. Chemother.* **55**: 645–54.

Coste, A. T., Karababa, M., Ischer, F., Bille, J. and Sanglard, D. (2004) TAC1, transcriptional activator of CDR genes, is a new transcription factor involved in the regulation of *Candida albicans* ABC transporters CDR1 and CDR2. *Eukaryot. Cell.* **3**: 1639–52.

Cowen, L. E., Nantel, A., Whiteway, M. S., Thomas, D. Y. *et al.* (2002) Population genomics of drug resistance in *Candida albicans*. *Proc. Natl. Acad. Sci. USA.* **99**: 9284–9.

De Backer, M. D., Ilyina, T., Ma, X. J., Vandoninck, S. *et al.* (2001) Genomic profiling of the response of *Candida albicans* to itraconazole treatment using a DNA microarray. *Antimicrob. Agents. Chemother.* **45**: 1660–70.

DeRisi, J., Penland, L., Brown, P. O., Bittner, M. L. *et al.* (1996) Use of a cDNA microarray to analyse gene expression patterns in human cancer. *Nat. Genet.* **14**: 457–60.

Douglas, L. J. (2002) Medical importance of biofilms in *Candida* infections. *Rev. Iberoam. Micol.* **19**: 139–43.

Eisen, M. B., Spellman, P. T., Brown, P. O. and Botstein, D. (1998) Cluster analysis and display of genome-wide expression patterns. *Proc. Natl. Acad. Sci. USA*. **95**: 14863–8.

Enjalbert, B., Nantel, A. and Whiteway, M. (2003) Stress-induced gene expression in *Candida albicans*: Absence of a general stress response. *Mol. Biol. Cell*. **14**: 1460–7.

Enjalbert, B. and Whiteway, M. (2005) Release from quorum-sensing molecules triggers hyphal formation during *Candida albicans* resumption of growth. *Eukaryot. Cell*. **4**: 1203–10.

Fradin, C., De Groot, P., MacCallum, D., Schaller, M. *et al.* (2005) Granulocytes govern the transcriptional response, morphology and proliferation of *Candida albicans* in human blood. *Mol. Microbiol.* **56**: 397–415.

Garcia-Sanchez, S., Aubert, S., Iraqui, I., Janbon, G. *et al.* (2004) *Candida albicans* biofilms: A developmental state associated with specific and stable gene expression patterns. *Eukaryot. Cell*. **3**: 536–45.

Garcia-Sanchez, S., Mavor, A. L., Russell, C. L., Argimon, S. *et al.* (2005) Global roles of Ssn6 in Tup1- and Nrg1-dependent gene regulation in the fungal pathogen *Candida albicans*. *Mol. Biol. Cell*. **16**: 2913–25.

Harcus, D., Nantel, A., Marcil, A., Rigby, T. and Whiteway, M. (2004) Transcription profiling of cyclic AMP signaling in *Candida albicans*. *Mol. Biol. Cell*. **15**: 4490–9.

Hochberg, Y. and Benjamini, Y. (1990) More powerful procedures for multiple significance testing. *Stat. Med.* **9**: 811–8.

Hromatka, B. S., Noble, S. M. and Johnson, A. D. (2005) Transcriptional response of *Candida albicans* to nitric oxide and the role of the YHB1 gene in nitrosative stress and virulence. *Mol. Biol. Cell*. **16**: 4814–26.

Jones, T., Federspiel, N. A., Chibana, H., Dungan, J. *et al.* (2004) The diploid genome sequence of *Candida albicans*. *Proc. Natl. Acad. Sci. USA*. **101**: 7329–34.

Kadosh, D., and Johnson, A. D. (2005) Induction of the *Candida albicans* filamentous growth program by relief of transcriptional repression: A genome-wide analysis. *Mol. Biol. Cell*. **16**: 2903–12.

Kohrer, K. and Domdey, H. (1991) Preparation of high molecular weight RNA. *Methods. Enzymol.* **194**: 398–405.

Kumamoto, C. A. (2002) Candida biofilms. *Curr. Opin. Microbiol.* **5**: 608–11.

Lan, C.Y., Newport, G., Murillo, L. A., Jones, T. *et al.* (2002) Metabolic specialization associated with phenotypic switching in *Candida albicans*. *Proc. Natl. Acad. Sci. USA*. **99**: 14907–12.

Lan, C.Y., Rodarte, G., Murillo, L. A., Jones, T. *et al.* (2004) Regulatory networks affected by iron availability in *Candida albicans*. *Mol. Microbiol.* **53**: 1451–69.

Lane, S., Zhou, S., Pan, T., Dai, Q. and Liu, H. (2001) The basic helix-loop-helix transcription factor Cph2 regulates hyphal development in *Candida albicans* partly via TEC1. *Mol. Cell. Biol.* **21**: 6418–28.

Lee, C.M., Nantel, A., Jiang, L., Whiteway, M. and Shen, S. H. (2004) The serine/threonine protein phosphatase SIT4 modulates yeast-to-hypha morphogenesis and virulence in *Candida albicans*. *Mol. Microbiol.* **51**: 691–709.

Liu, T. T., Lee, R. E., Barker, K. S., Wei, L. *et al.* (2005) Genome wide expression profiling of the response to azole, polyene, echinocandin, and pyrimidine antifungal agents in *Candida albicans*. *Antimicrob. Agents. Chemother.* **49**: 2226–36.

Lo, H.J., Kohler, J. R., DiDomenico, B., Loebenberg, D. *et al.* (1997) Nonfilamentous *C. albicans* mutants are avirulent. *Cell.* **90**: 939–49.

Lorenz, M. C., Bender, J. A. and Fink, G. R. (2004) Transcriptional response of *Candida albicans* upon internalization by macrophages. *Eukaryot. Cell.* **3**: 1076–87.

Murad, A. M., d'Enfert, C., Gaillardin, C., Tournu, H. *et al.* (2001a) Transcript profiling in *Candida albicans* reveals new cellular functions for the transcriptional repressors CaTup1, CaMig1 and CaNrg1. *Mol. Microbiol.* **42**: 981–93.

Murad, A. M., Leng, P., Straffon, M., Wishart, J. *et al.* (2001b) NRG1 represses yeast-hypha morphogenesis and hypha-specific gene expression in *Candida albicans. Embo. J.* **20**: 4742–52.

Murillo, L. A., Newport, G., Lan, C. Y., Habelitz, S. *et al.* (2005) Genome-wide transcription profiling of the early phase of biofilm formation by *Candida albicans*. *Eukaryot. Cell.* **4**: 1562–73.

Nantel, A., Dignard, D., Bachewich, C., Harcus, D. *et al.* (2002) Transcription profiling of *Candida albicans* cells undergoing the yeast-to-hyphal transition. *Mol. Biol. Cell.* **13**: 3452–65.

Nicholls, S., Straffon, M., Enjalbert, B., Nantel, A. *et al.* (2004) Msn2- and Msn4-like transcription factors play no obvious roles in the stress responses of the fungal pathogen *Candida albicans*. *Eukaryot. Cell.* **3**: 1111–23.

Nikawa, H., Jin, C., Hamada, T., Makihira, S. *et al.* (2000) Interactions between thermal cycled resilient denture lining materials, salivary and serum pellicles and *Candida albicans in vitro.* Part II: Effects on fungal colonization. *J. Oral. Rehabil.* **27**: 124–30.

Nobile, C. J. and Mitchell, A. P. (2005) Regulation of cell-surface genes and biofilm formation by the *C. albicans* transcription factor Bcr1p. *Curr. Biol.* **15**: 1150–5.

Pappas, P. G., Rex, J. H., Lee, J., Hamill, R. J. *et al.* (2003) A prospective observational study of candidemia: Epidemiology, therapy, and influences on mortality in hospitalized adult and pediatric patients. *Clin. Infect. Dis.* **37**: 634–43.

Quackenbush, J. (2002) Microarray data normalization and transformation. *Nat. Genet.* **32** (Suppl): 496–501.

Ramage, G., Bachmann, S., Patterson, T. F., Wickes, B. L. and Lopez-Ribot, J. L. (2002) Investigation of multidrug efflux pumps in relation to fluconazole resistance in *Candida albicans* biofilms. *J. Antimicrob. Chemother.* **49**: 973–80.

Raychaudhuri, S., Stuart, J. M. and Altman, R. B. (2000) Principal components analysis to summarize microarray experiments: Application to sporulation time series. *Pac. Symp. Biocomput.*: 455–66.

Rocha, C. R., Schroppel, K., Harcus, D., Marcil, A. *et al.* (2001) Signaling through adenylyl cyclase is essential for hyphal growth and virulence in the pathogenic fungus *Candida albicans*. *Mol. Biol. Cell.* **12**: 3631–43.

Rogers, P. D. and Barker, K. S. (2002) Evaluation of differential gene expression in fluconazole-susceptible and -resistant isolates of *Candida albicans* by cDNA microarray analysis. *Antimicrob. Agents. Chemother.* **46**: 3412–7.

Rogers, P. D. and Barker, K. S. (2003) Genome-wide expression profile analysis reveals coordinately regulated genes associated with stepwise acquisition of azole resistance in *Candida albicans* clinical isolates. *Antimicrob. Agents. Chemother.* **47**: 1220–7.

Sanglard, D. and Odds, F. C. (2002) Resistance of *Candida* species to antifungal agents: Molecular mechanisms and clinical consequences. *Lancet. Infect. Dis.* **2**: 73–85.

Selmecki, A., Bergmann, S. and Berman, J. (2005) Comparative genome hybridization reveals widespread aneuploidy in *Candida albicans* laboratory strains. *Mol. Microbiol.* **55**: 1553–65.

Slonim, D. K. (2002) From patterns to pathways: Gene expression data analysis comes of age. *Nat. Genet.* **32** (Suppl): 502–8.

Staab, J. F. and Sundstrom, P. (1998) Genetic organization and sequence analysis of the hypha-specific cell wall protein gene HWP1 of *Candida albicans*. *Yeast*. **14**: 681–6.

Tusher, V. G., Tibshirani, R. and Chu, G. (2001) Significance analysis of microarrays applied to the ionizing radiation response. *Proc. Natl. Acad. Sci. USA*. **98**: 5116–21.

Xu, Z., Cao, Y. B., Zhang, J. D., Cao, Y. *et al.* (2005) cDNA array analysis of the differential expression change in virulence-related genes during the development of resistance in *Candida albicans. Acta. Biochim. Biophys. Sin. (Shanghai)*. **37**: 463–72.

Zhao, R., Daniels, K. J., Lockhart, S. R., Yeater, K. M. *et al.* (2005a) Unique aspects of gene expression during *Candida albicans* mating and possible G(1) dependency. *Eukaryot. Cell*. **4**: 1175–90.

Zhao, X., Oh, S. H., Yeater, K. M. and Hoyer, L. L. (2005b) Analysis of the *Candida albicans* Als2p and Als4p adhesins suggests the potential for compensatory function within the Als family. *Microbiology*. **151**: 1619–30.

10

Molecular techniques for application with *Aspergillus fumigatus*

Nir Osherov and **Jacob Romano**

10.1 Introduction

Aspergillus fumigatus is the most commonly encountered mould pathogen of humans. Diseases caused by *A. fumigatus* can be divided into three categories: (A) allergic responses to inhaled conidia in individuals with a hyperactive immune response, (B) chronic aspergillosis in generally immunocompetent persons, including aspergillomas or 'fungal ball' and (C) invasive aspergillosis (IA) in immunocompromised patients. IA is associated with high mortality rates of between 30% to 90%, despite antifungal treatment (Vogeser *et al.*, 1999).

A. fumigatus is an opportunistic pathogen of humans and farm animals. It is normally found in the soil, particularly in decomposing leaf litter, where it plays an important role in recycling nutrients (Latge, 1999). It produces only a small proportion of all aerial spores, yet is responsible for approximately 90% of systemic *Aspergillus* infections, suggesting that it may possess intrinsic virulence factors that allow it to preferably infect immunocompromised patients.

A. fumigatus is a particularly tenacious mould, able to withstand a wide variety of growth conditions (Latge, 1999). It grows rapidly on minimal agar plates containing a wide variety of carbon sources, a nitrogen source and trace elements. It can grow at temperatures of up to 50 °C, and conidia can survive temperatures of 70 °C. It produces prodigious numbers of minute greenish conidia (2–3 μm in diameter), which are easily aerosolized and dispersed.

Medical Mycology: Cellular and Molecular Techniques. Edited by Kevin Kavanagh

Surprisingly, despite its rising prominence as an important human pathogen, the molecular study of *A. fumigatus* has lagged significantly behind that of the model organism *Aspergillus nidulans* and the biotechnological 'workhorse' *Aspergillus niger.*

This disparity is being rapidly narrowed: transformation techniques enabling gene knockout and complementation, selectable markers and auxotrophic strains, reporter genes and proteins have all been developed for *A. fumigatus* (Brakhage and Langfelder, 2002). Most importantly, the *A. fumigatus* genome has been sequenced and annotated (Tekaia and Latge, 2005). Researchers in the field can now perform computational analyses of the *A. fumigatus* genome to identify novel genes and pathways and compare the *A. fumigatus* genome with the sequenced genomes of other *Aspergilli* to identify genes and pathways unique to this organism (Nierman *et al.*, 2005). They can prepare microarrays to measure transcriptional changes at the whole-genome level. The bottleneck to this potential flood of novel information depends on the ability of the researcher to rapidly identify, clone and accurately delete potentially interesting new genes. However, many genes may prove non-essential or display a subtle phenotype when deleted. In this chapter, we provide methods to rapidly prepare knockout vectors (Protocol 10.2), transform them efficiently and accurately (Protocols 10.3 and 10.4) and assess the resulting mutant phenotype (Protocol 10.5). A wide variety of growth assays and microscopic techniques are described (Protocol 10.5) and a list of specific inhibitory compounds commonly used to analyse the phenotypes of *Aspergillus* mutants is provided (Protocol 10.5).

10.2 Preparation of knockout vectors for gene disruption and deletion in *A. fumigatus*

Molecular gene disruption techniques are widely used in the *A. fumigatus* field to identify and investigate the role of specific genes, to understand mechanisms involved in virulence and identify novel drug targets.

Several gene disruption methods have been described. (1) Cloning the target gene of interest with 5′ and 3′ flanking regions by PCR into a plasmid vector, and replacement of all or part of the gene with a selectable marker utilizing suitable restriction sites found in the target gene (Mouyna *et al.*, 2005; Bhabhra *et al.*, 2004; Mellado *et al.*, 2005). (2) PCR cloning the 5′ and 3′ flanking sequences of the target gene and ligation of a selectable marker between them, utilizing restriction sites engineered by PCR into the flanking regions (Xue *et al.*, 2004). (3) Integrating a selectable marker by homologous recombination into a target gene using an *Escherichia coli* strain expressing the phage λ Red (gam, bet, exo) functions (Chaveroche *et al.*, 2000). (4) RNA interference: constructing small portions of RNA homologous to a gene of interest. The small fragments of RNA will hybridize to the specific mRNA, generating double-stranded RNA and causing degradation (Mouyna *et al.*, 2004). This method enables down-regulation of several genes simultaneously.

All methods, aside from RNA interference, require long flanking regions from both sides of the targeted region which are necessary for successful homologous recombination into the genomic DNA. The first method depends upon finding suitable restriction sites in the genetic sequence, which can be problematic. The second method uses PCR-cloned flanking regions from both sides with restriction sites engineered at the ends. This enables a ligation of any chosen marker gene engineered with the same restriction site, and generates a whole-gene deletion, not leaving a partial target gene region. It does, however, entail finding suitable unique restriction sites and performing a 4-fragment ligation. The third method is based on constructing the deletion fragment *in vivo* utilizing an *E. coli* strain expressing the phage λ Red cosmid.

Here we describe two additional approaches for disruption and deletion of genes. One approach (Protocol 10.1.1) utilizes a modified transposon containing the *N. crassa pyr4* gene, which is randomly inserted *in vitro* into a target sequence of interest. Clones in which the gene of interest has been disrupted are identified by PCR and used to transform a *pyrG*-deficient strain of *A. fumigatus* (Jadoun *et al.*, 2004). The second approach (Protocol 10.1.2), known as double-jointed PCR, is a wholly PCR-based technique for DNA assembly to fuse long flanking regions of the target gene to an antibiotic-resistant gene or promoter, constructing recombinant DNA for deletion or gene control respectively (Yu *et al.*, 2004). Using this method eliminates the use of intermediate cloning and the transformation of competent *E. coli* bacteria. This method utilizes multiple rounds of PCR and long-PCR techniques and, thus, may result in undesired mutations in the genomic sequence.

Protocol 10.1
Preparation of knockout vectors for gene disruption and deletion in *A. fumigatus*

Equipment, materials and reagents

A. fumigatus genomic DNA prepared with the MasterPure Yeast DNA Purification Kit™ (Epicenter, Madison, WI, USA) (Jin *et al.*, 2004).

PCR reagents

Expand high-fidelity PCR system (Roche Mannheim, Germany)

PCR thermal block

pGEM T/A cloning vector system™ (Promega Corp., Madison, WI)

Competent *E. coli* bacteria strain

Nucleic acid gel purification kit

Mini- and midiprep kit

GPS-1 Genome Priming System™ (New England Biolabs, Beverly, MA)

Modified GPS-1/*pyr4* transposon™ (Available at the FGSC, USA)

A. fumigatus pyrG-deficient strain AF293.1 (Available at the FGSC, USA)

Method

Transposon-based disruption method

1. Extract *A. fumigatus* genomic DNA using the MasterPure Yeast DNA Purification Kit, with minor protocol modifications described by Jin *et al.* (2004).
2. Design primers to contain an *Asc*I restriction site at their 5′ end to generate a ~4 kb fragment of the desired gene (i.e. ~2 kb upstream to the ATG start codon and ~2 kb downstream to the TAG stop codon of the gene).
3. Amplify the ~4 kb fragment using the expand high-fidelity PCR system as recommended by the manufacturer.
4. Cut and purify the PCR product by gel electrophoresis.
5. Clone the fragment containing the desired gene of interest into the pGEM T/A cloning vector according to the manufacturer's instructions.
6. Transform the plasmid into an *E. coli* strain (any competent strain will suffice) and grow overnight at 37 °C on LB agar plates containing 100 μg/ml ampicillin for selection.
7. Extract the plasmid using a standard miniprep protocol.
8. Perform inactivation of the selected gene using the GPS-1 genome priming system with the modified GPS-1/*pyr4* transposon according to the manufacturer's instructions (Figure 10.1).
9. Transform the transposed plasmid into an *E. coli* strain and grow on LB agar plates containing 100 μg/ml ampicillin and 50 μg/ml of kanamycin for selection.
10. Identify clones carrying transposon-disrupted genes by toothpick PCR with the forward and reverse primers of each gene.

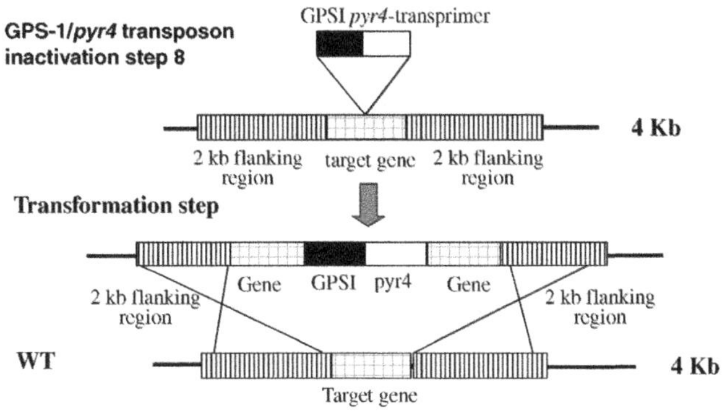

Figure 10.1 Disruption of a target gene in *A. fumigatus* by transposon mutagenesis. Schematic representation of the PCR-cloned wild-type locus and the GPS-*pyr4* transprimer used to disrupt the target gene. The disruption construct is subsequently transformed into *A. fumigatus* strain 293.1 according to Protocol 10.3

11. Grow several positive colonies for plasmid miniprep extraction. Select a plasmid in which the *pyr4*-transposon has integrated near the 5′ start of the target gene.

12. Check orientation and position of the GPS-1 transposon by PCR with the transprimer 1 inverted primers N and S and the forward and reverse primers of each gene.

13. Cleave the fragment, with the transprimer 1-*pyr4* disruption and its flanking sequence with *Asc*I restriction enzyme.

14. Cut and purify the fragment following gel electrophoresis using any DNA cleanup kit.

15. Transform the fragment into *A. fumigatus* strain AF293.1 according to Protocol 10.3.

Double-jointed PCR based gene-deletion method

This protocol corresponds to the method described by Yu *et al.* (2004).

1. For first-round PCR design six primers: (1) 5′-forward, (2) 5′-reverse +flanking marker gene region (chimeric primer), (3) 3′-forward +flanking marker gene region (chimeric primer), (4) 3′-reverse, (5) 5′-nested and (6) 3′-nested (Figure 10.2 below). The nested primers should start just after a T to avoid any mismatch mutations caused by the *Taq* or long-expand polymerases adding an additional A at the 3′-ends.

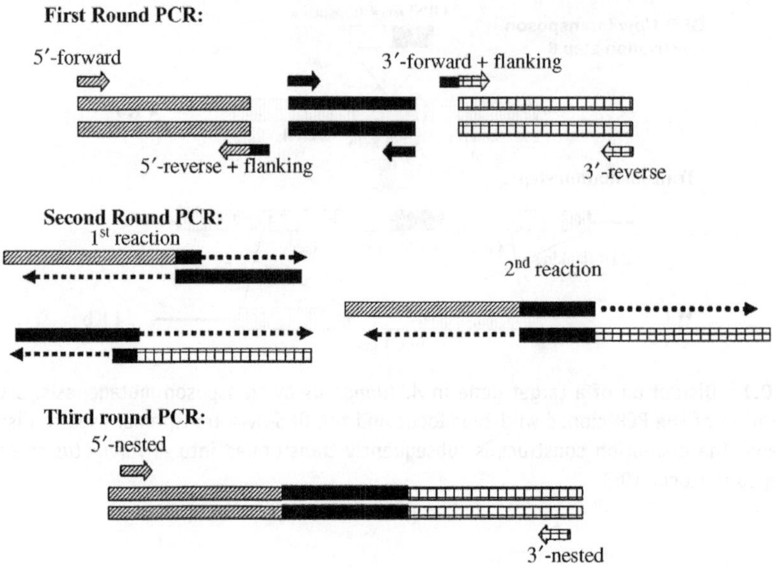

Figure 10.2 Schematic representation of the assembly of a gene-deletion construct using 'double-jointed PCR. First round PCR: amplification of the components using specific primers 5′-forward, 3′-reverse, forward and reverse marker gene primers (in black). Chimeric primers, e.g. 5′-reverse+flanking and 3′-forward+flanking should carry 25–30 bases of homologous sequence overlapping with the ends of the selectable marker gene. Second round PCR: fusion of the three fragments is executed without primers; the overhanging extensions act as primers. Third round PCR: amplification of the final product using the nested primers

2. Perform two separate PCR reactions using the 5′-forward and 5′-reverse +flanking marker gene region primers in one reaction and the 3′-forward +flanking marker gene region with the 3′-reverse primers in the second reaction.

3. Cut and purify the fragment by gel electrophoresis and any PCR-cleanup kit.

4. For the second-round PCR, which involves a fusion of the three fragments, e.g. the 5′-flanking region, a selectable marker gene and the 3′-flanking region, perform a PCR reaction (Table 10.1) without primers (Figure 10.2 above).

5. Third-round PCR should be performed in a total volume of 100 μl with less than 1 μl of the second-round product using the nested primers (Figure 10.2).

6. Run PCR product by gel electrophoresis for amplicon size conformation and cut with selected restriction enzymes for PCR product verification.

7. Purify the double-jointed final PCR product and directly use it for transformation.

Table 10.1 Second-round PCR ingredients

PCR mixture (final 25 µl)	PCR
1 µl of purified 5′-flanking amplicon	Step 1: 94 °C 2 min
1 µl of purified 3′-flanking amplicon	Step 2: 94 °C 30 sec
3 µl of purified selected marker gene*	Step 3: 58 °C 10 min
2 µl of dNTP's (2.5 mM each)	Step 4: 72 °C 5 min
2.5 µl of 10 x PCR buffer	Step 5: go to step 2 10–15 cycles
15.25 µl of sterile double distilled water	Step 6: 72 °C 10 min
0.25 µl of *Taq* polymerase	

*Use 1:3:1 molar ratio for the 5′-flanking amplicon: selectable marker gene: 3′-flanking amplicon. The total amount of DNA should be between 100 and 1000 ng. Annealing time can be reduced to 2 min.

10.3 Transformation of *A. fumigatus*

Several transformation systems have been developed for *A. fumigatus*, based on methods developed earlier for *A. nidulans* and *A. niger*. They include chemical transformation (described in this Chapter), electroporation (Weidner *et al.*, 1998) and *Agrobacterium tumefaciens*-based gene insertion (Sugui *et al.*, 2005).

Chemical transformation involves the removal of the cell wall from freshly grown germlings using a cocktail of cell-wall degrading enzymes. Glucanex (Novo Nordisk, UK), driselase/β-glucoronidase (Interspex, CA, USA) and Panzym® (Begerow, Mainz, Germany) have been used successfully. Protoplasts are maintained in an osmotically stabilized medium containing Ca^{+2} and PEG to facilitate uptake of DNA into the cells. Subsequently, protoplasts are plated in osmotically stabilized top-agar containing the relevant antibiotic or lacking the necessary nutritional supplement. Transformation efficiencies vary from 10–100 transformants/µg DNA, which can be a problem when attempting to complement mutant alleles. For complementation, an AMA1-based self-replicating plasmid library originally developed for *A. nidulans* (Xue *et al.*, 2004) has been successfully used and has achieved transformation efficiencies of up to 500 transformants/µg DNA.

Transformation by electroporation can also be used to transform *A. fumigatus*. Although transformation efficiencies are generally high (~100 transformants/µg DNA), the percentage of ectopic non-homologous recombination into the genome is significantly higher than in chemical mutagenesis (d'Enfert, 1996). Therefore, for targeted gene disruption, chemical mutagenesis is preferred.

Transformation using *Agrobacterium* has been demonstrated recently in *A. fumigatus* (Sugui *et al.*, 2005). This method involves the random integration of a single copy of the Ti-plasmid in the genome and can achieve high transformation efficiencies. Although this system cannot be used for targeted gene disruption, it may prove useful for the generation of insertional mutant libraries.

Selectable markers used to drive transformation include both dominant selectable markers and auxotrophic markers. The main dominant selectable marker used to transform *A. fumigatus* is the hygromycin B phosphotransferase (*hph*) gene of *E. coli*

Table 10.2 Auxotrophic markers for transformation of *A. fumigatus*

Auxotrophic Marker	Enzymatic activity	Selection used	Reference
pyrG	orotidine-5-monophosphate decarboxylase	Absence of uracil and uridine	Weidner *et al.* (1998)
sC	ATP sulphurylase	Absence of sulphur	De Lucas *et al.* (2001)
argB	ornithine carbamoyltransferase	Absence of arginine	Xue *et al.* (2004)
lysB	homoisocitrate dehydrogenase	Absence of lysine	Xue *et al.* (2004)

whose protein product phosphorylates and inactivates the antibiotic Hygromycin B. To express this gene in *A. fumigatus*, expression cassettes were constructed containing strong fungal promoters (i.e. the *A. nidulans gpdA* promoter) and terminators. Several auxotrophic markers and strains have been developed for transformation in *A. fumigatus* (see Table 10.2). Recently, a collection of auxotrophic strains of the AF293 reference strain has been prepared (Xue *et al.*, 2004).

The *pyrG* system is particularly useful in that it exhibits low background levels of non-integrative transformation and can be utilized for multiple rounds of transformation by counter-selection with 5-fluoroorotic acid (FOA) (d'Enfert, 1996).

We provide an improved protocol for chemical transformation of *A. fumigatus*. Changes include exposure of protoplasts to DTT prior to transformation (Dawe *et al.*, 2000) and their incubation at room temperature following transformation (Jadoun *et al.*, 2004).

Protocol 10.2
Chemical transformation of *A. fumigatus*

Equipment, materials and reagents

Freshly harvested *A. fumigatus* conidia (~10^9/ml) in double distilled water (DDW)

Water bath set at 47 °C

Clinical centrifuge

Transformation solutions:

- 1.1 M KCl, 0.1M citric acid pH 5.8 with KOH
- 1% yeast extract, 2% sucrose, 40 mM glucose, 2 x vitamin mix

- 0.6 M KCl, 0.05 M citric acid pH 5.8, 1% (w/v) glucose
- 25% PEG 8000, 100 mM $CaCl_2$, 0.6 M KCl, 10 mM Tris-HCl pH 7.5
- 0.6 M KCl, 100 mM $CaCl_2$, 10 mM Tris-HCl pH 7.5

Note: Transformation solutions are prepared with ultrapure DDW. Glassware is washed thoroughly in ultrapure DDW before use.

BSA (fraction V)

1 M $MgSO_4$

Driselase (Interspex, Products Incorporated, San Mateo, CA, USA)

β-glucoronidase (Interspex)

0.1 M DTT

Media and top agar plates: standard media are used with 0.6 M KCl to stabilize the transformed protoplasts. Alternative osmotic stabilizers such as NH_4SO_4 (0.8M) or sucrose (0.4 M for plates, 1 M for top agar) may be used. Top agar contains 1% agar only. For transformation of auxotrophic strains, the nutritional supplement is not added to the media.

Method

Note: Protoplast preparation (steps 1–6), transformation (steps 7–8).

1. Inoculate 5×10^8 conidia into 40 ml rich liquid media and incubate 32 °C for 5.5–6 h or until $>70\%$ of conidia have formed a germ tube.
2. Harvest the conidia by centrifugation in a clinical centrifuge for 5 min at 2000 x *g*.
3. Resuspend the conidia in freshly prepared lytic-mix (10 ml solution A, 10 ml solution B, 0.2 ml 1 M $MgSO_4$, 200 mg BSA, 0.2 gram driselase, 0.1 gram β-glucoronidase, mixed 5 min, centrifuged at 2000 x *g* before use and filter sterilized).
4. Incubate for ~2 h at 32 °C (incubation time varies between isolates and should be calibrated individually).
5. Harvest conidia by centrifugation for 5 min at 2000 x *g* and wash twice in 20 ml solution C.

6. Resuspend conidia in 0.5 ml solution E and 5 μl 0.1 M DTT. Store overnight at 4 °C.

7. Transformation: add 25 μl of protoplasts to a microcentrifuge tube with 1–2 μg of plasmid DNA in a volume not exceeding 5 μl. Add 12.5 μl of solution D and incubate on ice for 15 min.

8. Add 150 μl of solution D and incubate at room temperature for 15 min.

9. Aliquot the entire mix into 3 ml of 0.6 M KCl top agar media at 47 °C, mix gently and pour onto 0.6 M KCl media plate pre-warmed to 42 °C. Tilt and shake the plate gently to facilitate even spreading of the top agar onto the plate.

10. Incubate the plates for 24 h at room temperature.

11. Incubate further (24–48 h) at 37 °C.

12. For transformation with an integrating vector expect about 50–100 colonies/plate. For transformation with the AMA1 self-replicating vector expect ~1000 colonies/plate. Always include a no-DNA control in your transformation. This plate should be empty of transformants at this stage.

10.4 Molecular verification of correct single integration (PCR-based)

When preparing a knockout or disrupted *A. fumigatus* strain only a small and variable percentage of transformants carry the correctly integrated construct. It is crucial to verify that the construct has been inserted in the correct orientation and site in the genomic DNA. Here, we describe a method developed for the initial rapid screening of primary *A. fumigatus* directly from conidia.

As a first step, conidia are collected from primary transformant colonies and snap frozen. PCR using primers spanning the deleted region is performed. This enables the rapid identification of putative positive disrupted mutants without the need to prepare genomic DNA from a large number of transformants.

A second step, to verify correct insertion and orientation, involves a primer designed for the 5′-upstream region to the inserted construct, the second, a reverse primer from within the construct, preferably from the transposon or another selected marker gene region.

Finally, Southern blot analysis is performed to verify that a single copy of the knockout/disruption construct has been inserted into the nucleus.

Protocol 10.3
Molecular verification of correct integration by PCR

Equipment, materials and reagents

Liquid nitrogen

PCR reagents

PCR thermal block

A. fumigatus genomic DNA (prepared with the MasterPure Yeast DNA Purification Kit™ (Epicenter, Madison, WI, USA) (Jin *et al.*, 2004)

Miracloth™ (Calbiochem, Darmstadt, Germany)

Paper towels

0.2% (v/v) Tween 80

YAG agar pates and liquid (8.4)

Method

Rapid primary analysis of transformants by PCR

1. Collect conidia from transformed *A. fumigatus* colonies and transfer to 500 µl of DDW to reach a conidial concentration of $\sim 10^7$ conidia/ml.

2. Snap-freeze tubes in liquid nitrogen for 10 min.

3. Perform a PCR reaction with 10 µl of the conidial extract and forward and reverse gene primers; Initiate PCR reaction with a 95 °C for 5 min to release the conidial genomic DNA.

Note: It is recommended to amplify in parallel the *gpdA* housekeeping gene as a positive control (*AfugpdA* forward 5′-TCTCCAACGTTCTTGCACC-3′ and *AfugpdA* reverse 5′-CCACTCGTTGTCGTACCAGG-3′). Choose colonies negative for gene amplification but positive for *gpdA* amplification for further analysis (Figure 10.3). Select several isolates containing the disrupted gene for further analysis.

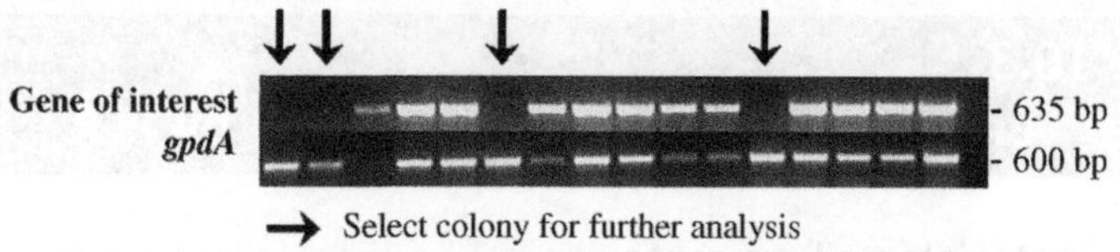

Figure 10.3 Analysis of transformants by PCR. PCR using target-gene specific primers is performed directly on snap-frozen and heated conidial extracts. *gpdA*-specific primers are used as a positive control. Isolates containing the disrupted gene (negative for the WT target gene and positive for the *gpdA* gene) are selected for further analysis

Secondary PCR analysis of transformants to verify correct integration of the transforming plasmid

1. Grow colonies carrying the disrupted gene for 2 days at 37 °C on YAG agar plates.
2. Scrape conidia with 0.2% Tween 80 and wash 3 times with sterile DDW.
3. Grow conidia 16–24 h at 37 °C in 25 ml liquid YAG medium at a concentration of 10^6 conidia/ml.
4. Collect hyphae onto Miracloth, rinse with DDW and press between paper towels. Transfer into a microtube and puncture the lid several times. Snap-freeze in liquid nitrogen.
5. Lyophilize and extract genomic DNA using the MasterPure Yeast DNA Purification Kit, with minor protocol modifications as described by Jin *et al.* (2004).
6. Validate the correct orientation of insertion in the genomic DNA by a PCR reaction with a primer designed for the 5′ upstream region to the inserted construct and the transprimer 1 inverted primers N (5′-CAGTTTAAGACTTTATTGTCCGCC-3′) or S (5′-CAGTTCCCAACTATTTGTCCGC-3′) from within the transposon (Figure 10.4). Figure 10.5 demonstrates a knockout PCR

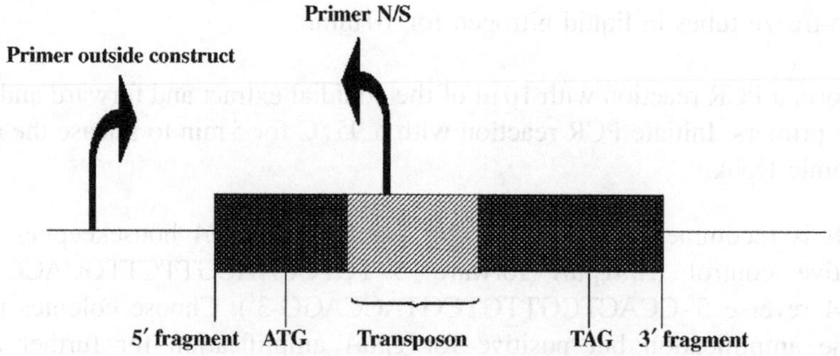

Figure 10.4 A schematic representation of the Secondary PCR analysis of transformants to verify correct integration of the transforming plasmid. A primer 5′-upstream to the construct is designed and used with primer N or S from within the transprimer to verify whether the construct has integrated into the correct site in the genomic DNA

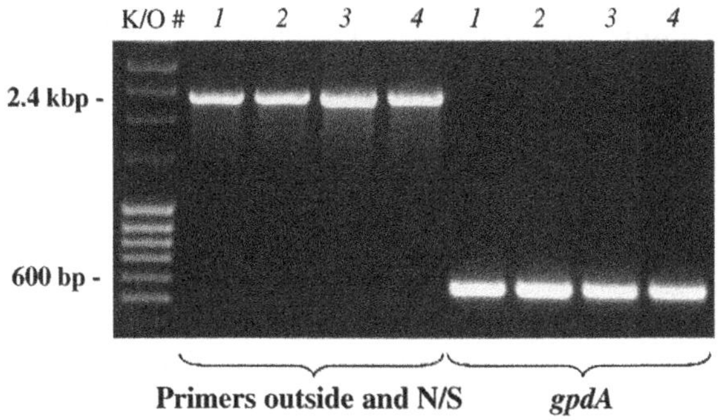

Figure 10.5 Secondary PCR analysis. Verification of correct integration of the disruption construct into the genomic DNA. Four transformed isolates (K/01-4) initially mapped by PCR using protocol 10.3.1 were subsequently analysed by an additional PCR reaction using the 5′-upstream to the construct and transprimer N/S primers as represented schematically in Figure 10.4. Amplification of the *gpdA* housekeeping gene was used as a positive PCR control

validation on 4 mutant isolates performed with a primer outside the construct and a primer from within the transposon region.

10.5 General strategies for the phenotypic characterization of *A. fumigatus* mutant strains

Following transformation and verification that the target gene has been deleted, several possibilities should be taken into account. (i) Gene deletion results in obvious phenotypic changes such as reduced growth rate, conidiation etc. (ii) gene deletion results in no obvious phenotypic changes. The latter possibility dictates a careful analysis of the mutant phenotype under a wide variety of conditions. The following protocols provide a comprehensive approach for identifying both obvious and subtle phenotypic changes in *A. fumigatus* mutant strains. They include a description of a wide range of techniques used to assess growth rates, undertake a detailed morphological characterization, analyse growth under stress and measure altered susceptibility to growth-perturbing compounds.

Protocol 10.4 General strategies for the phenotypic characterization of mutants

Media

YAG-rich medium: 0.5% (w/v) yeast extract, 1% (w/v) glucose, 10 mM $MgCl_2$, supplemented with trace elements, vitamins and 1.5% (w/v) agar for plates.

MM minimal medium: 70 mM $NaNO_3$, 1% glucose, 12 mM KPO_4 pH 6.8, 4 mM $MgSO_4$, 7 mM KCl, vitamin mix, trace elements and 1.5% (w/v) agar for plates.

Trace elements (x1000): $Na_2B_4O_7$:H_2O 40 mg/litre, $CuSO_4$:5 H_2O 400 mg/litre, FeCl 800 mg/litre, $MnSO_4$:H_2O 800 mg/litre, $NaMo_4$:H_2O 800 mg/litre, $ZnSO_4$:7 H_2O 8 g/litre.

Vitamin mix (x500): p-aminobenzoic acid 1 gram/litre, niacin 1 gram/litre, pyridoxine HCl 1 gram/litre, riboflavin 1 gram/litre, thiamine HCl 1 gram/litre, choline HCl 1 gram/litre, d-biotin 2 mg/litre.

Radial growth rate on solid medium

Point inoculation

1. Prepare a stock of 10^8/ml freshly harvested mutant and control wild-type conidia in DDW.
2. Gently dip a sterile toothpick approximately 5 mm into the conidial stock. Lightly prick the toothpick into the centre of the agar plate.
3. Incubate at 37 °C for 24–72 h. Measure radial growth diameter periodically. Measure at least 5 individual colonies for statistical analysis.

Streaked inoculation

1. Prepare a stock of 10^8/ml freshly harvested mutant and control wild-type conidia in DDW.
2. Streak with the aid of a sterile loop onto the agar plates.
3. Incubate at 37 °C for 24–72 h. Measure radial growth of individual colonies periodically. Do not measure the growth of colonies that are in physical contact with their neighbours. Measure at least 20 individual colonies for statistical analysis.

Growth rate on liquid medium

1. Place 13 mm sterile circular glass coverslips onto the bottom of 24-well plates.
2. Inoculate between 100–10 000/ml freshly harvested mutant and wild-type conidia into 24-well plate (1 ml/medium/well).

3. Incubate at 37 °C for up to 24 h. At 2, 4, 6, 8, 12, 16 and 24 h gently remove the coverslips, flip onto slides and view microscopically. Using a microscope grid, measure the hyphal length of 20–50 germinating conidia per time point.

Gross morphological characterization

Changes in conidiophore development and conidiation

Microscopic visualization of conidiophores: pour ~200 µl YAG-rich agar medium on top of a sterile 22 mm coverslip. Place the cooled coverslip in a sterile Petri dish containing solidified agar. Add 50 µl of 10^6/ml conidia on top of the agar-coated coverslip. Add another coverslip on top of the inoculated agar square. Incubate for 24–48 h at 37 °C. For observation of conidiophore structure, dip the coverslips with aerial hyphae and conidiophores attached into 100% ethanol, air dry, mount on slides and observe microscopically. Note the shape of the phialides and the length of the conidial chains.

Conidiation in submerged culture: germinate conidia in YAG-rich medium on coverslips as described above. After 24–48 h of growth at 37 °C, gently discard the floating mycelial mat. Carefully take out the coverslip, place face down onto a slide and view microscopically, looking for the presence of conidiophores. Conidiophore development and conidiogenesis do not normally occur under liquid. Their presence is suggestive of a deregulation in the conidiation pathway.

Changes in conidial density: carefully spread 200 µl of 10^6 freshly harvested conidia/ml onto agar plates (5 plates/mutant). After 2, 3, 4 and 5 days of incubation at 37 °C thoroughly collect conidia in 5 ml/plate PBS+ 0.2% Tween 80 and count using a haemocytometer. Perform at least 4 independent counts from each plate.

Changes in conidial morphology

Gross morphological changes: assess the hydrophobicity ('wet-ability') of the conidia in PBS containing 0.2% Tween 80 detergent and their ability to disperse when the plate is shaken or scraped. Note the colour of the conidia in solution. Assess the stability of conidia during storage (1–60 days) at room temperature. Store conidia in DDW, PBS or PBS containing 0.2% (v/v) Tween 80. Analyse conidial stability by microscopic examination and plating.

Calcofluor staining and microscopy: resuspend freshly harvested conidia in 0.1–1 µg/ml calcofluor white and view with a fluorescence microscope under x400–x1000 magnification. Measure the size and shape of the conidia, as well as the intensity and localization of the calcofluor staining. Assess the degree of conidial clumping by counting the percentage of single vs. clumped conidia and the number of conidia per clump.

Conidial adhesion: coat 96-well plate wells overnight at 4 °C with 10 µg/well of either laminin, BSA, collagen or fibrinogen (diluted in 100 µl DDW). Aspirate wells and wash 3 times with PBS. Add 200 µl *A. fumigatus* conidia at a concentration of 10^8/ml to each well. Incubate for 1 h at 37 °C. Gently wash the plates 3 times with 200 µl of PBS containing 0.05% Tween 20. Add 100 µl PBS and count adherent conidia under the microscope.

Conidial germination in the absence of a carbon source: prepare a fresh conidial stock and wash it 5 times in DDW to thoroughly remove possible carbon-nutrient sources. Germinate the conidia in MM-lacking glucose on coverslips as described above. Assess the degree of conidial swelling, adherence and hyphal tube emergence after incubation for 2–24 h at 37 °C.

Changes in hyphal branching and morphology

Germinate conidia on coverslips in liquid MM for 1–12 h at 37 °C. Gently remove the coverslip, dry off excess liquid and place face down on a microscope slide containing 10 µl of calcofluor white. Measure the time of emergence of the first and second germ tube and the angle between them (an angle of 180° is seen during normal growth). Measure the time of emergence of the first and second branches. Measure the angle between the branch and the main germ tube (an angle of 45° is seen during normal branching). Measure the length, width and shape of each hyphal compartment (a compartment is a portion of hyphal tube divided between two septae). Measure the angle and shape of the septae (the septum is normally at a right angle to the cell wall). Analyse the shape of the hyphal tips (abnormal tips may be swollen or pinched instead of crescent-shaped) and see whether abnormal bifurcations occur at the hyphal tip. For quantitation, measure the percentage of cells showing abnormal branching patterns at the corresponding times ($n \geq 100$ cells).

Changes in nuclear morphology

DAPI staining of nuclei: germinate conidia on coverslips in liquid MM for 1–24 h at 37 °C. Gently remove the coverslip, dip in DDW and place face down on a microscope slide containing 100 µl of DAPI fixative (50 mM KPO4, pH 6.8, 0.2% Triton X-100, 5% EM-grade-1 gluteraldehyde, 0.05 µg/ml DAPI). Incubate for 20 min. Dip/wash 3 times in DDW. Blot excess liquid and place face down on a microscope slide. View stained nuclei by fluorescence microscopy. Measure the number of nuclei per cell ($n \geq 50$ cells) during germination and early hyphal growth (2–12 h). Count the number of highly condensed mitotic nuclei and decondensed interphasic nuclei and calculate the ratio between them. Look for abnormal fragmented or extended nuclei indicative of defects in mitosis.

Changes in the actin cytoskeleton, endocytosis and vacuoles

Actin staining: germinate conidia on coverslips in liquid MM for 8–12 h at 37 °C. Gently remove the coverslip and place face down on a microscope slide containing a 200 µl drop of fixing solution (3.7% formaldehyde, 50 mM HEPES (pH 6.8), 5 mM $MgSO_4$, 25 mM EGTA (pH 7.0)). Incubate for 45 min at room temperature. Perform three 5 min washes in HEM solution (50 mM HEPES (pH 6.8), 25 mM EGTA (pH 7.0), 5 mM $MgSO_4$). Digest the cell wall by placing the coverslip in 500 µl lytic mix (at 32 °C for 1–2 h. Wash for 10 min in HEM solution, 5 min in extraction solution (50 mM HEPES (pH 6.8), 0.1% Nonidet P-40, 25 mM EGTA (pH 7.0)) and 5 min in HEM solution. Immerse the coverslip in absolute methanol for 10 min at −20 °C. Wash twice for 5 min in HEM solution, then once for 5 min in PBS containing 0.1% BSA. Place the coverslip on 0.5 ml primary antibody solution (PBS, 0.1% BSA, Sigma-Aldrich rabbit anti-actin antibody at a dilution of 1:500) and incubate for 1 h at room temperature in a sealed humid box. Wash 3 times for 5 min in PBS containing 0.1% Nonidet P-40. Hybridize with the secondary antibody (Cy3-conjugated anti-rabbit IgG antibody at a dilution of 1:500) exactly as described for the primary antibody. View by fluorescence microscopy.

Endocytosis: germinate conidia on coverslips in liquid MM for 2–12 h at 37 °C. Gently remove the coverslip and place face down on a pre-chilled (4 °C) slide containing a 10 µl droplet of FM4-64 (32 µM in DDW). (A stock solution of FM4-64 is prepared at 1000 x in DMSO and stored in small aliquots at −20 °C). View the uptake of fluorescent dye into the cells over ~30 min using a fluorescence microscope. Compare the rate of dye uptake between the mutant and wild-type strains.

Vacuole staining: germinate conidia on coverslips in liquid MM for 12 h at 37 °C. Add 1/10 and 1/1000 volumes of 500 mM citrate-NaOH (pH 3.0) and 10 mM CFDA (5-(and 6-) carboxy-2′,7′-dichlorofluorescein diacetate) in dimethylformamide, respectively. Incubate the cells for 30 min at 37 °C. Gently dip and wash the coverslips in DDW and compare the morphology and fluorescence of the cells using a fluorescence microscope.

Changes in morphology following growth under stress

Temperature: assess growth at 25–50 °C for 24–72 h (solid media) or 2–16 h (liquid media).

pH: for media at pH 9.0 add 50 mM BIS=1,3-bis-[tris(hydroxymethyl)methylamino]-propane to MM and titrate with NaOH to pH 9.0. For media at pH 5.0: add 50 mM citrate to MM and titrate to pH 5.0 with NaOH.

Table 10.3 Compounds commonly used in the analysis of mutant *Aspergillus* strains

Compound	Mode of action	Effect	Concentration range
Congo red	Binds cell-wall polysaccharides	Cell-wall destabilizing	1–200 μg/ml
Calcofluor white	Binds cell-wall polysaccharides	Cell-wall destabilizing	10–500 μg/ml
Nikkomycin Z	Chitin-synthase inhibitor	Cell-wall destabilizing	50–500 μg/ml
Caspofungin	Glucan-synthase inhibitor	Cell-wall destabilizing	0.01–100 μg/ml
Glucanex	β-glucanase activity	Cell-wall destabilizing	0.1–5 mg/ml
Sodium dodecyl sulphate (SDS)	Extracts cell-wall proteins, dissolves plasma membrane	Cell-wall and membrane destabilizing	10–500 μg/ml
Amphotericin B	Ergosterol-binding, pore-forming	Membrane destabilizing	0.125–20 μg/ml
Itraconazole	Ergosterol biosynthesis inhibitor	Membrane destabilizing	0.125–10 μg/ml
Terbinafine	Ergosterol biosynthesis inhibitor	Membrane destabilizing	0.03–10 μg/ml
Staurosporine	Protein kinase C inhibitor	Cell-wall repair inhibition	0.1–10 μg/ml
Tunicamycin	N-glycosylation inhibitor	Endoplasmatic reticulum function, secretion	1–80 μg/ml
Hydroxyurea	DNA synthesis inhibitor	DNA damage	1–100 mM
4-nitroquinolone oxide (4-NQO)	Mutagen	DNA damage	10–200 μg/ml
Camptothecin	DNA topoisomerase I inhibitor	DNA damage	0.1–2 μg/ml
Methylmethane sulphonate (MMS)	DNA-alkylating agent	DNA damage	10–100 μg/ml
Trifluoroperazine (TFP)	Ca^{+2}/Calmodulin pathway inhibitor	Inhibition of vacuole function, secretion, cell cycle	10–160 μg/ml
EGTA	Ca^{+2}/Calmodulin pathway inhibitor	Inhibition of vacuole function, secretion, cell cycle	0.01–5 mM
Tacrolimus (FK-506)	Calcineurin (Ca^{+2}-dependent phosphatase) inhibitor	Amino acid starvation, high osmolarity stress	0.1–10 μg/ml

Table 10.3 (*Continued*)

Compound	Mode of action	Effect	Concentration range
Cytochalasin A	F-actin disruption	Cytoskeletal damage	0.1–20 μg/ml
Nocodazole	Microtubule disruption	Cytoskeletal damage	0.1–10 μg/ml
Benomyl	Microtubule disruption	Cytoskeletal damage	0.1–5 μg/ml
H_2O_2	Oxidant	Oxidative stress	0.01–1 mM
Dithiothreitol (DTT)	Reduction of disulphide bonds	Reductive stress	1–25 mM
Cycloheximide	Protein-synthesis inhibitor	Protein synthesis inhibition	1–100 μg/ml

Carbon source: prepare MM without glucose. Add one of the alternative carbon sources listed: sucrose, sorbitol, lactose, maltose, trehalose, raffinose, soluble starch, glycerol, propionate, sodium acetate (1%, w/v), glutamate, albumin (0.2%, w/v), ethanol (1.5% v/v).

Nitrogen source: prepare MM without $NaNO_3$. Replace with ammonium sulphate $(NH_4)_2SO_4$ (70 mM) or urea $(NH_2)_2CO$ (70 mM). For nitrogen starvation prepare MM without a nitrogen source.

Phosphates: for phosphate starvation, grow in MM with 1 mM KPO_4, pH 6.8.

Vitamins: assess growth in media in the absence and presence of vitamin mix.

Susceptibility to growth-perturbing compounds

The susceptibility of mutant strains to a variety of compounds can provide important clues to the function of the deleted gene. Susceptibility can be tested either on MM agar plates or in multiwell plates containing liquid MM. Table 10.3 lists compounds commonly used in the analysis of mutant *Aspergillus* strains.

10.6 References

Bhabhra, R., Miley, M. D., Mylonakis, E., Boettner, D. *et al.* (2004) Disruption of the *Aspergillus fumigatus* gene encoding nucleolar protein CgrA impairs thermotolerant growth and reduced virulence. *Infect. Immun.* **72**(8): 4731–740.

Brakhage, A. A. and Langfelder, K. (2002) Menacing mold: The molecular biology of *Aspergillus fumigatus*. *Annu. Rev. Microbiol.* **56**: 433–55.

Chaveroche, M. K., Ghigo, J. M. and d'Enfert, C. (2000) A rapid method for efficient gene replacement in the filamentous fungus *Aspergillus nidulans*. *Nucleic Acids Res.* **28**(22): E97.

Dawe, A. L., Willins, D. A. and Morris, N. R. (2000) Increased transformation efficiency of *Aspergillus nidulans* protoplasts in the presence of dithiothreitol. *Anal. Biochem.* **283**(1): 111–2.

De Lucas, J. R., Dominguez, A. I., Higuero, Y., Martinez, O. *et al.* (2001) Development of a homologous transformation system for the opportunistic human pathogen *Aspergillus fumigatus* based on the *sC* gene encoding ATP sulfurylase. *Arch. Microbiol.* **176**(1–2): 106–13.

d'Enfert, C. (1996) Selection of multiple disruption events in *Aspergillus fumigatus* using the orotidine-5′-decarboxylase gene, *pyrG*, as a unique transformation marker. *Curr. Genet.* **30**(1): 76–82.

Jadoun, J., Shadkchan, Y. and Osherov, N. (2004) Disruption of the *Aspergillus fumigatus argB* gene using a novel *in vitro* transposon-based mutagenesis approach. *Curr. Genet.* **45**(4): 235–241.

Jin, J., Lee, Y. K. and Wickes, B. L. (2004) Simple chemical extraction method for DNA isolation from *Aspergillus fumigatus* and other *Aspergillus* species. *J. Chem. Microbiol.* **42**(9): 4293–4296.

Latge, J. P. (1999) *Aspergillus fumigatus* and aspergillosis. *Clin. Microbiol. Rev.* **12**(2): 310–50.

Mellado, E., Garcia-Effron, G., Buitrago, M. J., Alcazar-Fouli, L. *et al.* (2005) Targeted gene disruption of the 14 α-sterol demethylase (*cyp*51A) in *Aspergillus fumigatus* and its role in azole drug susceptibility. *Antimicrob. Agents. Chemother.* **49**(6): 2536–2538.

Mouyna, I., Henry, C., Doering, T. L. and Latge, J. P. (2004) Gene silencing with RNA interference in the human pathogenic fungus *Aspergillus fumigatus*. *Fems. Microbiol. Lett.* **237**(2): 317–324.

Mouyna, I., Morelle, W., Vai, M., Michel, M. *et al.* (2005) Deletion of GEL2 encoding for a β(1-3)glucanosyltransferase affects morphogenesis and virulence in *Aspergillus fumigatus*. *Mol Microbiol* **56**(6): 1675–1688.

Nierman, W. C., May, G., Kim, H. S., Anderson, M. J. *et al.* (2005) What the *Aspergillus* genomes have told us. *Med. Mycol.* **43**(1): S3–5.

Sugui, J. A., Chang, Y. C. and Kwon-Chung, K. J. (2005) *Agrobacterium tumefaciens*-mediated transformation of *Aspergillus fumigatus*: An efficient tool for insertional mutagenesis and targeted gene disruption. *Appl. Environ. Microbiol.* **71**(4): 1798–802.

Tekaia, F. and Latge, J. P. (2005) *Aspergillus fumigatus*: Saprophyte or pathogen? *Curr. Opin. Microbiol.* **8**(4): 385–92.

Vogeser, M., Wanders, A., Haas, A. and Ruckdeschel, G. (1999) A four-year review of fatal aspergillosis. *Eur. J. Clin. Microbiol. Infect. Dis.* **18**(1): 42–5.

Weidner, G., d'Enfert, C., Koch, A., Mol, P. C. and Brakhage, A. A. (1998) Development of a homologous transformation system for the human pathogenic fungus *Aspergillus fumigatus* based on the *pyrG* gene encoding orotidine 5′-monophosphate decarboxylase. *Curr. Genet.* **33**(5): 378–85.

Xue, T., Nguyen, C. K., Romans, A., Kontoyiannis, D. P. and May, G. S. (2004) Isogenic auxotrophic mutant strains in the *Aspergillus fumigatus* genome reference strain AF293. *Arch. Microbiol.* **182**(5): 346–53.

Xue, T., Nguyen, C. K., Romans, A. and May, G. S. (2004) A mitogen-activated protein kinase that senses nitrogen regulates conidial germination and growth in *Aspergillus fumigatus*. *Euk. Cell.* **3**(2): 557–560.

Yu, J. H., Hamari, Z., Han, K. H., Seo, J. A. *et al.* (2004) Double-joint PCR: A PCR-based molecular tool for gene manipulations in filamentous fungi. *Fungal Genet. Biol.* **41**(11): 973–981.

11

Promoter analysis and generation of knock-out mutants in *Aspergillus fumigatus*

Matthias Brock, Alexander Gehrke, Venelina Sugareva and **Axel A. Brakhage**

11.1 Introduction

Reporter genes are a useful tool to study promoter activity of certain genes or the localization of proteins in different cellular compartments. Different reporters are available and the choice is dependent on the experiment to be performed.

Fluorescent reporters like the green fluorescent protein (GFP) and its derivatives are generally used for localization studies. These reporters are suitable for monitoring localization of proteins under *in vivo* conditions. However, fluorescent reporters are not suited to determine the activity of a specific promoter under different growth and stress conditions since quantification of fluorescence is quite difficult and easily produces saturation of the detection system. Therefore, another commonly used reporter is the *lacZ* gene (coding for β-galactosidase) from *Escherichia coli*. The use of a *lacZ*-fusion allows comparison of promoter activity of a gene of interest under changing growth conditions or induced stresses. Furthermore, fusions with the *lacZ* gene are also used to monitor the change of promoter activity after the mutation of specific sites within a promoter region. The procedure to introduce a site-directed mutation in a promoter element will be explained in detail in a later section of this chapter. When comparing a wild-type promoter element with a mutated one, it is crucial that the reporter construct is integrated at the same genomic locus to avoid positioning effects. Therefore, a system was developed that allows the integration of a gene-fusion construct at a specific site. In our lab, we routinely use the *pyrG* locus (*pyrG* gene coding for orotidine

Medical Mycology: Cellular and Molecular Techniques. Edited by Kevin Kavanagh

5′-monophosphate decarboxylase) of strain CEA17, which contains a mutated *pyrG1* allele. By using a mutated *pyrG2* allele in the reporter plasmid, which carries a mutation in a different region of *pyrG* than in the *pyrG1* allele, homologous recombination is forced to yield a functional *pyrG* gene and thus the uracil prototrophy of transformants. This strategy ensures the integration at the same chromosomal locus in strains containing different reporter constructs (Weidner *et al.*, 1998).

In addition to the investigation of promoter elements, *lacZ* gene fusions can be used to look for the production of specific proteins in tissues by staining with X-Gal™ (American Bioanalytical, Natick, MA, USA) (5-bromo-4-chloro-3-indolyl-β-D-galactopyranoside). This compound is colourless but acts as an artificial substrate of β-galactosidases and can be cleaved into galactose and 5-bromo-4-chloroindigo, which is strongly blue-coloured. Smith *et al.* (1994) have used a fusion of the alkaline protease gene with *lacZ* to monitor the production of alkaline protease in lung tissues infected with *A. fumigatus.*

In order to study the importance of a specific gene for metabolism or the development and maintenance of an infection with *A. fumigatus*, gene deletions are required. Deletion of specific genes in *A. fumigatus* is hampered by several problems. Only a few selection markers are available and, additionally, owing to the lack of a sexual crossing system (*A. fumigatus* belongs to the deuteromycetes) the introduction of more than one deletion in a single clone is difficult. Markers, which are quite often used for transformant selection, are uracil auxotrophy (*pyrG*⁻ strains like CEA17) and antimycotics such as hygromycin B. Hygromycin B is an aminoglycoside antibiotic from *Streptomyces hygroscopicus* that acts on bacteria, fungi and higher eukaryotes by inhibiting protein synthesis (Cabanas *et al.*, 1978). Resistance is obtained via a kinase that phosphorylates hygromycin B. The encoding *hph* gene was isolated from *E. coli*. Both markers can be used in the same strain either to produce double mutations or for the reconstitution of a strain in which a deletion was formerly introduced.

Wild-type strains of *A. fumigatus* have a rather low rate of homologous recombination. Flanking regions of 800–1000 bp are generally needed to enforce a double cross-over. Therefore, in a deletion construct the resistance or auxotrophic marker replaces the gene of interest and is flanked by long up- and downstream-regions from the chromosomal locus of the gene to be deleted. When a *pyrG*⁻ strain is used for transformation, generally a gene coding for orotidine 5′-monophosphate decarboxylase from another fungus, like *pyrG* from *Aspergillus niger*, is used, because it avoids the integration into the endogenous *pyrG* locus. An elegant method to employ the uracil auxotrophy for more than one deletion is the use of the so-called URA-blaster. In this case, the *pyrG* gene is additionally flanked by direct repeats of the neomycin resistance cassette, which allows a later out-recombination of the *pyrG* gene by a counter selection against a functional *pyrG* gene, for example by use of 5-FOA (5-fluoroorotic acid or 5-fluorouracil-6-carboxylic acid). Strains with an intact orotidine 5′-monophosphate decarboxylase are able to decarboxylate 5-FOA and produce 5-fluorouracil, which is toxic to the cells. A defective gene avoids conversion of 5-FOA and therefore enables the screening of strains, which have lost the URA-blaster and therefore are uracil auxotrophic.

The use of the hygromycin B resistance system has at least one tremendous advantage over the URA-blaster. The URA-blaster can only be used in strains which already carry a defective *pyrG* gene. New isolates of *A. fumigatus* do not encode such a defective gene and therefore cannot be used with the URA system but are sensitive to hygromycin B. Nevertheless, no fixed concentrations for the use of hygromycin B can be given. The sensitivity of different strains varies greatly. Therefore, they need to be tested before a transformation is performed. For using the hygromycin resistance gene in *A. fumigatus* it has to be cloned under the control of a strong promoter. In our laboratory, the promoter from the *gpd* gene from *A. nidulans* (coding for glyceraldehyde-3-phosphate dehydrogenase) was found to be suitable to confer hygromycin B resistance to transformants. However, other promoters may also work.

11.2 Site-directed mutagenesis of promoter elements

Site-directed mutagenesis is an important technology developed to study the effect of a mutation on the function of the gene or promoter in which the mutation was introduced. It can be used to obtain detailed information on the importance of an individual base pair for promoter (or protein) activity.

Generally, the sequence possessing the intended mutation is fused in the correct reading frame with a reporter gene like *lacZ* for determination of the effect of the introduced mutation. An oligonucleotide based method for the generation of point mutations in promoter elements by use of PCR is described by Higuchi *et al.* (1988). A slightly modified version, which omits the use of Klenow fragment, is described here.

Protocol 11.1
Site-directed mutation of promoter elements

Equipment, materials and reagents

Two sets of oligonucleotides (Figure 11.1):

Set one: oligonucleotide P1 binding upstream of the mutated region and reading downstream to the site of mutation. Oligonucleotide P2, which contains the mutated site and reads upstream in direction of P1.

Set two: oligonucleotide P4 binding downstream of the mutated region and reading upstream to the site of mutation. Oligonucleotide P3, which contains the mutated site, reads downstream in direction of P4. (Generally, oligonucleotide P3 is reverse-complementary to P2.)

Set of automatic pipettes and sterile tips

Template DNA

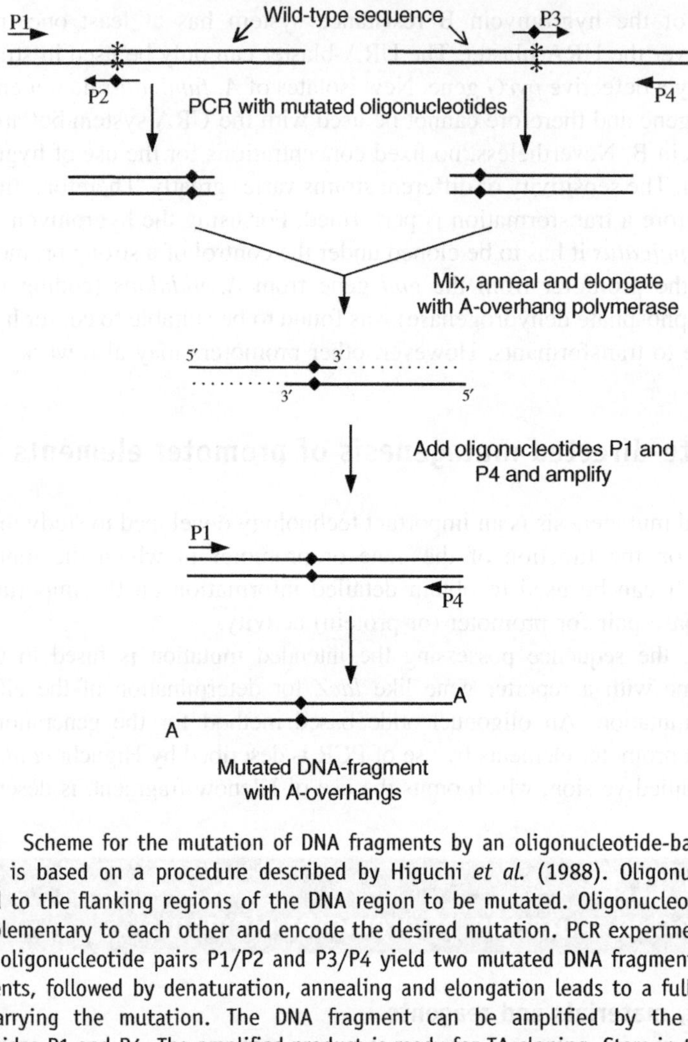

Figure 11.1 Scheme for the mutation of DNA fragments by an oligonucleotide-based method. The method is based on a procedure described by Higuchi *et al.* (1988). Oligonucleotides P1 and P4 bind to the flanking regions of the DNA region to be mutated. Oligonucleotides P2 and P3 are complementary to each other and encode the desired mutation. PCR experiments with the two sets of oligonucleotide pairs P1/P2 and P3/P4 yield two mutated DNA fragments. Mixing of both fragments, followed by denaturation, annealing and elongation leads to a full length DNA fragment carrying the mutation. The DNA fragment can be amplified by the addition of oligonucleotides P1 and P4. The amplified product is ready for TA-cloning. Stars in the wild-type sequence represent the desired site of mutation, whereas the closed diamonds represent the mutated site

dNTPs

PCR buffers containing Mg^{2+} ions

Proof-reading polymerase (no A-overhangs)

High-fidelity polymerase, which adds A-overhangs

PCR tubes

PCR thermocycler

Competent *E. coli* cells

Water bath or electroporator for transformation of *E. coli* cells

PCR cloning vector (for A-overhangs)

Plates for streaking transformed *E. coli* cells with respective antibiotic

Equipment for agarose gel-electrophoresis

Agarose gel for purification of PCR products

Gel-extraction kit for isolation of DNA fragments

(If possible, sequencing facility)

Method

1. Carry out two independent PCR reactions with your template DNA and oligonucleotide sets 1 and 2, respectively. Use the proof-reading polymerase for high fidelity and to avoid addition of A-overhangs. Follow the manufacturer's instructions for cycling. Make sure that the final extension time at 72 °C is <u>not less</u> than 15 min.

2. Load the whole PCR samples on a 1–2% agarose gel and excise the independent bands for the upstream and downstream DNA fragments from the gel. Do not mix the fragments at this stage.

3. Gel-purify the single fragments and elute them in a suitable buffer allowing downstream PCR and quantify the amount. (Alternatively, 0.2–1 μl of the respective elutes can be used.)

4. Set up another PCR in a final volume of 50 μl (after addition of oligonucleotides). Use the two PCR fragments from above (5–20 ng each) as templates. Leave out oligonucleotides at this stage and use a high-fidelity polymerase, which is able to add A-overhangs. Heat the reaction to 95 °C for 1 min. This step leads to single-stranded DNA fragments. Annealing at a temperature near to the T_m of the oligonucleotides P2 and P3 should be used and the annealing time should be around 30–40 sec. At this point, the two ssDNA fragments have annealed and chain elongation can be performed at 72 °C. Choose at least 1 min/kb of elongation time at 72 °C to ensure that all DNA products are of full length. Repeat this cycling 4–5 times and pause at 95 °C.

5. Mix oligonucleotides P1 and P4 (these are the outer upstream and downstream oligonucleotides) and add them to the reaction paused at 95 °C.

6. Continue with a standard PCR programme for template amplification. In that programme a full-length DNA fragment encoding the mutation is amplified.

7. Check the purity of the PCR product by agarose gel-electrophoresis.

8. Proceed with TA cloning of the PCR product. When required, gel-purify the fragment before cloning. (Generally, a purification step is not required because fragments from the first PCR approach do not contain an A-overhang and do not ligate in TA cloning.)

9. Transform competent *E. coli* cells by standard procedures and streak cells onto plates.

10. Analyse plasmids from transformants by restriction analysis or PCR.

11. To ensure that only mutations are present, which are introduced by oligonucleotides P2 and P3, it is recommended to sequence the entire DNA fragment.

Additional information

In order to mutate a promoter element and to fuse the mutated DNA fragment with a reporter, oligonucleotides P1 and P4 may directly contain a restriction site, allowing subsequent in frame fusion with the reporter gene.

To compare the promoter activity of a wild-type and a mutated DNA fragment, integration of the reporter construct at the same chromosomal locus is required. This can be achieved by integration into the *pyrG* locus. Weidner *et al.* (1998) developed a system ensuring the integration at the *pyrG* locus of *A. fumigatus*. For this purpose, two mutant alleles of the *pyrG* gene are used: *pyrG1* is a mutated allele of strain CEA17 and *pyrG2* is a mutated allele used in the transformation plasmid. Neither allele encodes for a functional protein. The mutations in *pyrG1* and *pyrG2* are located at the opposite ends of the gene. The uracil-auxotrophic *A. fumigatus* strain bearing the *pyrG1* allele is transformed with the plasmid bearing the *pyrG2* allele. Transformants are uracil prototrophic. Such strains most likely have a functional *pyrG* gene and the reporter fusions located at identical sites.

11.3 *lacZ* as a reporter gene

The *lacZ* gene from *E. coli* codes for *β*-galactosidase and is often used as a reporter to study gene expression. The gene was shown to be functional in several eukaryotic systems (Arnone *et al.*, 2004; Liebmann *et al.*, 2003; Trainor *et al.*, 1999). Besides the ability to convert the artificial substrate X-Gal™ (5-bromo-4-chloro-3-indolyl-β-D-galactopyranoside) to 5-bromo-4-chloroindigo, which enables, for example, the detection of hyphae in tissue, it can be also employed to measure promoter activity by use of another artificial substrate, which is *ortho*-nitrophenyl-β-D-galactopyranoside (ONPG). The cleavage of ONPG to *ortho*-nitrophenol and galactose leads to a change in colour to bright yellow, which can be easily measured spectrophotometrically. Two different methods have been described for the determination of β-galactosidase activity. One is a discontinuous method, which is generally used, when low activities are expected (e.g. promoters of genes involved in secondary metabolism) and the other is a continuous method, generally applied when activities are quite high (e.g. promoters of genes from primary metabolism). Both methods will be described in the following section.

Protocol 11.2
lacZ as a reporter gene: discontinuous determination of β-galactosidase activity

Equipment, materials and reagents

Harvested and shock-frozen mycelium containing the reporter construct grown under inducing/non-inducing conditions

Mortar and pestle

Liquid nitrogen

Ice

Stopwatch

Set of automatic pipettes and tips

2 ml microcentrifuge tubes

Microcentrifuge

Wassermann tubes

Spectrophotometer (wavelength 420 nm)

1 ml cuvettes

Water bath at 37 °C

Z-buffer:

$Na_2HPO_4 \times 2\ H_2O$	10.681 gram/litre (0,06 M)
$NaH_2PO_4 \times H_2O$	6.24 gram/litre (0,04 M)
KCl	0.75 gram/litre (0,01 M)
$MgSO_4 \times 7\ H_2O$	0.246 gram/litre (0,001 M)
β-Mercaptoethanol	0.7 ml/l
pH 7.0	

Phosphate buffer:

(100 mM)	pH 7.0
Na_2HPO_4	17.6 gram/litre (100 mM)
NaH_2PO_4	15.6 gram/litre (100 mM)

ONPG-solution (*ortho*-nitrophenyl-β-D-galactopyranoside)

4 mg ONPG/ml in 100 mM phosphate buffer

Na_2CO_3-solution (1M)

Reagents for determination of protein concentrations including standard protein

Method

1. Grind approximately 100–200 mg of frozen mycelium to a fine powder and transfer it to a 2 ml tube. Add 0.5–1 ml of Z-buffer and mix by pipetting.

2. Centrifuge at full speed for 5 min in the microcentrifuge. Take the supernatant and transfer it to a new 2 ml tube. Store the supernatants on ice.

3. Equilibrate 1 ml of Z-buffer in Wassermann tubes to 37 °C. Add 20–200 μl of protein extract and start the reaction with 200 μl of the ONPG solution and note the time. Perform a parallel reaction (blank) in which 2 ml of sodium carbonate solution are added prior to ONPG.

4. During incubation in the water bath, try to follow the change of the colour in the reaction mix. When a bright yellow colour becomes visible (which can take up to 1 h), the reaction is stopped by addition of 2 ml of sodium carbonate solution. Stop

the time and write it down. Take both the blank and the assay out of the water bath.

5. Determine the absorption of an aliquot of the assay against the blank on a spectrophotometer at 420 nm.

6. Calculate the change of absorbance per minute.

7. Determine protein concentration of sample (generally 2–10 mg/ml).

8. Calculate specific activity using the following formula:

$$\text{Specific activity} = V \times (\Delta E_{420})/[V_a \times \text{Prot}\,(\text{mg/ml}) \times t \times d \times \varepsilon]$$

Abbreviations

ΔE_{420} = change in absorbance (calculated per min)

V = reaction volume (ml) (e.g. 3.1 ml, when 100 μl of protein extract were used)

V_a = volume of protein extract in the assay (e.g. 0.1 ml)

Prot = protein concentration (in mg/ml)

t = incubation time (min)

d = layer thickness of cuvette (cm) (generally 1 cm)

ε = extinction coefficient ($0.0045\ \text{mM}^{-1} \times \text{cm}^{-1}$)

Note: The extinction coefficient used here corresponds to the μM extinction coefficient for *ortho*-nitrophenol at alkaline pH. It is therefore not directly comparable with the units obtained in the continuous method described below (Protocol 11.3) and varies by a factor of approximately 1000. That means that 100 units determined in this assay correspond to 0.1 U in the continuous assay.

Protocol 11.3
lacZ as a reporter gene: continuous determination of β-galactosidase activity

Equipment, materials and reagents

Harvested and shock-frozen mycelium containing the reporter construct grown under inducing/non-inducing conditions

Mortar and pestle

Liquid nitrogen

Ice

Set of automatic pipettes and tips

2 ml microcentrifuge tubes

Microcentrifuge

Spectrophotometer (wavelength 410 nm; optional heatable to 37 °C)

1 ml cuvettes

Water bath at 37 °C

Reagents for determination of protein concentrations including standard protein

13.5 mM *ortho*-nitrophenyl-*β*-D-galactopyranoside (ONPG) in 50 mM MOPS buffer pH 7.5

Reaction buffer:

100 mM MOPS buffer

4 mM $MgCl_2$

20 mM *β*-mercaptoethanol

0.5 x Reaction buffer: reaction buffer diluted by factor 2 with water

Water

Method

1. Grind ~100–200 mg of frozen mycelium to a fine powder and transfer it to a 2 ml tube. Add 0.5–1 ml of 0.5 x reaction buffer and mix by pipetting.

2. Centrifuge at full speed for 5 min in the microcentrifuge. Transfer the supernatant to a new 2 ml tube. Store the supernatants on ice.

3. Set up the following assay mixture in a 1 ml cuvette and place it in the spectrophotometer set to 410 nm: 10–100 µl protein extract, 500 µl reaction buffer and H_2O to give a final volume of 800 µl.

Note: This mixture can be either equilibrated to 37 °C or room temperature, the temperature needs to be kept constant for comparability of different measurements.

4. Start the reaction by adding 200 µl ONPG solution.

5. Follow the change in absorbance for 1–20 min and calculate the change/min.

6. Determine protein concentration of sample (generally 2–10 mg/ml).

7. Calculate specific activity by the following formula:

$$\text{Specific activity} = V \times (\Delta E_{420})/[V_a \times \text{Prot(mg/ml)} \times t \times d \times \varepsilon]$$

Abbreviations

ΔE_{420} = change in absorbance (calculated per min)

V = reaction volume (ml) (fixed to 1 ml)

V_a = volume of protein extract in the assay (e.g. 0.1 ml)

Prot = protein concentration (in mg/ml)

t = incubation time (min)

d = layer thickness of cuvette (cm) (generally 1 cm)

ε = extinction coefficient (3.5 $mM^{-1} \times cm^{-1}$ (Hoyoux *et al.*, 2001))

Note: This calculation uses the milimolar extinction coefficient of *ortho*-nitrophenol at pH 7.5. Therefore, values determined in this continuous assay are at least by factor 1000 lower than that determined in the discontinuous method (see also note from Protocol 11.2).

11.4 Transformation of *A. fumigatus*

The transformation of filamentous fungi is an essential tool for studying function of genes and their corresponding proteins. Transformation is used to introduce DNA fragments or plasmids into the genome. The method is essential for the directed expression of genes, localization studies via reporter constructs by fluorescence and for the deletion or disruption of genes (Brakhage and Langfelder, 2002). In addition, transformation is required for the introduction of reporter constructs carrying, for example, a *lacZ* gene fusion to determine promoter activity of certain genes. DNA fragments are integrated into the fungal genome either via the mechanism of non-homologous end-joining, leading to ectopic integrations, or via homologous recombination. The latter is required when a gene deletion or disruption is desired.

In general, for transformation, mycelium is treated with an enzyme mix to degrade the fungal cell wall. Care has to be taken to avoid osmotic shock during the procedure, because protoplasts are highly sensitive towards plasmolysis. Protoplasts are mixed with DNA, and fusion of the cells is facilitated by the addition of polyethylene glycol (PEG) resulting in the formation of syncytia. This is a fusion of several protoplasts, which therefore contains multiple nuclei from independent cells. During the fusion of protoplasts, DNA is incorporated and introduced into the genome of single nuclei by the mechanisms mentioned above, thereby leading to the formation of heterokaryons. Outgrowing mycelium is still heterokaryotic, that is transformed and untransformed nuclei within one cell. Nevertheless, during the asexual process of sporulation, haploid conidia are formed containing only a single nucleus (monokaryon). By repeated streaking of spore suspensions on selective agar plates homokaryons can be isolated exclusively containing transformed nuclei.

In *A. fumigatus* only a few selection markers are available. Since no sexual life cycle has been reported, auxotrophies cannot be exchanged between different strains by sexual crossing. Only a few selection systems can be considered as well established. In our laboratory, we use the dominant marker hygromycin B or uracil-auxotrophic strains for transformation. Transformation procedures will be described in detail in the following sections.

Protocol 11.4
Transformation of *A. fumigatus*

Equipment, materials and reagents

solution I: 22.37 gram KCl in 500 ml H_2O with 5 ml of 1 M $NaPO_4$ buffer pH 5.8

solution II: 22.37 gram KCl in 500 ml H_2O with 100 ml 0.5 M Tris-HCl pH 7.0

solution III: 22.37 gram KCl, 0.555 g $CaCl_2$ in 500 ml H_2O with 5 ml 1 M Tris-HCl 7.5

PEG solution: 25% PEG 8000, 50 mM $CaCl_2$, 10 mM Tris-HCl pH 7.5 (**Note:** *Do not* autoclave; filter-sterilize!)

Aspergillus Minimal Medium (AMM) for pre-culture

For 1 litre, add $NaNO_3$ 6.0 gram, KCl 0.52 gram, $MgSO_4 \cdot 7\ H_2O$ 0.52 gram, KH_2PO_4 1.52 gram to 800 ml H_2O, adjust pH to 6.5 with 10 N KOH and make up to 1 litre (optionally add 15 gram agar for solid media)

After autoclaving add 1% (w/v) final concentration of sterile glucose and 1 ml of sterile filtered Hutner's trace elements

For transformation agar, supplement AMM with 0.6 M KCl (top and bottom agar)

Note: This medium must not contain supplements, which are used as auxotrophic markers.

Hutner's trace elements: to 80 ml H_2O add the following: $ZnSO_4 \cdot 7\ H_2O$ 2.2 gram, H_3BO_3 1.1 gram, $MnCl_2 \cdot 4\ H_2O$ 0.5 gram, $FeSO_4 \cdot 7\ H_2O$ 0.5 gram, $CoCl_2 \cdot 6\ H_2O$ 0.16 gram, $CuSO_4 \cdot 5H_2O$ 0.16 gram, $Na_2MoO_4 \cdot 2\ H_2O$ 0.11 gram, Na_4EDTA 5 gram. Add the compounds in given order and boil. Cool to 60 °C and adjust pH to 6.5 with KOH. The pH needs to be carefully adjusted. Cool to room temperature. Adjust volume to 100 ml with distilled water. With storage the colour of the solution turns from green to purple, which may contain a rusty-brown precipitate that should be resuspended before use.

Sterile Erlenmeyer flask (250 ml)

2 sterile funnels with sterile filter gauze (Miracloth™, Calbiochem Darmstadt, Germany)

Rotary shaker with adjustable temperature (30 °C and 37 °C) and rotary speed (60–220 rpm)

Thoma chamber for eukaryotic cells or equivalent

Cooling centrifuge

Heatable water bath (to at least 55 °C)

Incubator (at 37 °C)

Spatula

50 ml sterile reaction tube, suitable for centrifugation

15 ml sterile reaction tube

Ice

Sterile filter devices (0.2 μm pore size, optionally equipped with Luer lock™, e.g. Minisart, Sartorius AG, Goettingen, Germany)

20 ml syringe (optionally equipped with Luer lock)

2 ml sterile reaction tubes

Scissors for cutting tips

Set of automatic pipettes and tips

Petri dishes

Method

1. Inoculate 100–200 ml of AMM liquid medium with approx. 1×10^6 conidia/ml final concentration. The medium should contain all supplements to support growth of the strain used. Incubate culture overnight (14–16 h is usual but this depends upon the overall growth of the strain) at 37 °C and vigorous shaking (220 rpm).

   Note: Mycelium should just show first branches, which facilitates the following protoplasting procedure. Alternatively, a complete medium can be used but incubation times have to be reduced.

2. Prepare the protoplast mix by solving 0.4 gram of lysing enzyme (Sigma-Aldrich) in 20 ml of solution I and sterile filter it into the sterile 250 ml Erlenmeyer flask.

3. Harvest mycelium aseptically by using sterile filter gauze in a funnel and wash the mycelium extensively with solution I. This is important, because residual glucose may inhibit the formation of protoplasts.

4. Transfer a small amount of washed mycelium (bean-size) with a sterile spatula into the Erlenmeyer flask containing the protoplast formation mix and incubate at 28–30 °C at 60–70 rpm for several hours (depends on the strain used and the age of the culture). Take 20 μl aliquots and examine microscopically several times to follow the process of protoplast formation.

5. While the protoplast formation proceeds, pour the bottom agar plates with approximately 20 ml of transformation agar (contains 0.6 M KCl) and prepare 15 ml sterile reaction tubes with 10 ml of transformation agar each (called top agar) and store in a water bath at 55 °C to maintain the agar in the liquid state.

   Note: Prepare one plate with bottom agar and one tube with top agar, containing all supplements for a positive control of protoplast viability.

6. Protoplast formation is complete when no more mycelia can be found or the amount of protoplasts does not increase anymore.

7. Harvest the protoplasts over sterile filter gauze into a sterile 50 ml reaction tube.

   Note: The protoplasts are in the flow-through! For all following steps cells have to be kept on ice (cooled centrifuge, pre-cooled solutions II/III).

8. Centrifuge cells in a pre-cooled centrifuge at 4 °C and 3300 x *g* for 10 min.

9. Pour off the supernatant and add 1 ml of solution II. Carefully resuspend protoplasts with a 1 ml pipette and a cut 1 ml tip (cut tip is required to reduce the mechanical pressure on the protoplasts). After resuspension make up to 20 ml with solution II and centrifuge as described in step 8.

10. Pour off supernatant and add 1 ml of solution III and treat as described in step 9. Make up to 10 ml with solution III. Take an aliquot for counting of the protoplasts using a Thoma chamber. While determining the number of protoplasts, centrifuge the suspension as described in step 8.

11. Discard the supernatant and carefully resuspend the protoplasts in solution III to reach a final density of 5×10^6 to 2×10^7 protoplasts/100 µl.

12. Prepare 2 ml sterile reaction tubes with 5–10 µl of DNA fragment/plasmid (representing 5–10 µg of DNA) and 10 µl sterile water as control. Then add 50 µl of protoplast solution to the tube and mix by gently inverting.

13. Add 13 µl of PEG solution and gently mix by stirring with the tip. Incubate on ice for 20–30 min.

14. Add another 500 µl of PEG solution and incubate further on ice for 5 min. Then add 1 ml of chilled solution III.

15. After the final incubation on ice the protoplast/DNA mix is ready for plating. Divide the mix in three portions (e. g. 200, 400 and 900 µl) by adding the protoplasts to the 15 ml reaction tubes containing 10 ml of top agar (stored at 55 °C).

16. Mix by inverting the reaction tube and quickly pour the top agar evenly on the bottom agar plates.

    Note: For control of your transformation mix one aliquot with the top agar for the negative control and another aliquot with all supplements, for example uridine for *pyrG*$^-$ strains for the positive control. By this means the viability of the protoplasts can be checked. Furthermore, the negative control should leave a blank plate, showing that no contamination occurred during the procedure.

17. Incubate the agar plates in a 37 °C incubator for several days. When colonies have sporulated, carefully transfer conidia to a fresh selective agar plate in order to obtain colonies deriving from single spores.

Note: To verify the result of the transformation it is necessary to isolate genomic DNA from the corresponding mutants and to check the integration or deletion by PCR and Southern blot.

11.5 Hygromycin B as a selection marker for transformation

Hygromycin B is an aminocyclitol antibiotic (belonging to the family of aminoglycoside antibiotics), which has a broad spectrum of activity against prokaryotes and eukaryotes. The hygromycin phosphotransferase gene (*hph*) confers resistance by catalysing a phosphorylation reaction at the 4-hydroxyl group at the cyclitol ring, thereby inactivating hygromycin B (Figure 11.2). It can be used to transform *A. fumigatus* strains sensitive to the antimycotic drug. Initially, for each strain the level of sensitivity has to be determined. Usually, concentrations of 100 µg per ml and higher are necessary to prevent growth of the wild type but resistance is also dependent upon the inoculation amount. Inoculation with a high cell density requires higher concentrations of hygromycin B.

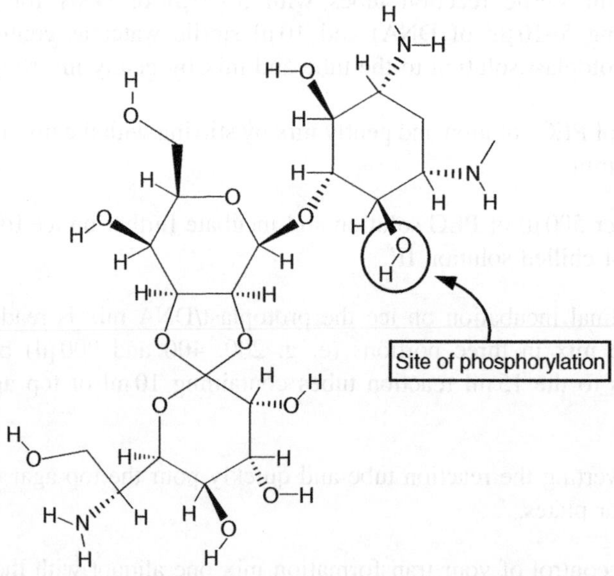

Figure 11.2 Structure of hygromycin B with site of phosphorylation by hygromycin B phosphotransferase (*hph* gene product) indicated. Phosphorylation leads to the inactive form 7′-O-phosphohygromycin

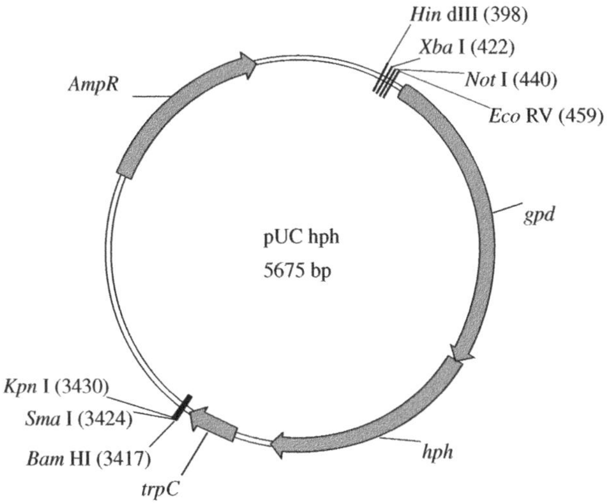

Figure 11.3 Schematic drawing of plasmid pUChph, which contains the hygromycin B resistance cassette. *gpd* = promoter of the glyceraldehyde-3-phosphate dehydrogenase from *Aspergillus nidulans*; *hph* = gene coding for the hygromycin B phosphotransferase from *Escherichia coli*; *trpC* = terminator sequence from *A. nidulans* to avoid a read-through from the *hph*-gene. Some restriction sites in front and behind the resistance cassette are shown

A gene cassette used for transformation consists of the constitutively active *A. nidulans gpd* promoter, the *hph* gene from *E. coli* and the *trpC* terminator from *A. nidulans*. For a scheme of the *hph* resistance cassette in the pUChph vector see Figure 11.3.

Protocol 11.5
Hygromycin B as a selection marker for transformation

Equipment, materials and reagents

Hygromycin B (purity >80%; good results were obtained with hygromycin B from Roche Diagnostics)

AMM agar plates (see Protocol 11.4) containing varying amounts of hygromycin B (50 to 500 μg/ml)

Petri dishes

Transformation agar containing a slightly higher concentration (~50 μg) of hygromycin B than the experimentally determined concentration

Inoculating loop

1.5 ml reaction tubes

Sterile water or PBS

Set of automatic pipettes and tips

Incubator at 37 °C

Method

1. Choose the desired strain for transformation. When hygromycin B is used as a selection marker, the transformation strain may be any wild-type strain without further auxotrophic markers.

2. To determine the minimum inhibitory concentration (MIC) required to prevent sporulation, make up a diluted conidia suspension (50 conidia/μl) and point inoculate with 1–2 μl of conidia suspensions AMM agar plates containing the different hygromycin B concentrations.

3. Incubate at 37 °C for at least 3 days and monitor development of colonies.

   Note: It is not required to prevent vegetative growth completely, but sporulation of colonies needs to be inhibited. This procedure determines the minimal concentration of hygromycin B necessary for transformation of the desired *A. fumigatus* strain.

4. Proceed with the transformation procedure described in Protocol 11.4.

   Note: You have to add hygromycin B at step 5 of the transformation protocol in both top and bottom agar. Generally, slightly higher concentrations than the MIC of hygromycin B are added because the high density of protoplasts in the transformation mix may lead to a higher level of resistance. For the positive control, leave out hygromycin B from both top and bottom agar.

5. Transformants appearing 3–6 days after transformation have to be purified several times. This step is essential to get rid of any contaminating un-transformed nuclei. Carefully pick some conidia with the inoculating loop and transfer them into 500 μl of sterile water or PBS. Vortex vigorously to separate conidia.

6. Plate different amounts of the conidia suspension on agar plates containing the MIC of hygromycin B. Incubate at 37 °C for 3–4 days and pick conidia from colonies deriving from single conidia.

7. Strains should be propagated in the presence of hygromycin B, especially when a circular plasmid was introduced by transformation, which can easily recombine out in the absence of selective pressure.

11.6 *pyrG* as a selection marker for transformation

Utilizing uracil-auxotrophic strains for transformation, *pyrG* genes from different *Aspergillus* species can be used as selection marker genes. The *pyrG* gene coding for orotidine 5′-monophosphate decarboxylase (OMP decarboxylase) is required for *de novo* synthesis of uracil. Recipient strains, for example *Aspergillus fumigatus* CEA17, either contain a point mutation in the *pyrG* locus or carry a deletion of this gene. Generally, a *pyrG* gene from a closely related species is used to avoid integration of the construct into the endogenous *pyrG* locus of *A. fumigatus*, when deletion of a specific gene of interest is desired.

One important fact has to be considered. Deletion mutants derived from uracil auxotrophic *A. fumigatus* strains have a clear drawback. It has been demonstrated that *A. fumigatus* strains devoid of a functional *pyrG* were completely avirulent (D'Enfert *et al.*, 1996). In *Candida albicans*, problems have been reported using this marker in virulence tests. The reduction of virulence could not be ascribed exclusively to the deleted gene, but, in some cases, also resulted from an insufficient complementation of the uracil auxotrophy (Brand *et al.*, 2004; Lay *et al.*, 1998). However, for both – gene expression studies and promoter analyses – this possible insufficiency may be of minor importance. For a comparison of promoter activity of wild-type and mutated sequences, the integration of the different reporter constructs at the same chromosomal locus is necessary. To achieve this site-specific integration, we use the genomic *pyrG* locus of strain CEA17, which contains a truncated *pyrG1* allele due to a point mutation leading to an unexpected stop codon. A mutated *pyrG2* allele, carrying a mutation upstream of that present in the *pyrG1* allele, is used as a selection marker in the reporter plasmid. Ectopic integration of the plasmid carrying the mutated *pyrG2* allele results in uracil-auxotrophic strains, which are unable to grow under selective conditions. Only a homologous recombination of the *pyrG1* allele with the *pyrG2* allele yields a functional *pyrG* gene and restores uracil prototrophy. By employing the two mutated *pyrG* alleles, one can ensure that transformants have integrated the reporter constructs at the *pyrG* locus (Weidner *et al.*, 1998).

Protocol 11.6
pyrG as a selection marker for transformation

Equipment, materials and reagents

Uracil (added to the medium before autoclaving at a final concentration of 10 mM)

Uridine 0.5 M sterile filtered stock solution (added to the medium after autoclaving at a final concentration of 5 mM; some protocols only use uridine in the given concentration)

AMM medium (see Protocol 11.4) containing 5–10 mM uracil/uridine for pre-culturing of uracil-auxotrophic strains

AMM agar plates (see Protocol 11.4)

AMM transformation agar (see Protocol 11.4)

Note: This medium must not contain uracil/uridine, except the plates and top agar for the positive control

Petri dishes

Inoculating loop

1.5 ml reaction tubes

Sterile water or PBS

Set of automatic pipettes and tips

Incubator at 37 °C

Method

1. Choose the desired strain for transformation. Using *pyrG* as the selection marker, the strain **must be** uracil auxotrophic.

2. Follow the protocol for transformation as described in Protocol 9.4. For the introduction of a reporter construct carrying a *pyrG2* allele, it is necessary to use the circular and not a linearized plasmid for transformation (Figure 11.4).

   Note: Both the top and bottom agar do not contain uracil or uridine. However, for a positive control, one bottom agar plate and one top agar tube have to be prepared with the supplement. The negative control is essential to monitor the frequency of spontaneously appearing uracil prototrophic revertants.

3. Colonies appearing 3–6 days after transformation have to be purified. Carefully pick some conidia with an inoculating loop and transfer them to a sterile 1.5 ml

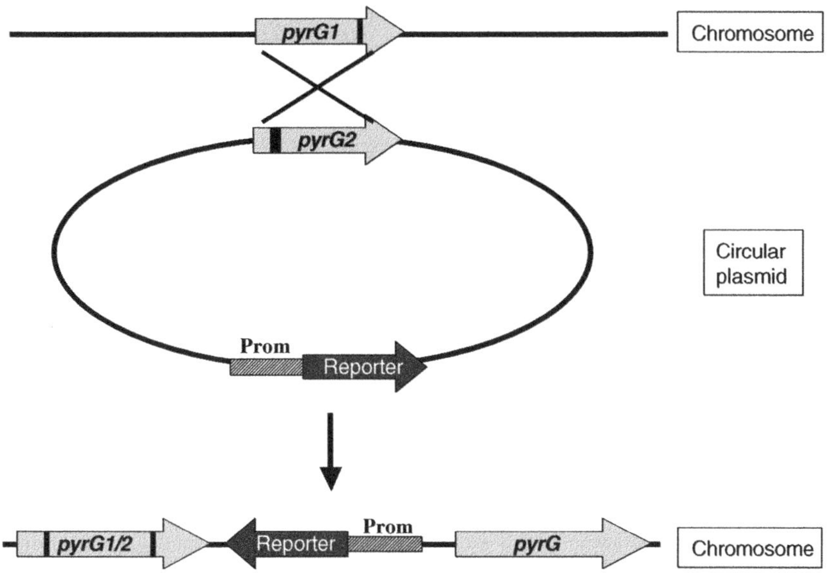

Figure 11.4 Schematic drawing of site-directed integration of a promoter-reporter gene-fusion by use of the *pyrG1* and *pyrG2* allele as selection marker. Light grey arrows represent the different alleles of *pyrG*. The black bars in the respective *pyrG* allele indicate the location of mutations. Therefore, none of the genes (in the fungal chromosome and on the plasmid) encodes a functional protein. Homologous recombination of the two defective alleles results in the following: a marker gene with two mutations (*pyrG1/2*), the integrated promoter-reporter (Prom = promoter) gene fusion and a functional marker gene (*pyrG*)

reaction tube containing 500 µl sterile water or PBS. Vortex vigorously to separate conidia.

4. Plate different aliquots of the conidia suspension on AMM agar plates (**without** uridine or uracil to maintain selective pressure) to obtain colonies derived from single conidia. Repeat several times (3–4 should be sufficient) until the colonies have a uniform morphology.

Note: When a reporter construct carrying a *lacZ* gene fusion is used, transformants can be pre-selected by a transfer of conidia to plates which contain 40 µg/ml of X-Gal (5-bromo-4-chloro-3-indolyl-β-D-galactopyranoside). Incubation of agar plates at respective inducing conditions should lead to colonies staining the agar blue.

11.7 URA-blaster (*A. niger pyrG*) as a reusable selection marker system for gene deletions/disruptions

Sometimes it is of importance to introduce reporter-gene constructs into formerly created deletion mutants. For example, after creating a signal transduction mutant,

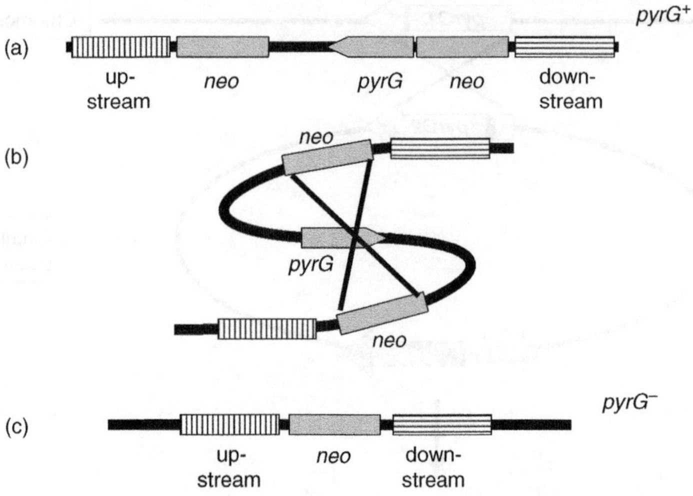

Figure 11.5 Schematic drawing of the excision of the URA-blaster from a gene deletion event under selective conditions. (a) Chromosomal situation: The URA-blaster is integrated into the genome flanked by the up-/downstream regions of the gene of interest. (b) With a frequency of 10^{-4} a homologous recombination of the consecutive *neo* genes occurs leading to the excision of the *Aspergillus niger pyrG*. (c) Genomic situation after excision of the URA blaster. The *pyrG* gene can be used again as a selection marker

one may want to integrate promoter-*lacZ* gene fusions to study the expression pattern of target genes downstream of the signal pathway. The required multiple transformations can only be achieved efficiently by employing the reusable URA-blaster (also called *pyrG*-blaster). The URA-blaster was first developed for yeast systems to move genetic markers back and forth between cloned yeast sequences on plasmids and the yeast genome (Alani *et al.*, 1987). Later on, the URA blaster was established in *Aspergillus* species (D'Enfert, 1996). The construct used for *A. fumigatus* consists of two consecutive bacterial *neo* genes as direct repeats flanking the OMP decarboxylase encoding *pyrG* from *Aspergillus niger.* (The *neo* gene confers resistance to kanamycin in bacteria.). For generation of deletion constructs, DNA regions flanking upstream and downstream (~800–1000 base pairs) the gene of interest have to be cloned 5′ and 3′ the *neo* genes (Figure 11.5). These flanking regions are required for the homologous integration of the deletion construct at the genomic locus by double cross-over. This recombination event replaces the gene of interest by the URA-blaster, thereby leading to the deletion of the gene of interest.

After the successful transformation with the URA-blaster, the prototrophic mutants can be incubated on a medium containing 5′ fluoroorotic acid (5-FOA) and uracil. 5-FOA serves as a substrate for OMP decarboxylase which converts 5-FOA in a cytotoxic product (5′-fluorouracil). With a frequency of $\sim 10^{-4}$ a mitotic recombination event occurs in which the *neo* repeats recombine and the *A. niger pyrG* is excised, thereby turning the transformed strain to uracil auxotrophy again. This allows selection of *pyrG*$^-$ mutants on media containing 5-FOA and uracil/uridine and provides a strain capable for a new transformation with *pyrG* as selection marker.

Protocol 11.7
URA-blaster (*A. niger pyrG*) as a reusable selection marker system for gene deletions/disruptions

Equipment, materials and reagents

Uracil (added to the medium before autoclaving at a final concentration of 10 mM)

Uridine 0.5 M sterile filtered stock solution (added to the medium after autoclaving at a final concentration of 5 mM; some protocols only use uridine in the given concentration)

1 gram 5-fluoroorotic acid (5-FOA) per litre

AMM medium (see Protocol 9.4) containing 5–10 mM uracil/uridine for preculturing of uracil-auxotrophic strains

Citrate-buffered AMM agar plates (see Protocol 11.4) containing 5–10 mM uracil/uridine and 1 mg/ml 5-FOA:

500 ml 2 x AMM agar

500 ml 50 mM citrate buffer pH 6.5

50 ml syringe with Luer lock

Aseptical filter devices (0.2 μm)

Incubator at 70 °C

Magnetic stirrer

AMM agar plates (see Protocol 11.4)

AMM transformation agar (see Protocol 11.4)

Note: This medium must not contain uracil/uridine, except the agar plates and top agar for the positive control.

AMM agar plates (see Protocol 11.4) containing 5–10 mM uracil/uridine and 1 mg/ml 5-FOA

Petri dishes

Inoculating loop

Thoma chamber for eukaryotic cells or equivalent

1.5 ml reaction tubes

Sterile water or PBS

Set of automatic pipettes and tips

Incubator at 37 °C

Method

See Protocols 11.4 and 11.6 for transformation and treatment of uracil-auxotrophic strains.

Note: The whole URA-blaster cassette must be part of the DNA fragment employed in the transformation. The construct should be linearized prior to transformation as plasmids are rarely integrated via double cross-overs, but rather by a single recombination event.

Strains have to be purified to obtain pure homokaryotic cultures.

A. Preparation of citrate-buffered AMM agar plates containing 5-FOA

Autoclaving of 5-FOA is not recommended. However, the solubility of 5-FOA in water is low. Although 50 mg/ml can be dissolved in 4 M ammonium hydroxide solution, the addition of that solution (20 ml/l) to the medium alters the pH. By contrast, addition of 5-FOA to the medium without further buffering leads to a strong acidification of AMM agar.

1. Prepare 2 x AMM agar by dissolving all components needed for 1 litre of AMM (see Protocol 9.4) in 500 ml of water. Sterilize by autoclaving and keep liquid at 70 °C.

2. Add 1 gram of 5-FOA to 500 ml of 50 mM citrate buffer pH 6.5. Stir until all 5-FOA has completely dissolved and sterile-filter the solution.

3. Heat up the sterile citrate buffered 5-FOA solution to 70 °C and mix 1 part with 1 part of the 2 x AMM agar from step 1.

4. Add 10 ml of 0.5 M uridine to 1 litre of citrate-buffered 5-FOA/AMM agar and pour into Petri dishes.

B. Selection of mitotic pop-out-events of the URA-blaster

1. With an inoculating loop take conidia from an already purified and characterized transformant and transfer them to PBS.
2. Calculate the number of conidia per ml and make up to 1×10^7 conidia/ml.
3. Plate 10 to 100 µl of the conidial suspension on the 5-FOA-containing agar plates from above and incubate at 37 °C for 3–5 days.
4. Take some conidia from developing colonies, transfer them again to 5-FOA-containing agar plates and incubate the agar plates at 37 °C for 3 days. Conidia from the developing colonies should be replica-plated on AMM agar plates with and without the addition of uracil/uridine to check the loss of uracil prototrophy.

11.8 References

Alani, E., Cao, L. and Kleckner, N. (1987) A method for gene disruption that allows repeated use of URA3 selection in the construction of multiply disrupted yeast strains. *Genetics* **116**: 541–545.

Arnone, M. I., Dmochowski, I. J. and Gache, C. (2004) Using reporter genes to study *cis*-regulatory elements. [B]Method*s Cell. Biol.* **74**: 621–652.

Brakhage, A. A. and Langfelder, K. (2002) Menacing mold: The molecular biology of *Aspergillus fumigatus. Annu. Rev. Microbiol.* **56**: 433–455.

Brand, A., MacCallum, D. M., Brown, A. J., Gow, N. A. and Odds, F. C. (2004) Ectopic expression of URA3 can influence the virulence, phenotypes and proteome of *Candida albicans* but can be overcome by targeted reintegration of URA3 at the RPS10 locus. *Eukaryot. Cell.* **3**: 900–909.

Cabanas, M. J., Vazquez, D. and Modolell, J. (1978) Dual interference of hygromycin B with ribosomal translocation and with aminoacyl-tRNA recognition. *Eur. J. Biochem.* **87**: 21–27.

D'Enfert, C. (1996) Selection of multiple disruption events in *Aspergillus fumigatus* using the orotidine-5′-decarboxylase gene, *pyrG*, as a unique transformation marker. *Curr. Genet.* **30**: 76–82.

D'Enfert, C., Diaquin, M., Delit, A., Wuscher, N. *et al.* (1996) Attenuated virulence of uridine-uracil auxotrophs of *Aspergillus fumigatus. Infect. Immun.* **64**: 4401–4405.

Higuchi, R., Krummel, B. and Saiki, R. K. (1988) A general method of *in vitro* preparation and specific mutagenesis of DNA fragments: Study of protein and DNA interactions. *Nucleic Acids Res.* **16**: 7351–7367.

Hoyoux, A., Jennes, I., Dubois, P., Genicot, S. *et al.* (2001) Cold-adapted beta-galactosidase from the Antarctic psychrophile *Pseudoalteromonas haloplanktis. Appl. Environ. Microbiol.* **67**: 1529–1535.

Lay, J., Henry, L. K., Clifford, J., Koltin, Y. *et al.* (1998) Altered expression of selectable marker URA3 in gene-disrupted *Candida albicans* strains complicates interpretation of virulence studies. *Infect. Immun.* **66**: 5301–5306.

Liebmann, B., Gattung, S., Jahn, B. and Brakhage, A. A. (2003) cAMP signaling in *Aspergillus fumigatus* is involved in the regulation of the virulence gene *pksP* and in defense against killing by macrophages. *Mol. Genet. Genomics* **269**: 420–435.

Smith, J. M., Tang, C. M., Van Noorden, S. and Holden, D. W. (1994) Virulence of *Aspergillus fumigatus* double mutants lacking restriction and an alkaline protease in a low-dose model of invasive pulmonary aspergillosis. *Infect. Immun.* **62**: 5247–5254.

Trainor, P. A., Zhou, S. X., Parameswaran, M., Quinlan, G. A. *et al.* (1999) Application of *lacZ* transgenic mice to cell lineage studies. [B]Method*s Mol. Biol.* **97**: 183–200.

Weidner, G., d'Enfert, C., Koch, A., Mol, P. C. and Brakhage, A. A. (1998) Development of a homologous transformation system for the human pathogenic fungus *Aspergillus fumigatus* based on the *pyrG* gene encoding orotidine 5′-monophosphate decarboxylase. *Curr. Genet.* **33**: 378–385.

12

Microarray technology for studying the virulence of *Aspergillus fumigatus*

Darius Armstrong-James and **Thomas Rogers**

12.1 Introduction

Members of the genus *Aspergillus* are ascomycetous mould fungi that are widespread in the environment including soil, decaying organic matter, plants and house dust (www.aspergillus.man.ac.uk).

Although there are close to 200 *Aspergillus* species, human disease is only commonly attributable to no more than six of these, of which by far the most important is *Aspergillus fumigatus* (Denning *et al.*, 2002). Less commonly found as pathogens are *A. flavus*, *A. terreus*, *A. nidulans* and *A. niger*, although *A. terreus* infections are reported to have recently increased in incidence (Lass-Florl *et al.*, 2005).

A. fumigatus is responsible for a diversity of human diseases ranging from exacerbations of allergic asthma, saprophytic colonization of lung cavities and locally invasive disease in individuals with chronic obstructive pulmonary disease to the most severe form: invasive pulmonary aspergillosis, which typically affects immunocompromised patients (Hope *et al.*, 2005).

Aspergillus conidia are constantly being inhaled but are normally killed by cells of the innate immune system, notably alveolar macrophages and neutrophils, although the adaptive immune response can also play a major role in protecting immunocompetent individuals from developing infection (Romani, 2004).

Invasive aspergillosis usually occurs in patients who are undergoing treatment for haematological malignancies, a typical example being an allogeneic stem cell transplant recipient with acute leukaemia.

Medical Mycology: Cellular and Molecular Techniques. Edited by Kevin Kavanagh

There has been considerable interest in gaining a better insight into why *A. fumigatus* in particular is such a successful opportunistic pathogen in this setting. What is known is that this is a versatile and highly adaptable fungus that is capable of growth across wide temperature and pH ranges and of surviving under adverse environmental conditions.

An initial approach was to identify putative virulence determinants, carry out targeted disruption of the relevant gene of interest and establish that the mutant had attenuated virulence compared to the background wild strain using an immunocompromised animal model (Latge, 2001). This has led to some success but, even though a number of virulence factors have been identified in this and other ways (Table 12.1), it is clear that there is no single one of these that is responsible for invasive disease.

How these virulence factors are expressed in the course of an invasive infection and how they interact with the host's immune cells is now the main area of research interest. To achieve this requires more versatile laboratory tools that will enable us to study which genes (both fungal and host) are either up- or down-regulated at different time points in the course of infection. This can be done by studying animal models of invasive pulmonary aspergillosis and ideally human *Aspergillus* infections.

The genome of strain *Af*-293, a pathogenic isolate of *A. fumigatus*, has been sequenced by Nierman *et al.* (2005) using a whole-genome random-sequencing

Table 12.1 Examples of genes and molecules identified as having a role in *A. fumigatus* virulence (Adapted from Rementeria *et al.* 2005)

Cell wall and cell surface related:
Beta(1-3)-glucan
Galactomannan
Galactomannoproteins (Afmp1, Afmp2)
Chitin synthetases
PacC regulatory system (environmental pH sensing)*
Resistance to host's innate immune response:
Hydrophobicity genes (*rodA*, *hyp1*)
Catalases (Cat 1p, Cat 2p)
Superoxide dismutases
Siderophore biosynthesis (*sidA* gene)**
Toxins
Gliotoxin, Fumagillin, Restrictocin
Allergens (associated with type 1 hypersensitivity)
Asp f1 (mitogillin), f2, f5, f6 etc
Extracellular enzymes
Alkaline serine protease
Elastase
Metalloprotease, Aspartic protease

*Bignell *et al.* (2005); **Schrettl *et al.* (2004).

method (Nierman *et al.*, 2001). It consists of eight chromosomes with a total size of 29.2 Mb containing over 9000 identified protein-coding sequences. However, the authors point to the incomplete reliability of their auto-annotation approach, especially in deducing gene function.

With reference to the construction of oligonucleotide microarrays for the purposes of performing expression-profiling experiments, they adopted a strategy to avoid the selected oligonucleotide being omitted from the expressed mRNA for a particular gene. Briefly, they selected regions immediately upstream and downstream of the target genes of interest for priming. In this way they designed an array comprising >99% of the predicted *A. fumigatus* gene regions from genomic DNA.

Using their whole-genome microarray Nierman *et al.* (2005) investigated the thermotolerance of *A. fumigatus* by comparing gene expression across a temperature range of 30 °C to 48 °C. Overall, more genes were upregulated at 37 °C compared to 48 °C. As they predicted, heat shock and stress response genes were expressed at higher levels at 48 °C, whereas more genes deemed to have a role in fungal pathogenicity had increased expression at 37 °C.

Another approach to unravelling the secrets behind why *A. fumigatus* is more pathogenic for humans than other *Aspergillus* species is to compare their genomes. However, even though the *A. nidulans* genome has been sequenced, it is viewed as too distantly related to *A. fumigatus* to allow meaningful comparisons to be made. By contrast, *Neosartorya fischeri* is closely related to *A. fumigatus* yet is regarded as a rare human pathogen. Nierman and colleagues took genomic DNA from these two fungi and, using their microarray, studied differences by comparative genomic hybridization; they found 700 genes of different function that were not present or diverged in *N fischeri* compared to *Af*-293. They propose to further study these by correlating their activities with phenotype differences.

Further applications of microarray technology to *Aspergillus* research include the study of gene expression relating to aflatoxin production in *A. flavus* and *A. parasiticus* (OBrian *et al.*, 2003), gene expression related to the activity of the glycolytic and tricarboxylic metabolic pathways in *A. oryzae* (Maeda *et al.*, 2004) and, similarly, the effects of glucose on gene expression in *A. nidulans* (Sims *et al.*, 2004).

A DNA microarray for the rapid identification of medically important fungal pathogens including four *Aspergillus* species has been developed by Leinberger *et al.* (2005), which may have valuable application in diagnostic mycology.

12.2 Isolation of RNA from *A. fumigatus*

TRIZOL® (Invitrogen Corporation, Carlsbad, CA, USA) is a solution containing both phenol and guanidine isothiocyanate, based on the original single-step RNA isolation method described by Chomczynski and Sacchi (Chomczynski and Sacchi, 1987). This method is used widely throughout the microarray community as it allows the rapid isolation of pure total RNA from a variety of organisms. Initial sample homogenization in TRIZOL results in the dissolution of cellular components, while preserving the integrity of total RNA. The subsequent addition of chloroform forces RNA exclusively into the aqueous phase, which is further isolated by centrifugation. RNA is then isolated from the aqueous phase by precipitation with isopropanol. Lithium chloride precipitation (Cathala *et al.*, 1983) offers the advantage of inefficiently precipitating contaminants such as DNA, protein and carbohydrate, and is excellent for the removal of inhibitors of reverse transcription.

Protocol 12.1
Isolation of total RNA from *A. fumigatus*

Equipment, materials and reagents

A. fumigatus mycelia to be processed

Liquid nitrogen

Pestle and mortar, at least 15 cm diameter

Absolve™ 2% (Du Pont, Wilmington, DE, USA)

Miracloth™ (Calbiochem, Darmstadt, Germany), autoclaved

TRIZOL® (Invitrogen Corporation, Carlsbad, CA, USA) reagent

Chloroform

Isopropanol

Diethylpyrocarbonate (DEPC)-treated water

70% (v/v) ethanol in DEPC-treated water on ice

4 molar lithium chloride in DEPC-treated water

Agarose

Ethidium bromide

RNA loading buffer:

80% (v/v) deionized formamide

1 mM EDTA, pH 8.0

0.1% w/v bromophenol blue

0.1% w/v xylene cyanol

TBE buffer:

89 mM Tris base

89 mM boric acid

2 mM EDTA

Spectrophotometer

Table-top centrifuge at 2–8 °C

Table-top centrifuge at 15–30 °C

Table-top vortex

Method

1. Soak a 15 cm diameter pestle and mortar overnight in 2% Absolve in DEPC-treated water. Rinse with DEPC-treated water. Air dry. Chill pestle and mortar with liquid nitrogen.

2. From liquid cultures, harvest mycelia by filtering through sterile Miracloth, ensure removal of excess medium and snap-freeze mycelia in liquid nitrogen.

3. Transfer biomass to pestle and mortar, and macerate to fine powder. Add TRIZOL reagent at a ratio of 1 ml TRIZOL/100 mg biomass, and further homogenize to smooth paste. The sample volume should not exceed 10% of the total volume of TRIZOL reagent used.

4. Allow paste to thaw to liquid, then quickly transfer to polypropylene tubes and centrifuge for 10 min at 12 000 x *g* at 2–8 °C. This step enables the removal of insoluble materials from the homogenate, such as protein and polysaccharides.

5. Remove supernatant and transfer to fresh polypropylene tube. Incubate at 15–30 °C for 5 min to allow dissociation of nucleoprotein complexes.

6. Add 0.3 volumes of chloroform per 1 ml of TRIZOL reagent initially used, securely close tube and shake vigorously by hand for 15 sec. Allow suspension to incubate for a further 2–3 min at 15–30 °C.

7. Centrifuge sample for 15 min at 2–8 °C at 12 000 x *g*. The sample will separate into a *red* lower chloroform/phenol phase, an inter-phase containing DNA and a *clear* upper-aqueous phase containing RNA.

8. Remove the *clear* upper-aqueous phase, which should be around 50% of the total volume of TRIZOL reagent initially used and transfer to a fresh polypropylene tube. Precipitate the RNA from the aqueous phase by adding an equal volume of isopropanol and incubating for a further 10 min at 15–30 °C.

9. Centrifuge sample for 10 min at 2–8 °C at 12 000 x *g*. The RNA forms a clear precipitate against the side of the tube.

10. Remove supernatant and wash pellet by vortexing at top speed in 1 ml ice-chilled 70% ethanol in DEPC-treated water. Centrifuge the sample for a further 10 min at 2–8 °C at 12 000 x *g*.

11. Remove supernatant, centrifuge briefly and remove residual liquid. Allow pellet to briefly air dry, then re-suspend in 500 μl of DEPC-treated water. Add 500 μl of 4 M lithium chloride and incubate sample at −20 °C for at least 1 h.

12. Centrifuge sample at 12 000 x *g* at 2–8 °C for 30 min, remove supernatant and wash in 1 ml 70% ethanol in DEPC-treated water by vortexing. Centrifuge the sample for a further 10 min at 2–8 °C at 12 000 x *g*.

13. Remove supernatant and wash pellet by vortexing at top speed in 1 ml ice-chilled 70% ethanol in DEPC-treated water. Centrifuge the sample for a further 10 min at 2–8 °C at 12 000 x *g*. Remove supernatant and centrifuge briefly. Remove residual liquid and allow to air dry briefly. Re-suspend pellet in 50 μl of DEPC-treated water.

14. Quantify RNA concentration by spectroscopy at 260 nm and 280 nm. The 260 nm:280 nm ratio should be greater than 1.8.

15. Analyse RNA by agarose gel electrophoresis. Wash gel tank in 2% absolve and rinse with DEPC-treated water. Cast a 1% agarose gel with ethidium bromide in TBE buffer. Mix 500 μg of RNA in equal volume of RNA loading buffer, and run

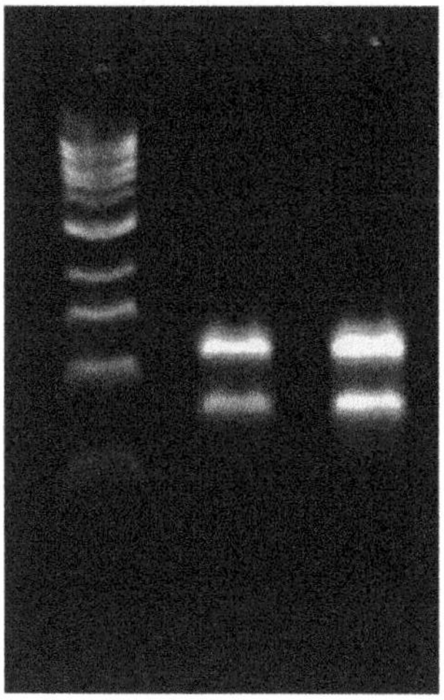

Figure 12.1 The 28s and 18s ribosomal bands are clearly visible, with no smearing. The TRIZOL method also results in isolation of 5s ribosomal RNA, which can sometimes be mistaken for degraded RNA. The ratio of 28s ribosomal RNA to 18s ribosomal RNA is roughly 2:1, indicating that the RNA has not been degraded

at 50 V for 30 min. Visualize gel with gel document system. The ideal image is shown in Figure 12.1.

12.3 Reverse transcription of RNA and fluorescent labelling of cDNA

Historically, fluorescently labelled cDNA probes used for microarray analysis have been generated by the direct incorporation (Yu *et al.*, 1994) of nucleotides attached to large fluorescent dyes such as Cy3 and Cy5. However, these dyes are inefficiently incorporated into cDNA molecules by reverse transcription owing to their bulky nature. Furthermore, direct labelling results in a differential incorporation of Cy3 and Cy5 dyes, which may result in signal-dye bias. A newer and more reliable approach is indirect dye incorporation (Hughes *et al.*, 2001). In this technique, amino-modified nucleotides are incorporated into cDNA molecules by reverse transcription, and fluorescent Cye dyes are then chemically incorporated into the molecules. This results in the generation of excellent quantities of cDNA, with high levels of dye incorporation. High levels of dye incorporation are essential for good-quality signals from subsequent microarray hybridization reactions. Indirect labelling takes longer than direct labelling; however, the greater reliability of this technique and reduced dye bias result in reduced costs and produce a far better result.

Protocol 12.2
Indirect labelling of cDNA with fluorescent dyes

Equipment, materials and reagents

Total RNA in DEPC-treated water

Sodium bicarbonate buffer: 0.1 M sodium bicarbonate adjusted to pH 9.0 with hydrochloric acid

Aminoallyl dNTP mix:

5 μl 100 mM dATP

5 μl 100 mM dCTP

5 μl 100 mM dGTP

3 μl 100 mM dTTP

2 μl 100 mM aminoallyl dUTP

Superscript™ III (Invitrogen Corporation, Carlsbad, CA, USA) reverse transcriptase 200 μl

Superscript™ III (Invitrogen) reverse transcriptase 5 x first strand buffer (Invitrogen)

0.1 M dithiothreitol (DTT)

Random hexamers 3 mg/ml

0.5 M EDTA

0.5 M sodium hydroxide

0.5 M hydrochloric acid

Ice

80% ethanol

CyScribe GFX Purification Kit™ (Amersham Biosciences, GE Healthcare, Little Chalfont, UK)

Cy3™ or Cy5™ mono-reactive dye pack (Amersham, cat. no. PA23001/PA25001)

Method

1. To 10 μg of total RNA in 10 μl of DEPC-treated water add 2 μl random hexamers (3 μg/μl) and make up to a total volume of 18.5 μl total with DEPC-treated water. Incubate at 70 °C for 10 min, then immediately chill on ice for 30 sec.

2. Centrifuge briefly and add:

   a. 6 μl first strand buffer

   b. 3 μl 0.1M DTT

   c. 0.6 μl aminoallyl dNTP mix

   d. 2 μl Superscript III.

3. Mix by pipetting, and incubate at 46 °C for 2 h at least (preferably overnight).

4. To hydrolyse RNA, add 2 μl 0.5 M sodium hydroxide and 2 μl 0.5 M EDTA. Incubate at 65 °C for 15 min. Add 2 μl 0.5 M hydrochloric acid to neutralize reaction.

5. Transfer reaction to a CyScribe column containing 500 μl capture buffer, and mix by pipetting 6 times. Centrifuge at 13 800 x *g* for 30 sec.

6. Remove eluate, and add 600 μl 80% ethanol to column. Centrifuge for 30 sec at 13 800 x *g*.

7. Remove eluate, and add 600 μl 80% ethanol to column. Centrifuge for 30 sec at 13 800 x *g*.

8. Remove eluate, and add 600 μl 80% ethanol to column. Centrifuge for 30 sec at 13 800 x *g*.

9. Spin for a further 10 sec at 13 800 x *g*, and place column in a fresh tube. Add 60 μl 0.1 M sodium bicarbonate, and wait 5 min.

10. Spin for 1 min at 13 800 x *g*. Add a further 60 μl 0.1 M sodium bicarbonate, and wait for a further 5 min.

11. Spin for 1 min at 13 800 x *g*, and quantify by spectroscopy at 260 nm.

12. Ideally at least 8 µg of cDNA is required for the current available *A. fumigatus* microarrays. If insufficient cDNA is available, perform further reactions using the same cDNA solution to elute.

13. Add cDNA solution to 1 vial of Cy3 or Cy5 mono-reactive dye. Incubate in the dark at room temperature for at least 2 h (preferably overnight).

14. Transfer reaction to a CyScribe column containing 500 µl capture buffer, and mix by pipetting 6 times. Centrifuge at 13 800 x *g* for 30 sec.

15. Remove eluate, and add 600 µl wash buffer to column. Centrifuge for 30 sec at 13 800 x *g*.

16. Remove eluate, and add 600 µl wash buffer to column. Centrifuge for 30 sec at 13 800 x *g*.

17. Remove eluate, and add 600 µl wash buffer to column. Centrifuge for 30 sec at 13 800 x *g*.

18. Spin for a further 10 sec at 13 800 x *g*, and place column in a fresh tube. Add 60 µl elution buffer, and wait 5 min.

19. Spin for 1 min at 13 800 x *g*. Add a further 60 µl elution buffer, and wait for a further 5 min.

20. Spin for 1 min at 13 800 x *g*. Quantify dye incorporation by spectroscopy at 550 nm (Cy3) or 650 nm (Cy5).

$$\text{pmol Cy3} = \text{OD550} \times \text{volume (in µl)}/0.15$$

$$\text{pmol Cy5} = \text{OD650} \times \text{volume(in µl)}/0.25$$

12.4 Hybridization of fluorescent probes to DNA microarrays and post-hybridization washing

In common with other hybridization procedures, the goal for microarray hybridization and washing is to achieve maximal signal in both channels with minimal background. The conditions used for microarray hybridization are very similar to those required for northern hybridization. The main requirements are for an excess of probe to target by at least ten fold, high ionic strength buffers to promote hybridization, blocking agents and detergents to reduce background and high stringency conditions to reduce cross-hybridization (Heller *et al.*, 1997).

Protocol 12.3 Hybridization of fluorescent probes to DNA microarrays and post-hybridization washing

Equipment, materials and reagents

Lyophilizer

Pre-hybridization solution:

25 ml of 20 x SSC

1 ml 10% SDS

1 gram BSA

Made up to 100 ml with double-distilled water

Hybridization solution:

5 ml deionized formamide

5 ml 10 x SSC

200 μl 10% SDS

10 ml double-distilled water

50 ml Coplin jar

Microarray hybridization chamber

Method

Note: Before beginning, it is important to note three important tips: (1) always try to work in a dust-free environment, (2) never touch the surface of the microarray and (3) never allow the microarray to dry out during the procedure; this would result in high background!

1. Lyophilize 800 pmol each fluorescently labelled cDNA sample, and re-suspend each in 30 μl hybridization solution.

2. Incubate 50 ml pre-hybridization solution at 42 °C for 30 min in a Coplin jar.

3. Fully submerge microarray in hybridization solution, and incubate at 42 °C for a further 1 h at least.

4. Remove microarray and immediately submerge in 50 ml of double-distilled water in a Coplin jar. Wash for 2 min at 100 rpm.

5. Transfer microarray to second Coplin jar containing 50 ml isopropanol, and wash for a further 2 min at 100 rpm.

6. Remove microarray and spin to dry by placing in a 50 ml plastic tube in a centrifuge and spinning at 3200 x *g* for 10 min.

7. While waiting, mix Cy3 and Cy5 probes, and denature cDNA by heating at 95 °C for 5 min, then spin briefly to pellet any debris.

8. Remove slide, and pipette cDNA directly onto middle of slide. Gently cover with a cover slip, taking care to avoid bubble formation.

9. Place in hybridization chamber, with sufficient water in wells, and incubate at 42 °C in a water tank for 16–20 h.

10. Remove slide, and submerge in Coplin jar containing 50 ml of 2 x SSC, 1% SDS. Leave for 1 min, gently remove slide. The cover slip should slip off.

11. Immediately transfer slide to Coplin jar containing 50 ml of 2 x SSC, 1% SDS at 55 °C and wash for 15 min at 100 rpm.

12. Transfer to a Coplin jar containing 50 ml of 1 x SSC, 0.2% SDS at 42 °C and wash for 8 min at 100 rpm.

13. Transfer to a Coplin jar containing 50 ml of 0.1 x SSC, 0.2% SDS at 42 °C and wash for 5 min at 100 rpm.

14. Transfer slide to Coplin jar containing 50 ml double-distilled water and wash for 2 min at 100 rpm.

15. Transfer slide to Coplin jar containing 50 ml isopropanol and wash for 2 min at 100 rpm.

16. Centrifuge slide in 50 ml plastic tube at 3200 x *g* for 10 min to dry.

12.5 Image acquisition from hybridized microarrays

The acquisition of microarray images is crucial, as it affects the quality of the data for subsequent analysis. The purpose of image analysis is to estimate the intensities of the signals for both red and green channels for each spot on the array. By convention, it is assumed that the overall red/green ratio for a given experiment is 1. The signal intensity for each channel corresponds to the degree of hybridization of the two competing probes to the target spot on the arrays, and therefore gives an estimate of the ratio of cDNA between the two conditions under analysis. The ideal microarray image should have high signal intensities, without signal saturation, with minimal background noise, and a red/green signal ratio of around 1. Signal saturation may lead to inaccurate measurements, that is falsely low signal. We currently use the Axon Instruments GenePix™ (Molecular Devices Corporation, Union City, CA) 4000b Dual Confocal Laser scanner, which is one of the industry-standard machines for DNA microarray image acquisition. Confocal lasers are used in this system to excite the flourophor-labelled targets. The resulting photons that are emitted from the flourophors are amplified by a photo-multiplier tube, resulting in a release of electrons which are measured as the signal. This scanner is easy to use, fast and reliable, with good resolution. Owing to its popularity, it also tends to be compatible with most downstream software applications.

Protocol 12.4
Image acquisition from hybridized microarrays

Equipment, materials and reagents

Slides to be scanned

Axon GenePix™ (Molecular Devices Corporation, Union City, CA) 4000b Dual Laser DNA Microarray Scanner

GenePix™ (Molecular Devices) Pro software

PC with at least:

1 GHz RAM

1 GHz processor

Windows® 2000 or XP (Microsoft® Corporation, Washington, USA)

Method

1. Turn on the scanner at least 15 min before use, in order to allow the lasers to warm up. The switch is located on the transformer box attached to the scanner.

2. Click on the GenePix software icon on the computer desktop to launch the software.

3. Insert slide face down into the slide holder by sliding the top door of the scanner to the left and opening the slide-holding mechanism. Snap the slide holder in place and close the top door of the scanner.

4. Set up the scan settings. Click on the hardware settings icon on the right of the window. Set the PMT voltages to 550 for Cy3 and 650 for Cy5 initially, with lines to average at 1 and pixel size at 10 μm.

5. Click the double-arrow icon for a 'preliminary scan'. This is a low-resolution scan which allows you to quickly check the channel gains are correct and to select the right area of the slide for formal scanning.

6. While scanning, the hardware settings can be adjusted to balance red/green ratio and ensure a good dynamic range of intensities. Clicking on the histogram tab allows the red/green ratio to be viewed. The ideal red/green ratio is 1. Use the cursor to run over the background area. The background intensity can then be seen in the left lower box. Ideally, the background intensity should be less than 100, and maximum spot intensity just under saturation, i.e. 65 000.

7. Click on the scan area icon and draw a box around the area to be imaged. Increase the lines to an average of 2 in the hardware settings and click on the single-arrow scan button. The smaller the scan area, the smaller the resulting image file. Scanning the slide should take at least 6 min.

8. Once scanned, save the file as a multi-image tiff. This can be done in the drop-down file menu on the right of the window.

12.6 Microarray image analysis

The way in which data are extracted from the scanned microarray image can have a significant impact on subsequent results and analysis. Fortunately, modern data-acquisition programs implement semi-automated methods to quickly and reproducibly extract data from microarray images. GenePix Array List (GAL) files are files containing all the data required to identify where each spot is located on the array and give the corresponding name for the related gene or DNA sequence. GAL files can be easily loaded onto image-analysis programs such as GenePix Pro to create grids over the array. Circles within the grids representing each spot are then lined up with the spots on the array, in a semi-automated fashion, allowing the subsequent measurement of median-spot intensities in red and green channels within circles, as well as the estimation of local background fluorescence around spots. It is essential to adjust for background fluorescence when estimating microarray spot fluorescence. This is because each spot intensity also partly results from non-specific hybridization and fluorescence from the chemical surface of the microarray. Local background is preferred to global background, as there is often substantial variation in the background across a hybridized microarray. We also advise that the median fluorescence is always used for subsequent analysis, rather than the mean. This is because artefacts such as dust can cause small areas of high-intensity fluorescence, which cause the mean values to be distorted but should have no effect on median fluorescence.

Protocol 12.5
Microarray image analysis

Equipment, materials and reagents

Scanned slide image

GenePix™ (Molecular Devices Corporation, Union City, CA) 4000b Dual Laser DNA Microarray Scanner

GenePix™ (Molecular Devices) Pro software

TIGR *A. fumigatus* GAL file

PC computer with at least:

- 1 GHz RAM
- 1 GHz Processor
- Windows® 2000 or XP (Microsoft® Corporation, Washington, USA)

Method

1. Open the scanned slide image by clicking the open/save icon on the right of the window and scrolling to the appropriate file. Display settings can be altered using the ratio colour mapping icon and brightness and contrast controls.

2. Zoom in on the slide using the zoom icon in order to survey the slide at higher magnification. The feature viewer can be used to check local background and spot intensities.

3. Open the TIGR *A. fumigatus* GAL file using the open/save icon. Line up the blocks from the GAL file with the blocks from the microarray in block mode using the mouse.

4. Once the blocks are lined up, precise alignments can be performed with the 'Find All Blocks, Align Selected Features' function in the top left of the window.

5. Use the zoom icon to zoom in, in order to fine-tune the alignment manually. Change to 'Feature' mode and use the mouse, arrow keys and control buttons to adjust the location and size of spots. Individual spots can also be flagged as good, bad or absent using the 'Feature Indicator' function. This is useful, as flagged spots will be identifiable for downstream analyses and data filtering.

6. After the spots are fully aligned and flagged as necessary, press the analyse button in order to fully analyse the image. Results will be sent to a *results spreadsheet* automatically after a few seconds. A JPEG of the scanned image can be saved at this stage, and is useful for image presentation. The raw data may be viewed in 'Scatter Plot' mode. Results can then be saved as a GPR (GenePix Results) file for subsequent analysis.

12.7 References

Bignell, E., Negrete-Urtasun, S., Calcagno, A. M., Haynes, K. *et al.* (2005) The *Aspergillus* pH-responsive transcription factor PacC regulates virulence. *Molecular Microbiology* **55**(4): 1072–84.

Cathala, G., Savouret, J. F., Mendez, B., West, B. L. *et al.* (1983) A method for isolation of intact, translationally active ribonucleic acid. *DNA* **2**(4): 329–35.

Chomczynski, P. and Sacchi, N. (1987) Single-step method of RNA isolation by acid guanidinium thiocyanate-phenol-chloroform extraction. *Analytical Biochemistry* **162**(1): 156–9.

Denning, D. W., Anderson, M. J., Turner, G., Latge, J-P. and Bennett, J. W. (2002) Sequencing the *Aspergillus fumigatus* genome. *Lancet Infectious Diseases*, **2**(4): 251–253.

Heller, R. A., Schena, M., Chai, A., Shalon, D. *et al.* (1997) Discovery and analysis of inflammatory disease-related genes using cDNA microarrays. *Proceedings of the National Academy of Sciences of the United States of America* **94**(6): 2150–5.

Hope, W. W., Walsh, T. J. and Denning, D. W. (2005) Laboratory diagnosis of invasive aspergillosis, *Lancet Infectious Diseases* **5**(10): 609–22.

Hughes, T. R. and Mao, M., Jones, A. R., Burchard, J. *et al.* (2001) Expression profiling using microarrays fabricated by an ink-jet oligonucleotide synthesizer. *Nature Biotechnology* **19**(4): 342–7.

Lass-Florl, C., Griff, K., Mayr, A., Petzer, A. *et al.* (2005) Epidemiology and outcome of infections due to *Aspergillus terreus*: 10 year single centre experience. *British Journal of Haematology*, **131**(2): 201–7.

Latge, J-P. (2001) The pathobiology of *Aspergillus fumigatus*. *Trends in Microbiology*, **9**(8): 382–389.

Leinberger, D. M., Schumacher, U., Autenrieth, I. B. and Bachmann, T. T. (2005) Development of a DNA microarray for detection and identification of fungal pathogens involved in invasive mycoses. *Journal of Clinical Microbiology*, **43**(10): 4943–53.

Maeda, H., Sano, M., Maruyama, Y., Tanno, T. *et al.* (2004) Transcriptional analysis of genes for energy catabolism and hydrolytic enzymes in the filamentous fungus *Aspergillus oryzae* using cDNA microarrays and expressed sequence tags. *Applied Microbiology and Biotechnology*, **65**(1): 74–83.

Nierman, W. C., Feldblyum, T. V., Laub, M. T., Paulsen, I. T. *et al.* (2001) Complete genome sequence of *Caulobacter crescentus*. *Proceedings of the National Academy of Sciences of the United States of America*, **98**(7): 4136–4141

Nierman, W. C., May, G., Kim, H. S., Anderson, M. J. *et al.* (2005) What the *Aspergillus* genomes have told us. *Medical Mycology*, **43**(Suppl 1): S3–S5.

OBrian, G. R., Fakhoury, A. M. and Payne, G. A. (2003) Identification of genes differentially expressed during aflatoxin biosynthesis in *Aspergillus flavus* and *Aspergillus parasiticus*. *Fungal Genetics and Biology*, **39**(2): 118–27.

Rementeria, A., Lopez-Molina, N., Ludwig, A., Vivanco, A. B. *et al.* (2005) Genes and molecules involved in *Aspergillus fumigatus* virulence. *Revista Iberoamericana de Micologia*, **22**(1): 1–23.

Romani, L. (2004) Immunity to fungal infections. *Nature Reviews Immunology*, **4**(1): 1–23.

Schrettl, M., Bignell, E., Kragl, C., Joechl, C. *et al.* (2004) Siderophore biosynthesis but not reductive iron assimilation is essential for *Aspergillus fumigatus* virulence. *Journal of Experimental Medicine*, **200**(9): 1213–19.

Sims, A. H., Robson, G. D., Hoyle, D. C., Oliver, S. G. *et al.* (2004) Use of expressed sequence tag analysis and cDNA microarrays of the filamentous fungus *Aspergillus nidulans*. *Fungal Genetics and Biology*, **41**(2): 199–212.

Yu, H., Chao, J., Patek, D., Mujumdar, R. *et al.* (1994) Cyanine dye dUTP analogs for enzymatic labeling of DNA probes. *Nucleic Acids Research* **22**(15): 3226–32.

13

Techniques and strategies for studying virulence factors in *Cryptococcus neoformans*

Nancy Lee and **Guilhem Janbon**

13.1 Introduction

Cryptococcus neoformans is an encapsulated pathogenic yeast that primarily infects immunocompromised patients (Figure 13.1). In severe cases, cryptococcal infection progresses to meningoencephalitis and is fatal if left untreated (Casadevall and Perfect, 1998). It is thought that *C. neoformans* is acquired early in life and remains dormant in the host until the appearance of an immunological defect (Garcia-Hermoso *et al.*, 1999). *C. neoformans* is then able to multiply, disseminate and, perhaps most interestingly, invade the central nervous system.

C. neoformans exists as four serotypes (A, B, C and D). Serotypes A and D (corresponding to the *grubii* and *neoformans* varieties respectively) are responsible for the vast majority of cryptococcosis cases worldwide, and thus account for more than 90% of clinical isolates. Serotypes B and C (corresponding to the variety *gattii*) affect patients with no apparent immunological defects, and are more common in tropical and subtropical regions. However, the recent outbreak on Vancouver Island (Canada) has demonstrated that these serotypes are also a threat in more northern regions (Kidd *et al.*, 2004).

In the recent years, a strong community effort in acquiring knowledge regarding the biology of *C. neoformans* has made this yeast a model for basidiomycete fungi (Idnurm *et al.*, 2005). Accordingly, *C. neoformans* was the first basiomycete for which a complete genome sequence was made publicly available (Loftus *et al.*, 2005). In fact,

Medical Mycology: Cellular and Molecular Techniques. Edited by Kevin Kavanagh

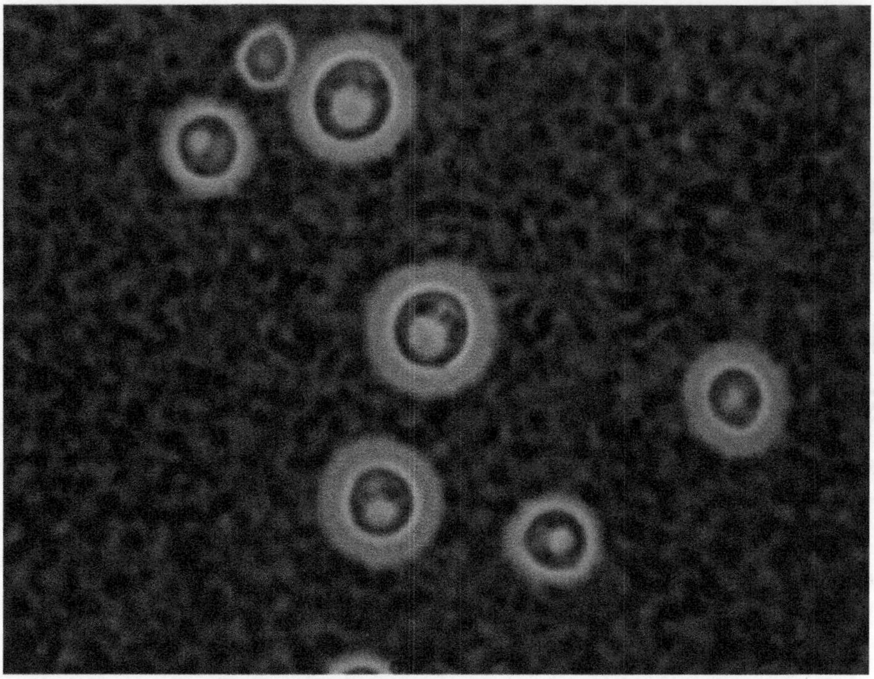

Figure 13.1 India ink negative staining of the capsule

four different genomes have already been sequenced (two serotype D strains, one serotype A strain and one serotype B strain), with another genome currently being sequenced (a serotype B strain from the Vancouver Island outbreak).

C. neoformans is an ideal micro-organism to study fungal virulence. Its sexual cycle and haploid genome combine to make this organism extremely genetically tractable, and to give the study of *C. neoformans* distinct advantages over two other notable human fungal pathogens, *Candida albicans* and *Aspergillus fumigatus*. The sexual cycle of *C. neoformans* described in the seventies (Kwon-Chung, 1975) is of interest both for basic research (Lengeler *et al.*, 2002) and for classical genetic analysis (Fox *et al.*, 2001; Janbon *et al.*, 2001; Marra *et al.*, 2004). Moreover, *Cryptococcus* haploid genome has been especially useful from a molecular genetics standpoint. Specifically, forward genetics has been widely used to understand the molecular mechanisms underlying *C. neoformans* pathogenesis. Numerous genes have been identified as important virulence factors using techniques such as functional complementation, multicopy suppression and insertional mutantagenesis (Moyrand *et al.*, 2004; Walton *et al.*, 2005). Reverse genetics has been also widely used both to confirm the involvement of a specific gene in *C. neoformans* virulence and to analyse the importance of a gene identified through bioinformatic analysis. The development of numerous tools and protocols to make mutants has been instrumental in advancing our knowledge of how fungi cause disease, and some of these protocols are detailed in this chapter.

The virulence factors involved in *C. neoformans* pathogenesis have been well defined (Buchanan and Murphy, 1998) and include the presence of a polysaccharide capsule, the production of melanin and the ability to grow at 37 °C (Chang and Kwon-Chung, 1994; Salas *et al.*, 1996; Odom *et al.*, 1997). In addition to these major virulence factors, different elements such as the production of phospholipase, urease and antioxidant enzymes have been clearly demonstrated to influence the pathophysiology of cryptococcosis (Liu *et al.*, 1999; Cox *et al.*, 2003; Olszewski *et al.*, 2004). These discoveries were made possible through elegant studies using cellular and animal models of cryptococcosis (Perfect *et al.*, 1980; Lortholary *et al.*, 1999; Goldman *et al.*, 2000; Steenbergen *et al.*, 2001; Tucker and Casadevall, 2002). Alternative models have also been recently described; these models are particularly useful in screening for mutant strains altered in virulence traits (Apidianakis *et al.*, 2004; Mylonakis *et al.*, 2004).

A number of molecular tools have been developed to study *C. neoformans* including genomic libraries, shuttle vectors, inducible promoters and GFP expression vectors (Hull and Heitman, 2002). The availability of fully annotated genome sequences for serotype D strains has added enormous value to the study *C. neoformans* biology, and, in the short period since their publication, the genome data have already proved to be fantastic tools (Loftus *et al.*, 2005). The recent construction of classical post-genomic tools, such as gene microarrays, serves as an attestation to the general utility of the genome sequencing projects (Fan *et al.*, 2005). In fact, the community of *Cryptococcus* researchers has developed considerably and can now be considered as one of the most dynamic communities in the field of fungal pathogenesis.

In this chapter, we describe strategies and methods to investigate gene function in *C. neoformans*. As an example we list relevant protocols to study a major virulence factor in *C. neoformans*, the capsule. The main constituent of the capsule is a large polymer of $(1\rightarrow3)$-α-mannosyl residues with xylose, glucuronic acid and O-acetyl branches. These polysaccharides, called glucuronoxylomannans (GXM), comprise 88% of the mass of the capsule. The two other constituents are a minor polysaccharide, galactoxylomannan (GalXM), and mannoproteins (MP) (Cherniak and Sundstrom, 1994).

Evidence of the relevance of the capsule in virulence lies in the facts that acapsular strains are avirulent and that *C. neoformans* serospecificity is based on capsular polysaccharide structure. Although there has not been a definitive demonstration of a causal relationship between capsule size and virulence, different environmental conditions have been shown to regulate the capsule size. For example, a change in capsule size has been observed *in vivo*, with differences corresponding to the infection of specific organs (Rivera *et al.*, 1998). Furthermore, the capsule structure was proven to vary *in vivo* (Charlier *et al.*, 2005) and influence the pathophysiology of cryptococcosis (Janbon *et al.*, 2001; Moyrand *et al.*, 2002). Thus, two phenotypes often used to characterize the *C. neoformans* capsule are the capsule size and the capsule structure.

The study of genes potentially involved in capsule biosynthesis or regulation starts with the identification of these putative capsule-related genes. There are many ways to select genes to characterize, including genetic screens and bioinformatics analysis. However, once selected, Your Favourite Gene (*YFG1*) must undergo scientific evaluation to determine its biological function and relevance to the question being studied, in this case whether *YFG1* has genetic influence on either capsule size or capsule structure. The way to definitively answer such a question relies on the process of gene deletion and the analysis of the resulting mutant's phenotypes. The following strategies and protocols serve as a guide to executing such a study in *C. neoformans*.

13.2 Construction of *C. neoformans* gene-disruption cassettes

Classical approaches to the construction of gene-disruption cassettes via restriction enzymes and subcloning can sometimes take months of preparation for the deletion of a single gene. A relatively new method using PCR overlap strategy has been widely adopted as one of the most efficient methods developed to construct deletion cassettes. This PCR fusion technique is similar in strategy to that used by Kuwayama and colleagues for the bacterium *Clostridium difficile* (Kuwayama *et al.*, 2002).

The disruption cassette consists of a marker gene (used to enable transformant selection) that is flanked by genomic sequences within the *YFG1* locus. The amount of flanking DNA required for an efficient transformation varies among organisms and even between different loci within the same organism. In general, the efficiency of homologous recombination is directly proportional to the length of the homologous sequence flanking the marker. In *C. neoformans*, under most circumstances, flanking sequences of 0.5–1 kb are sufficient to promote adequate rates of homologous recombination (1–50%) if biolistic technology is applied in combination (Toffaletti *et al.*, 1993).

Protocol 13.1
Construction of *C. neoformans* gene-disruption cassettes

Equipment, materials and reagents

Primers

DNA samples

0.2 ml PCR tubes

PCR machine

Method

A description of the primer sets used to construct the disruption cassette follows. The positions of the primers are indicated in Figure 13.2.

A. Primers YFG1-5′3 and YFG1-3′5 contain sequences recognized by the M13R and M13F primers, respectively. These primers also anneal to the gene of interest (Your Favourite Gene, or *YFG1*)

B. Similarly, the MKRYFG1f and MKRYFG1r primers are designed to anneal to M13R and M13F, as well as YFG1-5′3 and YFG1-3′5, respectively.

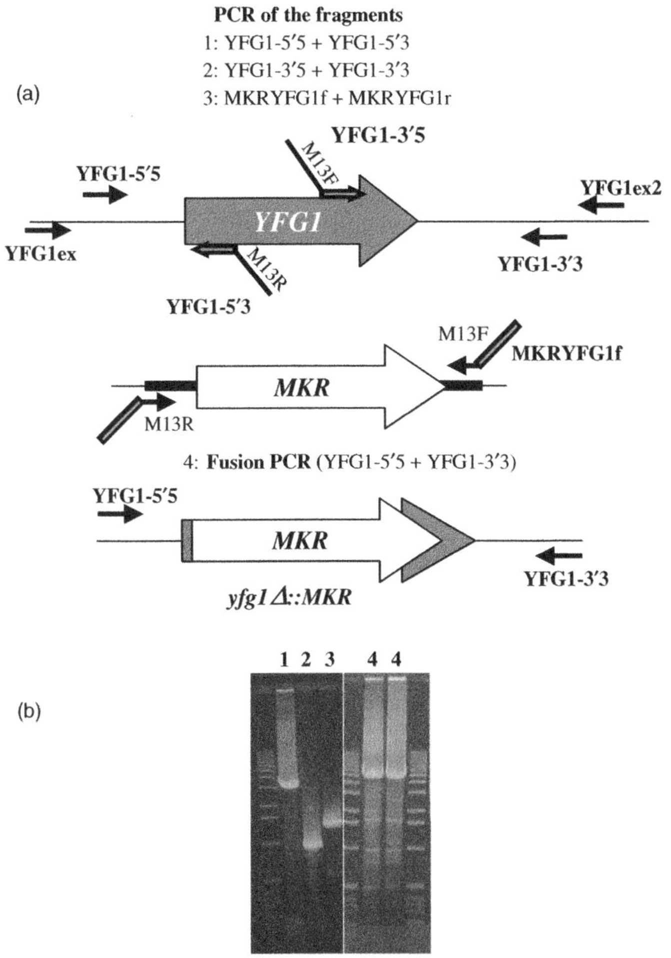

Figure 13.2 (a) Schematic representation of the strategy presented in this protocol to construct disruption cassettes. The positions of the different primers are indicated. (b) Example of the result that one can expect. 10 μl of each PCR reaction were loaded in each line

C. Consequently, YFG1-5′3 and YFG1-3′5 contain the reverse complements of MKRYFG1f and MKRYFG1r. The size, melting temperature (Tm), and percentage of GC content of all the primers must be very similar; ideally, the primers should be 21–23 bp in length, with a Tm of 60–63 °C and a GC content of >50%. The use of specialized software to aid in primer design is recommended.

D. Primers YFG1-5′5 and YFG1-3′3 are composed of sequences that are specific to *YFG1*. These primers anneal to the outermost extremities of the disruption construct.

E. Lastly, primers YFG1ex and YFG1ex2 lie outside of the disruption cassette and are used as controls to verify the sequences and orientations of the other primers involved in this experiment.

Note: While the primers YFG1ex and YFG1ex2 are used for screening of the transformants at the later stage (see Protocol 13.6), their compatibility should be checked in this experiment.

1. To start, a set of six PCRs must be performed:

   a. YFG1-5′5 + YFG1-5′3

   b. YFG1-3′5 + YFG1-3′3

   c. MKRYFG1f + MKRYFG1r

   d. YFG1-5′5 + YFG1-3′3

   e. YFG1ex + YFG1-5′3

   f. YFG1ex2 + YFG1-3′5

Note: To avoid the introduction of unwanted mutations in the disruption construct, a high-fidelity polymerase (e.g. HFPCR from Clontech Laboratories Inc., Palo Alto, CA, USA, K1914-1) should be used for PCR (a) and (b).

For all high-fidelity PCR experiments (PCR (a) and (b)), the reaction mix should contain:

32 μl of water

5 μl of *C. neoformans* genomic DNA (20 ng/μl)

5 μl of 10 x HF2 PCR buffer

5 μl of 10 x HF2 PCR dNTP mix

1 μl of primer1 (10 pmol/μl)

1 μl of primer2 (10 pmol/μl)

1 μl of 50 x Advantage HP2 polymerase mix (Clontech, cat. no. 639123)

Commonly used disruption markers include auxotrophic markers – for example *URA5* (Edman and Kwon-Chung, 1990) or *ADE2* (Sudarshan *et al.*, 1999) genes – and dominant markers (e.g. nourseothricin (McDade and Cox, 2001), hygromycin (Cox *et al.*, 1996) or geneticin (Hua *et al.*, 2000) resistance genes). The standard reaction mix contains:

38 μl of water

1 μl of linearized plasmid DNA containing the marker (100 ng/μl)

5 μl of 10 x Advantage 2 PCR buffer

1 μl of dNTP mix (10 mM)

1 μl of primer1 (10 pmol/μl)

1 μl of primer2 (10 pmol/μl)

1 μl of 50 x Advantage 2 polymerase mix (Clontech, cat. no. 8430-1)

Note: When the GC-rich nourseothricin resistance gene is used, 2 μl of DMSO should be added to the reaction solution.

The PCR reactions (d), (e) and (f) are done to check the compatibility of the primers. The standard reaction mix contains:

36 μl of water

5 μl of *C. neoformans* genomic DNA (20 ng/μl)

5 μl of 10 x Advantage 2 PCR buffer

1 μl of dNTP mix 10 mM

1 μl of primer1 (10 pmol/μl)

1 μl of primer2 (10 pmol/μl)

1 μl of 50 x Advantage 2 polymerase mix (Clontech, cat. no. 8430-1)

The parameters for PCR amplification should be:

1 cycle 94 °C 15 sec

35 cycles 94 °C 10 sec, 68 °C 4 min

1 cycle 68 °C 7 min

1 cycle 4 °C hold

Note: Check to ensure that the sizes of the PCR products amplified are correct and that the correct sequences have been amplified. If not, the primers must be redesigned before moving on to the next step.

2. Purify PCR products (a), (b) and (c) by agarose gel electrophoresis and re-suspend the DNA in water to a final concentration of 5 ng/μl. All three PCR amplification products are then used for the final PCR fusion reaction.

The standard PCR fusion reaction mix contains:

11 μl of water

10 μl of each PCR product (a), (b) and (c)

1 μl of YFG1-5′5 (10 pmol/μl)

1 μl of YFG1-3′3 (10 pmol/μl)

5 μl of 10 x Advantage 2 PCR buffer

1 μl of dNTP mix 10 mM

1 μl of 50 x Advantage 2 polymerase mix (Clontech, cat. no. 8430-1)

Note: The actual concentration of DNA and primers may differ depending on the specific gene sequences involved. Thus, the final concentrations of DNA and primers must be optimized with each fusion reaction to maximize the efficiency of the PCR amplification and the quality of PCR product.

The parameters for PCR amplification should be:

1 cycle 94 °C 15 sec

35 cycles 94 °C 10 sec, 68 °C 4 min,

1 cycle 68 °C 7 min

1 cycle 4 °C hold

3. Purify PCR fusion product by electrophoresis through an agarose gel. It may be necessary to combine PCR fusion products from several independent reactions to obtain enough DNA to transform *C. neoformans*.

13.3 Genetic transformation of *C. neoformans*

The introduction of nucleic acids by electroporation and through biolistic delivery are the two most commonly used methods of genetically modifying *C. neoformans*. A third method involving the use of *Agrobacterium* has been recently developed, and readers requiring further information on this protocol are directed to the publication in which it is described (Idnurn *et al.*, 2004). Historically, electroporation was the first method adapted for use in *C. neoformans*. Transformation by electroporation predominantly leads to episomal replication of the transforming DNA, with integration events occurring rarely (Edman and Kwon-Chung, 1990). Traditionally, this method is used to transform *C. neoformans* serotype D strains with genomic libraries in order to clone genes by functional complementation or to screen for multicopy supressors (Chang and Kwon-Chung, 1994; Fox *et al.*, 2003). Conversely, biolistic transformation results mostly in the genomic integration of the transforming DNA. This approach was first developed to transform serotype A strains of *C. neoformans* since this serotype is more recalcitrant than the D serotype (Toffaletti *et al.*, 1993). A protocol for a biolistic transformation of serotype D strains was later developed (Davidson *et al.*, 2000). Currently, transformation using biolistics is the method of choice for gene deletion.

Protocol 13.2
Biolistic transformation of *C. neoformans*

Equipment, materials and reagents

Cells to be transformed

DNA sample

Gold beads™ (Bio-Rad Laboratories, Hercules, CA, USA, 165-2262)

Microcarrier

Rupture disks 1100 psi or 1350 psi

Biolistic® Particle Delivery System (PDS) (Bio-Rad PDS 1000/He)

Method

Preparation of the gold beads

1. Wash 60 mg of 0.6 μm gold beads (Bio-Rad cat. no. 165-2262) with 1 ml of 100% ethanol by vortexing for 3 min.

2. Centrifuge the mixture at high speed in a microcentrifuge and aspirate the supernatant.

3. Wash the beads with 1 ml of sterile water.

4. Centrifuge the beads at high speed in a microcentrifuge and aspirate the supernatant.

5. Resuspend the beads in 1 ml of sterile water. The beads can then be stored at room temperature until use.

Preparation of the samples

6. For each DNA sample mix successively:

   a. 50 μl of gold beads

   b. 10 μl of DNA sample (~500 ng/μl)

   c. 50 μl of ice cold $CaCl_2$ (2.5 M)

   d. 10 μl ice cold spermidine (1 M; Sigma-Aldrich cat. no. S-7902).

7. Vortex the suspension for 10 min at room temperature and then let sit for 3 min.

8. Spin down the beads for 15 sec at high speed in a microcentrifuge and carefully aspirate the supernatant.

9. Wash the beads with 500 µl of 70% ethanol and vigorously agitate the mix (the beads should be completely resuspended).

10. Spin down the beads for 15 sec at high speed in a microcentrifuge and carefully aspirate the supernatant.

11. Wash the beads with 500 µl of 100% ethanol and vortex until the beads are completely resuspended.

12. Spin down the beads for 15 sec at high speed in a microcentrifuge and carefully aspirate the supernatant.

13. Resuspend the cells in 50 µl of 100% ethanol.

Preparation of the cells

14. Inoculate 50 ml of YPD with a loopful of cells. Allow the cells to grow overnight at 30 °C, with constant shaking at 150 rpm.

15. Spin down the cells at 3000 x *g* for 10 min at 4 °C.

16. Resuspend the cells in 5 ml of regeneration buffer.

17. Plate 200 µl of the cell suspension on agar plates and allow the plates to dry.

Note: Plate cells directly onto selective media if the selection is based on an auxotrophic marker. If selection is based on antibiotic resistance, plate the cells first on non-selective media. Allow cells to regenerate for 4 h after transformation on the non-selective plates before transferring the newly transformed cells to plates containing the antibiotic.

Biolistic transformation (using a Bio-Rad PDS 1000/He Biolistic® PDS)

18. Spot 15 µl of DNA onto a microcarrier and allow it to dry in a Petri dish containing desiccant.

19. Place a plate with cells to be transformed on the second level from the bottom of the transforming chamber of the Biolistic PDS apparatus (Bio-Rad).

20. Place the microcarrier on the top level of the chamber.

21. Transform the cells using a 1350 psi rupture disk (Bio-Rad cat. no. 165-2330).

22. Incubate the cells at 30 °C. The number of transformants one can expect from this method varies from strain to strain, and transformation conditions should be optimized for each strain, e.g. the use of a 1100 psi rupture disk (Bio-Rad cat. no. 165-2329) can sometimes increase transformation efficiency. Furthermore, optimal conditions of transformation are not identical for different machines even of the same model.

Regeneration medium

D-sorbitol (1M)

D-mannitol (1M)

YNB without amino acid (0.9% w/v)

Glucose (2.6% w/v)

Bactopeptone (0.054% w/v)

Gelatin (0.133% w/v)

Autoclave and store at 4 °C

Protocol 13.3
Transformation via electroporation

Equipment, materials and reagents

Cells to be transformed

DNA sample

Electroporation cuvette

Electroporation device (Gene Pulser® II, Bio-Rad Laboratories, Hercules, CA, USA)

Method

1. Inoculate 10 ml of YPD with a loopful of cells. Allow the cells to grow overnight at 30 °C, with constant shaking at 150 rpm.
2. Add 90 ml of YPD and grow the cells up to a concentration of 4×10^7 cells/ml.
3. Spin down half of the volume of cell culture at 3000 x *g* for 5 min.

4. Wash the cells with 50 ml of ice-cold sterile water.
5. Resuspend the cells in 25 ml of EB buffer.
6. Add 100 μl of 1M DTT and incubate on ice for 15 min.
7. Centrifuge the cells at 7800 x *g* for 10 min.
8. Eliminate the supernatant by inverting the tube and resuspend the cells in the residual supernatent (typically 250–500 μl).
9. Incubate the cells (use 50 μl of cell suspension for each transformation) and the DNA (~100 ng) on ice for 1 min.
10. Transfer the DNA/cells suspension mix to an ice-cold electroporation cuve.
11. Shock the cells using an electroporation device (parameters: $\Omega = \infty$ Cap = 25 μF, V = 470V; Bio-Rad Gene Pulser II). The resulting τ value should be typically about 40 ms).
12. Allow the cells to recover by adding 500 ml of EB buffer.
13. Plate the cells on selective media plate (A telomeric plasmid will typically give 10^6 transformants/μg of DNA).

YPD

Yeast extracts 10 gram/litre

Bactopeptone 20 gram/litre

Glucose 20 gram/litre

EB buffer

10 mM Tris-HCl, pH 7.5

DTT 1 mM

Sucrose 270 mM

13.4 Extraction of genomic DNA from *C. neoformans*

Not only do the *C. neoformans* polysaccharide capsule and cell wall serve as interesting virulence factors, but the resilience of these components acts as an additional barrier to cell damage. Thus, DNA-extraction protocols for *C. neoformans* differ from standard nucleic acid analysis methods in that they inherently include additional steps to compensate for capsule and cell-wall disruption and removal. Two methods to extract DNA from *C. neoformans* that differ in the extent to which DNA is purified from cell contaminants are described below. The first protocol is a modification of the method previously published by Varma and Kwon-Chung (1991). This method was developed to isolate high-quality DNA for the purposes of genomic library construction and hybridization blot analysis. The final step in the process (purification from residual polyphosphates) is necessary for the elimination of most of the residual capsular polysaccharides that co-precipitate with DNA in ethanol. These polyphosphates are potent inhibitors of modification and restriction enzymes (Rodriguez, 1993). The second protocol was adapted from an *Saccharomyces cerevisiae* protocol (Holm *et al.*, 1986) and outlines a relatively quick and simple way to obtain very crude DNA used for experiments such as PCR analysis.

Protocol 13.4
DNA for use in library construction and hybridization analysis

Equipment, materials and reagents

Lysing enzyme (Sigma-Aldrich cat. no. L1412)

Optical microscope

Centrifuges

Method

Preparation of protoplasts

1. Inoculate 50 ml of YPD with a loopful of cells. Allow the cells to grow overnight at 30 °C, with constant shaking at 150 rpm.

2. Centrifuge the culture for 5 min at 7800 x *g* in Sorvall rotor SS34 or equivalent at 4 °C and aspirate the supernatant.

3. Wash the cells with 20 ml of SCS buffer and repeat the centrifugation step outlined above.

4. Resuspend the cells in 10 ml of SCS buffer containing 20 mg/ml lysing enzyme (Sigma-Aldrich cat. no. L1412) and incubate at 37 °C until most of the cells are spheroplasted (typically 10–30 min). Aliquot 10 μl of the cell suspension onto a glass slide to monitor spheroplast formation using a microscope.

5. Spin down the protoplasts at 3000 x *g* in a Sorvall rotor SS34 or equivalent at 4 °C. Remove the supernatant by pipetting.

Protoplast lysis

6. Re-suspend the spheroplasts in 9 ml of lysing buffer.

7. Add 1 ml of 10% laurylsarcosine and incubate the suspension for 15 min at 65 °C to lyse the protoplasts.

8. Allow the suspension to cool down to room temperature (RT) add 1 ml of 5 M potassium acetate and mix. Incubate on ice for a minimum of 15 min.

9. Centrifuge the cell lysate for 20 min at 10 000 x *g* (using a Sorvall rotor SS34 or equivalent) at 4 °C. Save the supernatant and transfer the DNA-containing supernatant to a new centrifugation tube.

10. Add 1 volume of isopropanol to the supernatant, mix gently and incubate for at least 10 min at 20 °C.

11. Centrifuge for 5 min at 10 000 x *g* (using a Sorvall rotor SS34 or equivalent) at 4 °C and aspirate the supernatant.

12. Let the pellet dry at RT for 10 min.

13. Add 1 ml of TE to the DNA pellet at the bottom of the centrifugation tube and transfer the DNA suspension in a new 2 ml Eppendorf tube.

Purification of the DNA

14. Add 300 μg of RNAse A and incubate at 37 °C for 30 min.

15. Add 400 μg of Proteinase K and incubate at 37 °C for 30 min.

16. Add 1 ml of phenol:chloroform:isoamyl alcohol (24:24:1) and mix well.

17. Centrifuge for 5 min at high speed in a microcentrifuge and carefully pipette out the top, aqueous phase. Be sure to avoid the white pad of cell debris suspended in the middle of the mixture. Pipette the aqueous phase into a new Eppendorf tube.

18. Add 1 ml of chloroform:isoamyl alcohol (24:1) and mix well.

19. Centrifuge for 5 min at high speed in a microcentrifuge and carefully pipette out the top, aqueous phase.

20. Aliquot the solution into two tubes, add 1 ml of 100% ethanol at each tube and mix.

21. Collect the DNA precipitate using a Pasteur pipette and wash the DNA in 1 ml of 70% ethanol.

22. Again, collect the DNA precipitate using a Pasteur pipette. Allow the DNA to dry at room temperature for 10 min on the Pasteur pipette.

23. Dissolve the DNA in 100–500 µl TE buffer.

Purification from residual polyphosphates

24. Add 1 volume of 5 M ammonium acetate, mix and set the tube on ice for 30 min.

25. Centrifuge the tube at 4 °C for 10 min at 11 000 x *g*.

26. Carefully collect the supernatant and transfer the DNA suspension into a new tube.

27. Precipitate the DNA by adding 2 volumes of 100% ethanol.

28. Collect the DNA precipitate using a Pasteur pipette and wash the DNA in 1 ml of 70% ethanol.

29. Again, collect the DNA precipitate using a Pasteur pipette. Allow the DNA to dry at RT for 10 min on the Pasteur pipette.

30. Dissolve the DNA in 50–200 µl TE buffer.

Buffer SCS

Sorbitol 1 M

Sodium citrate 20 mM, pH 5.8

Buffer TE

Tris-HCl 10 mM, pH 8

EDTA 1 mM, pH 8

Lysing buffer

EDTA 0.1 M, pH 8

Tris-HCl 10 mM, pH 8

Proteinase K 2 mg/ml

Protocol 13.5
DNA for use in PCR

Equipment, materials and reagents

Glass beads (Sigma-Aldrich cat. no. G8772)

Cell lysis apparatus (Fastprep FP120, Bio-101 Incorporated, Vista, CA, USA)

Bench centrifuge

Method

1. Pellet the cells of a 2 ml overnight culture in an Eppendorf tube.
2. Resuspend the pellet in 500 μl of lysing solution and transfer the cell suspension into a cryotube containing 500 mg of glass beads (Sigma-Aldrich cat. no. G8772).
3. Add 500 μl of phenol:chloroform:isoamyl alcohol (24:24:1).
4. Break the cells using a cell lysis apparatus (e.g. Bio-101 Fastprep FP120) for 1 min at maximum speed. Repeat this step to treat the cells for a total of 2 min in the machine.
5. Add 500 μl of TE and mix.
6. Centrifuge the sample for 5 min at high speed in a microcentrifuge.
7. Carefully recover 700 μl of the top, aqueous phase, transfer the suspension into a new tube and add 2 volumes of 100% ethanol. Mix well.
8. Centrifuge for 5 min at high speed in a microcentrifuge.
9. Aspirate the supernatant and allow the DNA pellet to dry.

10. Resuspend the DNA in 400 μl of TE.

11. Add 30 μg of RNAse A and incubate for 10 min at 37 °C.

12. Precipitate the DNA by adding 1 ml of 100% ethanol. Mix the tube well.

13. Centrifuge the sample for 5 min at high speed in a microcentrifuge.

14. Wash the pellet with 70% ethanol and let the DNA dry at RT.

15. Resuspend the DNA in 50 μl of TE.

Lysing solution:

Triton X-100 2% (w/v)

SDS 1% (w/v)

NaCl 100 mM

Tris 10 mM, pH 8

EDTA 1 mM, pH 8

13.5 Screening and identification of deletion strains

Before describing any strain as a mutant deleted for *YFG1* (Your Favourite Gene), transformants must be confirmed as containing homologously integrated disruption cassettes within their genome. This process involves screening the transformant pool using three rounds of PCR and definitively confirming strain genotype using DNA hybridization analysis. With recombination frequencies of 1–50%, the optimization of transformation conditions is key to obtaining enough transformants to screen. To maximize the efficiency of transformation, two important aspects of the experimental design should be considered: the size of each fragment of homologous DNA flanking the cassette should be of at least 0.5 kb and the transformation should be done using biolistic delivery of the disruption cassette (see Protocols 13.1 and 13.2).

Protocol 13.6
Screening and identification of deletion strains

Equipment, materials and reagents

Cells to be analysed

0.2 ml PCR tubes

PCR machine

Southern blot classical equipment

Method

Transformant preparation

1. After transformation, incubate the transformation plates for 3–5 days at a permissive temperature.
2. Pick 20 transformants and streak them onto selective media to obtain isolated colonies. Note that when a dominant selection system is used not all the transformants will re-grow on the selective medium; so a larger number of transformants should be selected from the original transformation plate.
3. Pick some cells from an isolated colony from each plate (originating from a single transformant) and inoculate the colony into 100 μl of YPD in a 96-well plate. Allow the cells to grow for 24 h at 30 °C and then add 1 volume of glycerol in each well. The 96-well plate can now be stored at −80 °C indefinitely.
4. With the remaining cells from the same isolated colony, streak out a SD plate and incubate the plate at 30 °C for 2 days.

Colony PCR to screen large numbers of transformants

The transformants are initially screened with PCR to quickly identify candidate deletion mutants that have integrated the disruption cassette at the correct genomic locus. Two successive rounds of PCR reactions are made using a primer set that includes one primer that localizes within the selection marker (MKRr or MKRf) and a second primer that anneals to sequences outside of the disruption cassette (YFG1ex or YFG1ex2) (Figure 13.3).

5. Use a toothpick or a stab to pick a small mass of cells from a single colony and re-suspend those cells in a single PCR tube containing all of the components necessary for the reaction. Vortex the PCR tube.

The standard PCR mix for this stage contains:

41 μl of water

5 μl of rTaq buffer 10 X

1 μl of dNTP 10 mM

1 μl of primer MKRr 10 pmol/μl

1 μl of primer YFG1ex 10 pmol/μl

1 μl of rTaq polymerase 5 U/μl (Amersham Biosciences, cat. no. 27-0798-05)

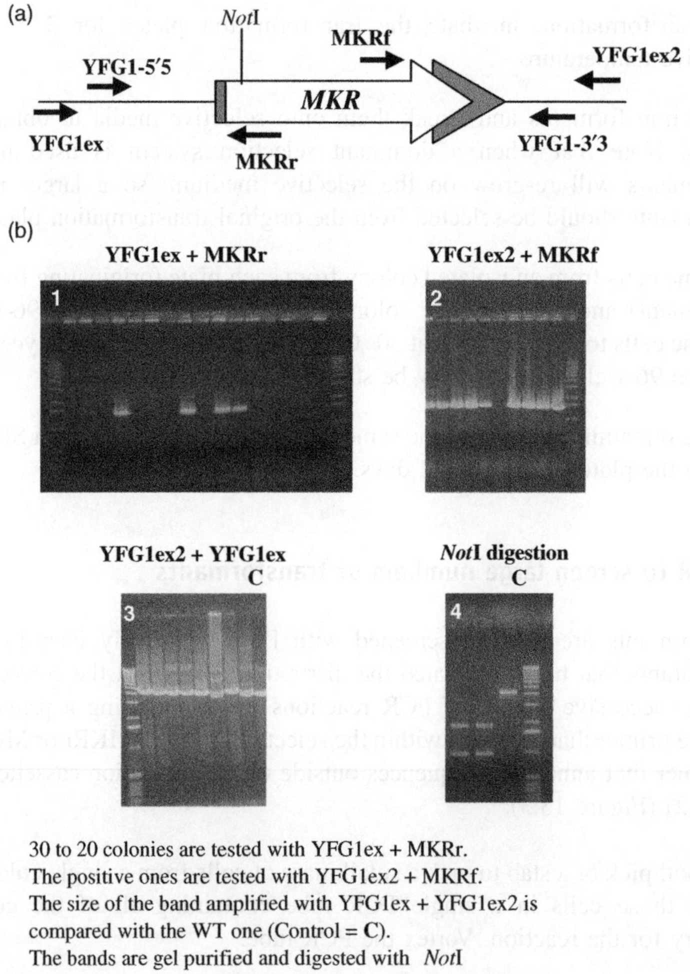
(a)
NotI
MKRf
YFG1-5′5
YFG1ex2
MKR
YFG1ex
YFG1-3′3
MKRr
(b)
YFG1ex + MKRr
YFG1ex2 + MKRf
1
2
YFG1ex2 + YFG1ex
C
NotI digestion
C
3
4
30 to 20 colonies are tested with YFG1ex + MKRr.
The positive ones are tested with YFG1ex2 + MKRf.
The size of the band amplified with YFG1ex + YFG1ex2 is
compared with the WT one (Control = C).
The bands are gel purified and digested with NotI

Figure 13.3 (a) Schematic representation of the strategy presented in this protocol to screen transformants. The positions of the different primers are indicated. (b) Example of the result that one can expect. 10 μl of each PCR reaction were loaded in each line

The parameters for the PCR program are:

1 cycle 95 °C, 5 min

35 cycles 95 °C, 15 sec; 68 °C 3 min

1 cycle 68 °C, 7 min

1 cycle 4 °C, hold

6. Analyse the results by running 30 µl of each PCR reaction in an agarose gel. Transformants with positive first-round PCR results can then be analysed by a second round of PCR screening using primers MKRf and YFG1ex2 (see Figure 13.3). These strains should also be stored at −80 °C for further analysis. Be sure to include the wild-type strain as a control for every set of PCR experiments performed.

Confirmation of the absence of *YFG1* in the putative deletion strains

The purpose of the first two rounds of PCR screens is to ensure that the disruption cassette has integrated into the correct spot in the genome. However, these screens do not demonstrate that *YFG1* is absent. Thus, an essential third round of screening by PCR focuses on determining the presence or the absence of the wild-type *YFG1* gene. Be sure to include the wild-type strain as a control for every set of PCR experiments performed.

7. Use glass beads (see Protocol 13.5) to extract genomic DNA from each transformant that successfully passed the first two PCR screens.

8. Amplify the entire genomic locus of interest using the YFG1ex and YFG1ex2 primers.

The standard PCR mix for this stage contains:

39 µl of water

2 µl of genomic DNA

5 µl of rTaq buffer 10 X

1 µl of dNTP 10 mM

1 µl of primer YFG1ex 10 pmol/µl

1 μl of primer YFG1ex2 10 pmol/μl

1 μl of rTaq polymerase 5 U/μl (Amersham Biosciences, cat. no. 27-0798-05)

The parameters for the PCR program are:

1 cycle 95 °C, 15 sec

35 cycles 95 °C, 15 sec; 68 °C 5 min

1 cycle 68 °C, 7 min

1 cycle 4 °C, hold

Note: The elongation time should be adjusted so that it is proportional to the size of the gene to be amplified. As a general guideline, 1 min for each kb of DNA works well.

9. Use agarose gel electrophoresis to visualize the results of the PCR screen. If the size of the cassette is too similar to the size of the wild-type gene, use restriction enzyme analysis to distinguish between the two PCR products.

   a. To check the PCR products by restriction enzyme analysis, isolate the PCR fragments by running the products of the reaction in a low melting point agarose gel and cutting out the bands.

   b. The fragments can then be purified from the agarose and digested using a restriction enzyme that results in restriction fragments that differentiates the two PCR products. For example, use an enzyme that cuts only the cassette and not the wild type such as *Not*1. The recognition sequence for this enzyme is frequently present in the marker cassette but relatively rare in genomic DNA.

   c. Separate the restriction fragments using agarose gel electrophoresis analysis to differentiate transformants that still contain an intact wild-type *YFG1* gene from those that do not.

Screening for ectopic integration of the disruption cassette by DNA hybridization

While screening transformants by PCR as described above demonstrates correct integration of the disruption cassette at the *YFG1* locus, these experiments do not reveal whether additional integration events occurred outside of the *YFG1* locus. Thus, to verify that additional ectopic integration at random loci within the genome did not occur in tandem with the homologous integration of the disruption construct, an examination of total genomic DNA by hybridization is very commonly used:

10. Extract total genomic DNA from the transformants using the adapted protocol (Protocol 13.4).

11. Digest the DNA samples using a restriction enzyme that cuts only once within the marker used to delete the gene.

12. Electrophoretically separate the DNA fragments and transfer them onto a nylon membrane using classical DNA blotting procedures.

13. Hybridize the fixed DNA with an appropriate probe.

14. If no additional ectopic integration has occurred, the autoradiograph should show only two bands that correspond to fragments of expected sizes given the specific genomic sequences around the *YFG1* locus. The corresponding strains can be then used to study the phenotypes associated with this gene deletion.

Note: The transformation process itself can be mutagenic. Thus, additional mutations can be introduced by the transformation process and it is possible that these mutations are responsible for observed phenotypes. Two strategies used to establish a link between an observed phenotype and a specific gene deletion are: (1) Reconstitution of the deletion strain with the wild-type gene (see Odom *et al.*, 1997) and (2) Back-crossing and analysis of progeny for co-segregation of the phenotype and the presence of disruption cassette (see Janbon *et al.*, 2001).

13.6 Measuring capsule size in *C. neoformans*

The importance of the *C. neoformans* polysaccharide capsule in virulence has prompted the need to quickly and easily assess this trait. The most obvious characteristic of the capsule is its size. Thus, measuring capsule size has been quickly adapted as a standard way to quantify and describe this key virulence factor. Although capsule size has not been directly linked to strain virulence, the expression of capsule-related genes is highly regulated and clearly inducible under a number of environmental conditions. For example, under conditions of low iron availability or high partial pressure of CO_2, wild-type *C. neoformans* strains are induced to produce larger capsules (Granger *et al.*, 1985; Vartivarian *et al.*, 1993). The capsule is also regulated to grow to varying sizes depending on the infected organ during cryptococcosis (Rivera *et al.*, 1998). The most common way to visualize the capsule and assess its size is to negatively stain *C. neoformans* using India ink. The ink particles are too large to penetrate the capsule structure, which results in the appearance of a white halo around each cell when observed under an optical microscope (Figure 13.1). This technique is also the most common approach to definitively diagnose cryptococcosis in patients.

Protocol 13.7
Measuring capsule size in *C. neoformans*

Equipment, materials and reagents

Cells to be analysed

Coverslips (22 × 22 mm)

Optical microscope equipped with a micrometer

Method

1. Grow each strain in 10 ml of YPD at 30 °C overnight.
2. Take 100 µl of the overnight culture, inoculate 10 ml of capsule-inducing medium and incubate the cells at 30 °C overnight.
3. Centrifuge 2 ml of overnight culture for 1 min at 12 000 x *g* in a bench centrifuge.
4. Wash the cells once with 2 ml of PBS and resuspend the cells in the same volume of PBS.
5. Mix 40 µl of the cell suspension with 20 µl of India ink.
6. Using a micrometer, visualize the cells under the highest magnification of an optical microscope and measure the diameter of 20 cells. For each cell, take two measurements: one that accounts for both the volume of the cell and the surrounding capsule, and a second that accounts for only the distance between cell walls and excludes the capsule size altogether.
7. Subtract one measurement from the other to calculate the size of the capsule:

   Capsule size = 1/2 (Mean of total cell sizes – Mean of cell size excluding capsule)

   (Figure 13.4)

13.7 Purification of glucuronoxylomannan (GXM)

Glucuronoxylomannan (GXM) is an essential component of the *C. neoformans* polysaccharide capsule. The structure of GXM plays an important role in the

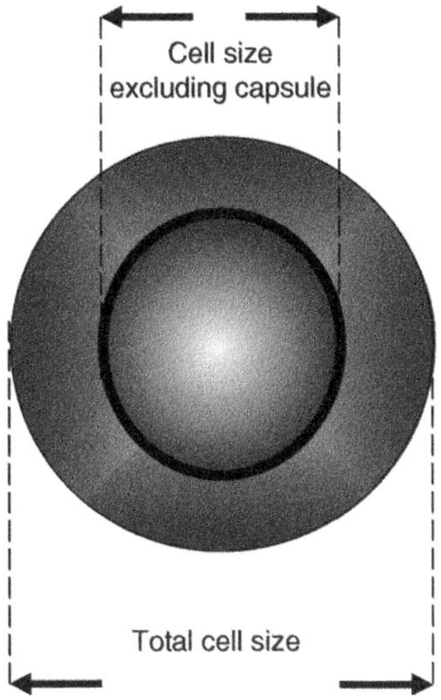

Figure 13.4 Schematic representation of a *C. neoformans* cell. The different measurements to estimate the capsule size are indicated

physiopathology of cryptococcosis. GXM structure varies amongst different serotypes, strains and even for the same strain depending upon the time of the analysis (Janbon, 2004). Characterization of GXM structure has been performed using non-commercially available anti-GXM antibodies and dot blot analysis (Cherniak *et al.*, 1995; Garcia-Hermoso *et al.*, 2004). However, these studies are limited in that they only reveal information regarding epitope recognition; it is impossible to glean information on possible GXM chemical structure modification using this approach. To obtain a more detailed understanding of the chemical composition of GXM, purification of GXM from the other components of the capsule is essential. Isolation of GXM is usually followed by nuclear magnetic resonance (NMR) analysis (Cleare *et al.*, 1999; Moyrand *et al.*, 2004). The protocol below is a modification of the protocol originally described by Cherniak *et al.* (1991).

Protocol 13.8
Purification of glucuronoxylomannan (GXM)

Equipment, materials and reagents

Cells to be analysed

Centrifuges

Hexadecyltrimethylammonium bromide (CTAB, Sigma-Aldrich cat. no. H-5882)

Dialysing MWCO3500 membrane (Spectrum Laboratories, Fort Lauderdale, FL, USA, cat. no. 132 725)

Lyophylizer

Method

1. Inoculate 1 litre of capsule induction medium (Janbon *et al.*, 2001) with a loopful of cells grown for 2 days at 30 °C on a YPD plate. Allow the cells to grow for 3 days at 30 °C, with constant shaking at 200 rpm.

2. Centrifuge the culture for 10 min at 10 000 rpm in Sorvall rotor GSA or equivalent at room temperature (RT) and recover the supernatant.

3. Repeat centrifugation at RT for 10 min, but at 10 000 x *g* this time, and again recover the supernatant.

4. Filter the solution using a 0.22 μm filter to eliminate the remaining cells. Save the supernatant.

5. Add three volumes of ethanol and store the supernatant at 4 °C overnight to promote precipitation of the polysaccharide (PS) within the supernatant.

6. Harvest the PS by centrifugation for 15 min at 12 000 x *g*. (**Note:** Weigh the centrifugation tube before centrifuging the sample to enable a calculation of the amount of PS precipitate within the supernatant.)

7. Leave the pellet to dry at RT and weigh the tubes again.

8. Dissolve the PS in 100 ml of 0.2 M NaCl and store the solution at 4 °C overnight. (**Note:** Use a magnetic stir bar to aid in dissolution.)

9. In a beaker, add 3 mg of hexadecyltrimethylammonium bromide (CTAB, Sigma-Aldrich cat. no. H-5882) for every 1 mg of PS and stir the solution until the CTAB is completely dissolved. Addition of CTAB purifies GXM from the polysaccharide mix.

10. Slowly add 1M NaCl to the solution until a precipitate appears.

11. To verify that the GXM is completely precipitated, set aside a 1 ml aliquot of the solution. Centrifuge the aliquot for 5 min at 12 000 x *g*, add 500 μl of CTAB to the supernatant and watch for a precipitate. If no precipitate appears, continue to the next step. If a precipitate does appear, add more CTAB to the solution.

12. Harvest the GXM precipitate at RT for 15 min at 10 000 x *g*.

13. Aspirate the supernatant and allow the pellet to dry at RT.

14. Dissolve the GXM in 100 ml of 1 M NaCl and store the solution at 4 °C overnight. (**Note:** Use a magnetic stir bar to aid in dissolution.)

15. Add 3 volumes (300 ml) of ethanol to the GXM solution and store at 4 °C overnight to permit GSM precipitation.

16. Harvest the GXM precipitate at RT for 15 min at 10 000 x *g*.

17. Aspirate the supernatant and allow the pellet to dry at RT.

18. Dissolve the GXM in 100 ml of 1 M NaCl and store the solution at 4 °C overnight. (**Note:** Use a magnetic stir bar to aid in dissolution.)

19. Dialyse (MWCO3500 membrane) the GXM solution for 2 days using tap water and for 3 days using distilled water. Be sure to change the water every day.

20. Filter the solution using a 0.22 μm filter.

21. Lyophilize the GXM.

Capsule induction medium (for 1 litre)

1.7 gram yeast nitrogen base (YNB) without amino acids and ammonium sulphate (BD Difco^{TM} 233520, BD, Franklin Lakes, NJ, USA)

1.5 g asparigine

20 g glucose

12 mM $NaHCO_3$

35 mM MOPS, pH 7.1

Filter sterilize the media before use

13.8 References

Apidianakis, Y., Rahme, L. G., Heitman, J., Ausubel, F. M. *et al.* (2004) Challenge of *Drosophila melanogaster* with *Cryptococcus neoformans* and the role of the innate immune responses. *Eukaryot Cell* **3**: 413–419.

Buchanan, K. L. and Murphy, J. W. (1998) What makes *Cryptococcus neoformans* a pathogen? *Emerg. Infect. Dis.* **4**: 71–83.

Casadevall, A. and Perfect, J. R. (1998) *Cryptococcus neoformans.* Washington: American Society for Microbiology Press.

Chang, Y. C. and Kwon-Chung, K. J. (1994) Complementation of a capsule-deficient mutation of *Cryptococcus neoformans* restores its virulence. *Mol. Cell. Biol.* **14**: 4912–4919.

Charlier, C., Chretien, F., Baudrimont, M., Mordelet, E. *et al.* (2005) Capsule structure changes associated with *Cryptococcus neoformans* crossing of the blood-brain barrier. *Am. J. Pathol.* **166**: 421–432.

Cherniak, R., Morris, L. C., Anderson, B. C. and Meyer, S. A. (1991) Facilitated isolation, purification and analysis of glucuronoxylomannan of *Cryptococcus neoformans. Infect. Immun.* **59**: 59–64.

Cherniak, R. and Sundstrom, J. B. (1994) Polysaccharide antigens of the capsule of *Cryptococcus neoformans. Infect. Immun.* **62**: 1507–1512.

Cherniak, R., Morris, L. C., Belay, T., Spitzer, E. D. and Casadevall, A. (1995) Variation in the structure of glucuronoxylomannan in isolates from patients with recurrent cryptococcal meningitis. *Infect. Immun.* **63**: 1899–1905.

Cleare, W., Cherniak, R. and Casadevall, A. (1999) *In vitro* and *in vivo* stability of *Cryptococcus neoformans* glucuronoxylomannan epitope that elicits protective antibodies. *Infect. Immun.* **67**: 3096–3107.

Cox, G. M., Toffaletti, D. L. and Perfect, J. R. (1996) Dominant selection system for use in *Cryptococcus neoformans. J. Med. Vet. Mycol.* **34**: 385–391.

Cox, G. M., Harrison, T. S., McDade, H. C., Taborda, C. P. *et al.* (2003) Superoxide dismutase influences the virulence of *Cryptococcus neoformans* by affecting growth within macrophages. *Infect. Immun.* **71**: 173–180.

Davidson, R. C., Cruz, M. C., Sia, R. A., Allen, B. *et al.* (2000) Gene disruption by biolistic transformation in serotype D strains of *Cryptococcus neoformans. Fungal Genet. Biol.* **29**: 38–48.

Edman, J. C. and Kwon-Chung, K. J. (1990) Isolation of the *URA5* gene from *Cryptococcus neoformans* var. *neoformans* and its use as a selective marker for transformation. *Mol. Cell. Biol.* **10**: 4538–4544.

Fan, W., Kraus, P. R., Boily, M. J. and Heitman, J. (2005) *Cryptococcus neoformans* gene expression during murine macrophage infection. *Eukaryot. Cell* **4**: 1420–1433.

Fox, D. S., Cruz, M. C., Sia, R. A. L., Ke, H. *et al.* (2001) Calcineurin regulatory subunit is essential for virulence and mediates interactions with FKBP12-FK506 in *Cryptococcus neoformans. Mol. Microbiol.* **39**: 845–849.

Fox, D. S., Cox, G. M. and Heitman, J. (2003) Phospholipid-binding protein Cts1 controls septation and functions coordinately with calcineurin in *Cryptococcus neoformans. Eukaryot. Cell* **2**: 1025–1035.

Garcia-Hermoso, D., Janbon, G. and Dromer, F. (1999) Epidemiological evidence for dormant *Cryptococcus neoformans* infection. *J. Clin. Microbiol.* **37**: 3204–3209.

Garcia-Hermoso, D., Dromer, F. and Janbon, G. (2004) *Cryptococcus neoformans* capsule structure evolution *in vitro* and during murine infection. *Infect. Immun.* **72**: 3359–3365.

Goldman, D. L., Lee, S. C., Mednick, A. J., Montella, L. and Casadevall, A. (2000) Persistent of *Cryptococcus neoformans* pulmonary infection in the rat is associated with intracellular parasitism, decreased inductible nitric oxide synthase expresion, and altered antibody responsiveness to cryptococcal polysaccharide. *Infect. Immun.* **68**: 832–838.

Granger, D. L., Perfect, J. R. and Durack, D. T. (1985) Virulence of *Cryptococcus neoformans*. Regulation of capsule synthesis by carbon dioxide. *J. Clin. Invest.* **76**: 508–516.

Holm, C., Meeks-Wagner, D. W., Fangman, W. L. and Botstein, D. (1986) A rapid, efficient method for isolating DNA from yeast. *Gene.* **42**: 169–173.

Hua, J. H., Meyer, J. D. and Lodge, J. K. (2000) Development of positive markers for the fungal pathogen, *Cryptococcus neoformans*. *Clin. Diagn. Lab. Immunol.* **7**: 125–128.

Hull, C. M. and Heitman, J. (2002) Genetics of *Cryptococcus neoformans*. *Annu. Rev. Genet.* **36**: 557–615.

Idnurm, A., Bahn, Y. S., Nielsen, K., Lin, X. *et al.* (2005) Deciphering the model pathogenic fungus *Cryptococcus neoformans*. *Nat. Rev. Microbiol.* **3**: 753–764.

Idnurn, A., Reedy, J. L., Nussbaum, J. C. and Heitman, J. (2004) *Cryptococcus neoformans* virulence gene discovery through insertional mutagenesis. *Eukaryot. Cell.* **3**: 420–429.

Janbon, G., Himmelreich, U., Moyrand, F., Improvisi, L. and Dromer, F. (2001) Cas1p is a membrane protein necessary for the O-acetylation of the *Cryptococcus neoformans* capsular polysaccharide. *Mol. Microbiol.* **42**: 453–469.

Janbon, G. (2004) *Cryptococcus neoformans* capsule biosynthesis and regulation. *FEMS Yeast Res.* **4/8**: 765–771.

Kidd, S. E., Hagen, F., Tscharke, R. L., Huynh, M. *et al.* (2004) A rare genotype of *Cryptococcus gattii* caused the cryptococcosis outbreak on Vancouver Island (British Columbia, Canada). *Proc. Natl. Acad. Sci. USA.* **101**: 17258–17263.

Kuwayama, H., Obara, S., Morio, T., Katoh, M. *et al.* (2002) PCR-mediated generation of a gene disruption construct without the use of DNA ligase and plasmid vectors. *Nucleic Acids Res.* **30**: e2.

Kwon-Chung, K. J. (1975) A new genus, *Filobasidiella*, the perfect state of *Cryptococcus neoformans*. *Mycopathologia.* **67**: 1197–1200.

Lengeler, K. B., Fox, D. S., Fraser, J. A., Allen, A. *et al.* (2002) Mating-type locus of Cryptococcus neoformans: A step in the evolution of sex chromosomes. *Eukaryot. Cell.* **1**: 704–718.

Liu, L., Tewari, R. P. and Williamson, P. R. (1999) Laccase protects *Cryptococcus neoformans* from antifungal activity of alveolar macrophages. *Infect. Immun.* **67**: 6034–6039.

Loftus, B., Fung, E., Roncaglia, P., Rowley, D. *et al.* (2005) The genome and transcriptome of *Cryptococcus neoformans*, a basidiomycetous fungal pathogen of humans. *Science* **307**: 1321–1324.

Lortholary, O., Improvisi, L., Rayhane, N., Gray, F. *et al.* (1999) Cytokine profiles of AIDS patients are similar to those of mice with disseminated *Cryptococcus neoformans* infection. *Infect. Immun.* **67**: 6314–6320.

Marra, R. E., Huang, J. C., Fung, E., Nielsen, K. *et al.* (2004) A genetic linkage map of *Cryptococcus neoformans* variety *neoformans* serotype D (*Filobasidiella neoformans*). *Genetics.* **167**: 619–631.

McDade, H. C. and Cox, G. M. (2001) A new dominant selectable marker for use in *Cryptococcus neoformans*. *Med. Mycol.* **39**: 151–154.

Moyrand, F., Klaproth, B., Himmelreich, U., Dromer, F. and Janbon, G. (2002) Isolation and characterization of capsule structure mutant strains of *Cryptococcus neoformans*. *Mol. Microbiol.* **45**: 837–849.

Moyrand, F., Chang, Y. C., Himmelreich, U., Kwon-Chung, K. J. and Janbon, G. (2004) Cas3p belongs to a seven-member family of capsule structure designer proteins. *Eukaryot. Cell.* **3**: 1513–1524.

Mylonakis, E., Idnurm, A., Moreno, R., El Khoury, J., Rottman, J. B. *et al.* (2004) *Cryptococcus neoformans* Kin1 protein kinase homologue, identified through a *Caenorhabditis elegans* screen, promotes virulence in mammals. *Mol. Microbiol.* **54**: 407–419.

Odom, A., Muir, S., Lim, E., Toffaletti, D. L. *et al.* (1997) Calcineurin is required for virulence of *Cryptococcus neoformans*. *EMBO J.* **16**: 2576–2589.

Olszewski, M. A., Noverr, M. C., Chen, G. H., Toews, G. B. *et al.* (2004) Urease expression by *Cryptococcus neoformans* promotes microvascular sequestration, thereby enhancing central nervous system invasion. *Am. J. of Pathol.* **164**: 1761–1771.

Perfect, J. R., Lang, S. D. R., and Durack, D. T. (1980) Chronic cryptococcal meningitis: A new experimental model in rabbits. *Am. J. of Pathol.* **101**: 177–194.

Rivera, J., Feldmesser, M., Cammer, M. and Casadevall, A. (1998) Organ-dependent variation of capsule thickness in *Cryptococcus neoformans* during experimental murine infection. *Infect. Immun.* **66**: 5027–5030.

Rodriguez, R. J. (1993) Polyphosphates present in DNA preparations from filamentous fungal species of *Colletotrichum* inhibits restriction endonucleases and other enzymes. *Ann. Biochem.* **209**: 291–297.

Salas, S. D., Bennett, J. E., Kwon-Chung, K. J., Perfect, J. R. and Williamson, P. R. (1996) Effect of the laccase gene, *CNLAC1*, on virulence of *Cryptococcus neoformans*. *J. Exp. Med.* **184**: 377–386.

Steenbergen, J. N., Shuman, H. A. and Casadevall, A. (2001) *Cryptococcus neoformans* interactions with amoebae suggest an explanation for its virulence and intracellular pathogenic strategy in macrophages. *Proc. Natl. Acad. Sci. USA.* **98**: 15245–15250.

Sudarshan, S., Davidson, R. C., Heitman, J. and Alspaugh, J. A. (1999) Molecular analysis of the *Cryptococcus neoformans ADE2* gene: A selectable marker for transformation and gene disruption. *Fungal Genet. Biol.* **27**: 36–48.

Toffaletti, D. L., Rude, T. H., Johnston, S. A., Durack, D. T. and Perfect, J. R. (1993) Gene transfer in *Cryptococcus neoformans* by use of biolistic delivery of DNA. *J. Bacteriol.* **175**: 1405–1411.

Tucker, S. C. and Casadevall, A. (2002) Replication of *Cryptococcus neoformans* in macrophages is accompanied by phagosomal permeabilization and accumulation of vesicles containing polysaccharide in the cytoplasm. *Proc. Natl. Acad. Sci. USA.* **99**: 3165–3170.

Varma, A. and Kwon-Chung, K. J. (1991) Rapid method to extract DNA from *Cryptococcus neoformans*. *J. Clin. Microbiol.* **29**: 810–812.

Vartivarian, S. E., Anaissie, E. J., Cowart, R. E., Sprigg, H. A. *et al.* (1993) Regulation of cryptococcal capsular polysaccharide by iron. *J. Infect. Dis.* **167**: 186–190.

Walton, F. J., Idnurm, A. and Heitman, J. (2005) Novel gene functions required for melanization of the human pathogen *Cryptococcus neoformans*. *Mol. Microbiol.* **57**: 1381–1396.

14

Genetic manipulation of zygomycetes

Ashraf S. Ibrahim and **Christopher D. Skory**

14.1 Introduction

Mucormycoses are life-threatening infections caused by fungi belonging to the order Mucorales of the class zygomycetes (Ribes *et al.*, 2000). Cutaneous and deep infections are angio-invasive in nature and can be caused by *Rhizopus*, *Mucor*, *Rhizomucor*, *Absidia*, *Cunninghamella*, *Apophysomyces*, and in rare causes by *Cokeromyces*, *Mortierella*, or *Saksenaea* (Ibrahim *et al.*, 2003). *Rhizopus oryzae* (syn. *Rhizopus arrhizus*) is the most common organism isolated from patients with mucormycosis (Ribes *et al.*, 2000). The major risk factors for mucormycosis include uncontrolled diabetes mellitus in ketoacidosis, other forms of metabolic acidosis, treatment with corticosteroids, organ or bone marrow transplantation, neutropoenia, trauma and burns, malignant haematological disorders and deferoxamine chelation therapy in patients receiving haemodialysis (Ibrahim *et al.*, 2003; Sugar, 1995). The underlying predisposing conditions influence the clinical manifestation of the disease. For example, about 70% of rhinocerebral mucormycosis cases are seen in patients with diabetic ketoacidosis (the most common form of the disease) (Pillsbury and Fischer, 1977; McNulty, 1982). Conversely, patients with malignancy, severe neutropoenia or a history of deferoxamine therapy usually suffer from pulmonary or disseminated mucormycosis (Ibrahim *et al.*, 2003). Owing to the rising prevalence of diabetes, cancer and organ transplantation in the ageing population of the United States, the number of patients at risk for this deadly infection is rising dramatically (Marr *et al.*, 2002). In fact, one centre has reported a staggering 1,300% increase in the incidence of this disease over the last 15 years (Kontoyiannis *et al.*, 2000, 2005).

Medical Mycology: Cellular and Molecular Techniques. Edited by Kevin Kavanagh

The standard therapy for invasive mucormycosis includes the reversal of predisposing factors (if possible), emergent, wide-spread surgical debridement and antifungal therapy (Ibrahim *et al.*, 2003; Sugar, 1995; Kwon-Chung and Bennett, 1992). Amphotericin B (AmB) remains the only antifungal agent approved for the treatment of invasive mucormycosis (Ibrahim *et al.*, 2003; Kwon-Chung and Bennett, 1992; Edwards, 1989). Because Mucorales are relatively resistant to AmB, high doses (1–1.5 mg/kg/d) are required, which frequently cause nephrotoxicity and other adverse effects (Ibrahim *et al.*, 2003; Edwards, 1989). Also, in the absence of surgical removal of the infected focus, antifungal therapy alone is rarely curative (Ibrahim *et al.*, 2003; Kwon-Chung and Bennett, 1992). Even when surgical debridement is combined with high-dose AmB, the mortality associated with mucormycosis exceeds 50% (Kwon-Chung and Bennett, 1992). In patients with prolonged neutropoenia, and in those with disseminated disease, mortality is 90–100% (Marr *et al.*, 2002; Kontoyiannis *et al.*, 2000; Husain *et al.*, 2003). Because of its increasing incidence, unacceptably high mortality and the extreme morbidity of highly disfiguring surgical therapy, new strategies to prevent and treat mucormycosis are urgently needed.

14.2 Genetic tools to manipulate mucorales

During the last decade, developments in molecular biology have greatly advanced our understanding of the pathogenesis of invasive fungal infections, especially candidosis and cryptococcosis (Kwon-Chung *et al.*, 1998), and most recently aspergillosis (Panepinto *et al.*, 2003) and blastomycosis (Klein, 2000). These advances were greatly facilitated by molecular genetic studies of these pathogens. However, the few published studies of the pathogenesis of mucormycosis have emphasized host-defence mechanisms (Waldorf *et al.*, 1984a, 1984b) without attempting to investigate specific virulence factors of the causative organism. The limited availability of genetic tools for the study of the Mucorales fungi is a major contributing factor to the paucity of studies on the organism's pathogenesis.

While several laboratories have described numerous genetic studies with *R. oryzae* (Skory, 2002; Michielse *et al.*, 2004a, 2004b), *R. delemar* (Horiuchi *et al.*, 1995), *R. niveus* (Liou *et al.*, 1992; Takaya *et al.*, 1996), *Mucor circinelloides* (Anaya and Roncero, 1991; Benito *et al.*, 1992; Ruiz-Hidalgo *et al.*, 1999; Roncero *et al.*, 1989), *Absidia glauca* (Burmester *et al.*, 1992), *Phycomyces blakesleeanus* (Arnau *et al.*, 1988) and *Rhizomucor pusillus* (Wada *et al.*, 1996; Yamazaki *et al.*, 1999), certain basic molecular biology procedures, such as stable transformation and double cross-over gene replacements are still problematic. Difficulties with the genetic modification of the Mucorales are partly due to the unique ability of these fungi to replicate the DNA used for transformation autonomously in a concatenated high molecular weight form. These transformants tend to lose the transformed plasmid(s) when cultured on non-selection media, resulting in a very low mitotic stability. Nevertheless, significant progress and understanding of the genetic mechanisms of the

Mucorales continues at an ever-increasing pace. An overview of these methods and techniques, with particular emphasis on *R. oryzae*, is presented in this chapter.

14.3 Selectable markers used with mucorales fungi

Selectable markers used with various Mucorales fungi are not uncommon compared to other filamentous fungi. They tend to fall into two general categories of auxotrophic markers and drug resistance markers. The first type of selection requires that the host organism contains a mutation that necessitates medium supplementation of a specific compound not required by the wild-type organism. Typically, this would involve a mutation preventing synthesis of a necessary component of a biosynthetic pathway. The vector used for transformation would then contain a gene or selectable marker capable of allowing the host organism to synthesize the missing component. Examples of this type of selection with the Mucorales fungi include leucine (Takaya *et al.*, 1996; Roncero *et al.*, 1989), methionine (Anaya and Rancero, 1991) and uracil biosynthesis (Skory, 2002; Benito *et al.*, 1992).

Selection based on drug resistance has the advantage that no mutations are necessary in the host organism. If the fungus is susceptible to a certain drug, it may be possible to provide resistance by introducing a gene capable of modifying the drug to a less toxic form, preventing import or increasing export. This type of selection is often preferred because no mutagenesis and selection of the host strain is necessary to obtain an auxotrophic mutation. However, Mucorales fungi tend to be recalcitrant to many of the typical drug markers used for transformation. It is sometimes possible to increase an organism's susceptibility to specific drugs by selection in submerged growth conditions, modifying the medium pH or following the addition of specific components to increase the permeability of the compound. Examples of drug resistance with Mucorales have primarily relied on expression of a neomycin phosphotransferase gene with selection using geneticin, kanamycin or neomycin (Revuelta and Jayaram, 1986; Appel *et al.*, 2004; Obraztsova *et al.*, 2004; Wostemeyer *et al.*, 1987).

A third category of selectable markers allows the host organism to utilize specific substrates, incapable of supporting growth with the parent strain. An example of this type of selection would be using acetamide as a sole nitrogen source for growth. Such methods have been quite successful with other fungi (Kelly and Hynes, 1985) but have not yet been demonstrated with Mucorales fungi.

The choice of selectable markers ultimately depends on the availability of options and the anticipated use of the final transformant strain. Studies investigating the physiology of the organism, such as pathogenic factors, would require that the selectable marker does not alter or interfere with the functions being examined. Additionally, the use of drug resistant markers may preclude the use of the transformant strain in certain applications. While auxotrophic markers may require more effort with host modification, they usually allow the researcher to bypass several of these pitfalls. However, the investigator must be careful to ensure that the mutagenesis and selection used to obtain the host strain does not result in other mutations that alter the physiology of the organism. The ideal method to overcome this possibility is the use of site-directed gene knockouts in the desired auxotrophic gene. However, genetic techniques with the Mucorales fungi have not yet advanced to allow such methods to be performed efficiently. Additionally, from work done with other fungi, it is also necessary to place the auxotrophic marker in its original locus to ensure a proper expression of the gene product which might alter, for example the pathogenicity of the organism being studied (Staab and Sundstrom, 2003).

14.4 Introduction of DNA used for transformation

A. Protoplast/polyethylene glycol methods

Regardless of the markers employed for selection, it is necessary to get the DNA used for transformation into the fungal cell and ultimately into the nucleus. Several different methods have been used successfully to accomplish this goal with the Mucorales. Like most filamentous fungi, methods employing protoplasting and polyethylene glycol (PEG) treatment tend to be the most common. Such techniques typically involve using a mixture of enzymes to strip away the cell wall of the germinating spore to allow the release of a protoplast in an osmotically stabilized buffer. PEG treatment is then used to facilitate transfer of the DNA into the cell. While this method does not require any specialized equipment, it does require considerable optimization and is prone to variability.

There are numerous lysing enzymes available for making protoplast with filamentous fungi, and the optimal mixture must be experimentally determined. However, Mucorales usually have cell walls rich in chitin and chitosan (Pochanavanich and Suntornsuk, 2002) and supplementation with chitosanase and chitinase is almost always necessary for generating protoplasts in reasonable yields. The amount of enzyme necessary for protoplasting must also be optimized for each fungal strain and can even vary from batch to batch. As an alternative to purchasing commercial chitosanase, which is costly, it can be generated by growing *Streptomyces* in media containing a chitin/chitosan mixture or purified cell-wall material from the fungal strain as the sole carbon source (van Heeswijck *et al.*, 1998). Spores that are still in the early stages of germination tend to be most susceptible to hydrolysis by the enzyme mixture (van Heeswijck *et al.*, 1998; van Heeswijck, 1984; Suarez *et al.*, 1985). Ungerminated spores are usually resistant to the action of the enzyme cocktail, whereas extended growth often aggregates into balls that resist enzyme action and decrease protoplast yield. To prevent the protoplasts from lysing, it is important to stabilize the turgor pressure, by increasing the osmolarity of the protoplasting buffer. The preferred compound and concentration to accomplish this goal must also be determined empirically. Factors such as protoplast yield, viability and, ultimately, transformation efficiency must all be taken into account. Sugar alcohols, such as sorbitol or mannitol (0.35–0.5 M) are effective for osmotic stabilization of the protoplasts. Other osmotic stabilizers that are frequently used for other organisms, including KCl, $(NH_4)_2SO_4$ or $MgSO_4$, are inappropriate because they inhibit the activity of chitosanase. A protocol that works well for producing protoplasts with *R. oryzae* is included as an example.

Protocol 14.1
Protoplasting of *R. Oryzae*

Equipment, materials and reagents

Spores from which protoplasts will be generated

Refrigerated centrifuge

Incubator shaker

Light microscope

Haemocytometer

Sonicator (Branson Sonifier 450, Branson, Rochester, MI, USA)

Phosphate-buffered saline (PBS, pH 7.2)

PBS containing 0.01% (v/v) Tween 80

Potato dextrose (PD) agar

YPD broth

0.5 M sorbitol

10 mM sodium phosphate buffer, pH 6.4 with 0.5 M sorbitol

Protoplasting enzymes:

Lysing enzyme (Sigma-Aldrich, Poole, Dorset, UK, cat. no. L-1412) stock concentration 312.5 mg/ml

Chitnase (Sigma-Aldrich cat. no. C-6137) stock concentration 30.0 mg/ml

Chitosanase (US Biological Incorporated, Swampscott, MA, USA, cat. no. C4163) stock concentration 10.0 mg/ml

Protoplasting solution (40 µl of lysing enzyme, 250 µl chitnase, 750 µl chitosinase) and 9 ml of 10 mM sodium phosphate buffer, pH 6.4 with 0.5 M sorbitol

All solutions must be sterilized, preferably by filtration through less than 0.45 µM membranes. It is important to use low protein binding filters (e.g. polyethylene sulphone) with enzyme solutions to prevent loss of activity.

Method

1. Spores are typically harvested from 4 PD plates (9 cm in diameter) inoculated with the desired organism and incubated until confluency (~3–5 days depending on the *Rhizopus* under study).

2. Spore harvesting is carried out normally by flooding the plate with 10 ml of 0.01% (v/v) Tween 80 to the plate and using a sterile glass rod to liberate the spores followed by using a pipette to collect the suspension.

3. Spores are spun down with a centrifuge at 500 x *g* for 5 min at room temperature, washed once with 5 ml of PBS and resuspended in 2 ml of PBS.

4. Spores are sonicated briefly at setting 4 for 5–10 sec to disrupt aggregation then diluted (1:200) in PBS and counted with a haemocytometer using a light microscope (more dilution might be required depending on the amount of spores collected).

5. Spores are inoculated into 50 ml of YPD broth (1% yeast extract, 2% peptone and 2% dextrose [w/v]) in 250 ml Erlenmeyer flask and incubated at 37 °C with shaking (200 rpm) until spores germinate.

6. For best protoplasting results the germ tube should be between 30 and 60 µm. This is usually achieved with *R. oryzae* after 5–6 h. Germ-tube length can be determined by the use of a micrometer lens on a light microscope.

7. Germlings are collected by centrifugation at 500 x *g* for 5 min at 4 °C with low break then washed twice with 10 ml 0.5 M sorbitol.

8. Germlings are resuspended in 100 ml Erlenmeyer flask containing 40 ml of 10 mM sodium phosphate buffer, pH 6.4 with 0.5 M sorbitol followed by the addition of 10 ml of freshly prepared protoplasting solution containing protoplasting enzymes.

9. Lysing the cell wall is allowed to take place by incubating the Erlenmeyer flask at 30 °C with shaking at 100 rpm.

10. After 1 h incubation, protoplast formation should be examined every 20–30 min by examining a sample under a light microscope. To check for the quality of protoplasts, apply a water sample with a pipette tip to the slide containing the sample. Protoplasts should quickly swell and explode because of the increased turgor pressure.

11. Protoplasts are collected by centrifugation for 5 min at 100 x *g* at room temperature with low brake. Wash twice with 10 ml of 0.5 M sorbitol with gentle shaking to resuspend the pellet then resuspend in 10 ml 0.5 M sorbitol.

12. The protoplasts are passed through a Miracloth to purify protoplasts from undigested hyphae.

13. Viability of the protoplast is determined by serially diluting protoplasts in 0.5 M sorbitol buffer and inoculating aliquots onto appropriate growth media containing 0.5 M sorbitol. The number of colonies obtained per ml of protoplasts is then compared back to the total number of protoplasts determined by haemocytometer counts.

Note: Certain fungal isolates tend to spread quite rapidly and so counts must be obtained early in growth. Using various growth-inhibition techniques such as lowering the pH of the medium, adding Triton X-100, Rose Bengal or oxgall can be useful, but may prevent the recovery of some protoplasts.

B. Biolistic delivery methods

Biolistic particle delivery system (Figure 14.1) is a method of transformation that utilizes helium pressure to introduce DNA-coated microcarriers into the host cell. It is accomplished by building a pressure differential between a helium and vacuum chamber separated by a membrane designed to rupture at a specific pressure. Upon breakage, the sudden high-pressure burst of helium pushes a macrocarrier membrane containing the DNA-coated particles, on the underside of the membrane, downward until it is stopped abruptly with a metal screen. The DNA-coated particles continue to travel downward towards the target plate, until a certain percentage is expected to collide and enter the fungal spores.

Microprojectile bombardment can transform a wide variety of organisms including bacterial, fungi, plants, insects and animal cells. *Mucor* was the first zygomycetes transformed by this method in 1997 (González-Hernández *et al.*, 1997). This method has the advantage that minimal preparation of the host is required, resulting in less variability and decreased likelihood for contamination. Because of the high cost associated with the biolistic delivery system, it may be prohibitive to many laboratories.

Considerable work with *R. oryzae* demonstrates the effectiveness of this technique, and strategies for optimization will likely apply to other Mucorales. Various factors

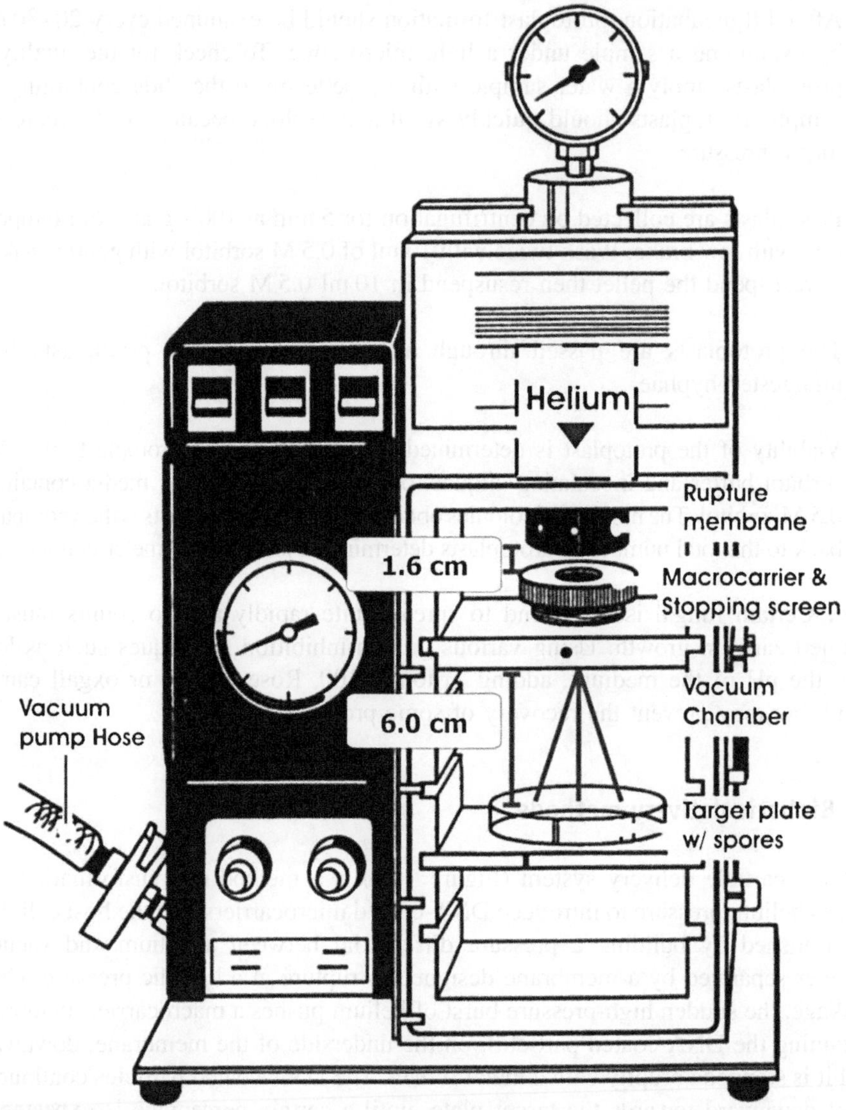

Figure 14.1 Biolistic particle delivery apparatus showing the recommended distance for *R. oryzae* bombardment

must be tested to achieve reasonably high transformation efficiencies. There are several choices of microcarriers (e.g. gold and tungsten particles of various sizes), rupture membranes that dictate the helium blast pressure used for particle acceleration and the distances between rupture membrane/macrocarrier and macrocarrier/host organism. Because of the limited number of Mucorales transformed with this method, each of the various factors should be optimized. However, work with *R. oryzae* did reveal some trends. Only tungsten particles less than 0.5 micrometers gave reproducible transformation. Additionally, greater rupture pressures generally gave higher transformation efficiencies, but were restricted by the physical limitations of stopping screen failure (i.e. stopping screen is unable to stop the macrocarrier membrane without being pushed into the target plate). Using freshly harvested spores as the recipient target has been the only option that reproducibly gave good transformation efficiencies. Any type of germination of the spores usually leads to higher variability and is probably the result of the organism's greater susceptibility to damage once it begins to grow and form intracellular turgor pressure. Not surprisingly, higher concentrations of spores on the target plate usually increase the efficiency of this technique. The success of biolistic transformation relies on the expectation of a DNA-coated particle bombarding and penetrating the target spore. Greater densities of spores increase the likelihood of this occurring.

While biolistic transformation is effective for introducing desired constructs into the host organism, the transformation efficiencies will likely never come close to the potential of protoplast/PEG methods. Protoplasting methods for filamentous fungi are frequently low as techniques are being developed. However, it is not uncommon to achieve efficiencies well above 10 000 transformants/μg DNA with continued optimization. Such efficiencies are probably unobtainable with biolistic delivery techniques, but this method still holds an important place in the arsenal of molecular tools for transforming fungi. A typical protocol for transformation of *R. oryzae* using an auxotrophic selectable marker is presented in Protocol 14.2.

Protocol 14.2
Biolistic delivery system transformation of *R. Oryzae*

Equipment, materials and reagents

Spores

Centrifuge

Light microscope

Haemocytometer

Sonicator (Branson Sonifier 450, Branson, Swampscott, MA, USA)

Vortex mixer

Biolistic® delivery system (PDS-1000/He, Bio-Rad) (see Figure 14.1)

Helium tank

Vacuum source capable of evacuating the bombardment chamber of $\geq$5 in. of mercury

PBS

PBS containing 0.01% (v/v) Tween 80

PD agar

70% and 100% (v/v) High-grade ethanol (HPLC or spectrophotometric grade)

2.5 M $CaCl_2$

0.1 M spermidine (free base, tissue culture grade)

Tungsten beads

Anhydrous calcium chloride

Method

1. Spores are typically harvested from 7 PD plates (9 cm in diameter) inoculated with the desired organism and incubated until confluency (~3–5 days depending on the *Rhizopus* understudy).

2. Spore harvesting is carried out normally by flooding the plate with 10 ml of 0.01% (v/v) Tween 80 to the plate and using a sterile glass spreader to liberate the spores followed by using a pipette to collect the suspension.

3. Spores are spun down with a centrifuge at 500 x *g* for 5 min at room temperature, washed once with 5 ml of PBS and resuspended in 2 ml of PBS.

4. Spores are sonicated at setting 4 for 5–10 sec briefly at setting 4 for 5–10 sec to disrupt aggregation (see above comments about sonication) then diluted (1:300) in PBS and counted with a haemocytometer using a light microscope (more dilution might be required depending on the amount of spores collected).

5. Spores (5×10^7/0.2 ml PBS) are spread on selection agar plates (e.g. minimal medium lacking uracil if *pyrG* or *pyrF* are used as selection markers) using a glass spreader and left to dry at room temperature.

6. Meanwhile the tungsten microcarriers will be prepared for bombardments (Skory 2002) as follows:

   i. 30 mg of tungsten particles are placed in a 1.5 ml microfuge tube containing 1 ml of 70% ethanol and vortexed vigorously for 3–5 min using a platform vortexer then soak for 15 min at room temperature (**Note:** In our experience, transforming *R. oryzae* using gold particles was found to be totally inefficient due to the high degree of agglomeration of the gold particles).

   ii. The particles are pelleted by spinning down in a microfuge at high speed for 5 sec and then washed in 70% ethanol 3 times.

   iii. After the third wash 500 µl of 50% glycerol is added and the particles are stored at −20 °C until used for coating with the DNA to be transformed.

   iv. 50 µl aliquot of tungsten particles is removed from the stock, while vortexing the tube containing the stock particles to maximize uniform sampling, then transformed to 1.5 ml microfuge tube.

   v. While vortexing the following are added in order:

      a. 5 µl of DNA (1 µg/µl)

      b. 50 µl of 2.5 M $CaCl_2$

      c. 20 µl of 0.1 M spermidine.

   vi. Vortexing is continued for another 2–3 min then the microcarriers are allowed to settle for 1 min followed by pelleting by spinning for 5 sec in a microfuge.

   vii. The liquid is discarded then the pellet is washed with 140 µl of 70% ethanol (high grade) followed by another 140 µl of 100% ethanol (high grade) and then resuspended in 48 µl of 100% ethanol by tapping the side of the tube several times, and then by vortexing at low speed for 2–3 sec.

7. The coated microcarrier tungsten particles (6 µl) are then spread on the macrocarrier placed within the macrocarrier holder. This step should be performed in a Petri dish containing anhydrous calcium chloride as a desiccant. The Petri dish lid should be placed immediately following application of the sample and should be used for bombardment within 2 h from the microcarrier application.

8. At a minimum, the chamber of PDS-1000/He should be disinfected with 70% ethanol prior to bombardment then left to dry. However, we prefer to use a series of washes with DNAZap (Ambion, Austin, TX) to remove residual plasmid from

the chamber, followed by a quaternary ammonium-based antimicrobial, and finally extensive rinsing with 70% ethanol.

9. Bombardment is then performed following the manufacturer's manual utilizing the following parameters (Skory, 2002):

    i. Distance from rupture disk to the top of the launch assembly should be 1.6 cm.

    ii. Distance from the bottom of the launch assembly to plate should be 6 cm.

    iii. Size of rupture disk used: ≥1100 psi

Note: Adjust helium tank so that the pressure reads approximately 200 psi above the size of the rupture disk used.

C. *Agrobacterium tumefaciens*-mediated transformation

A. tumefaciens-mediated transformation (AMT) of filamentous fungi is quickly gaining popularity, as it has been shown to be capable of transforming a broad range of hosts. While relatively little work has been done with AMT of Mucorales fungi, it has been used successfully with *R. oryzae* to integrate DNA used for transformation into the fungal genome (Michielse *et al.*, 2004a, 2004b). *A. tumefaciens* is a gram-negative plant pathogen that causes crown galls by transferring a part of its tumourinducing (Ti) plasmid DNA into the host genome (Abuodeh *et al.*, 2000). The transferred DNA (T-DNA) is located between two 24 bp border repeats on the binary vector of the bacterium into the genomic DNA of the host. A second region of the Tiplasmid known as the *vir* region is also essential for DNA transfer and encodes virulence proteins required for the transfer and integration of the foreign DNA. Induction of the *vir* genes by certain phenolic compounds, such as acetosyringone, initiates *A. tumefaciens* mechanism to transfer single-stranded DNA from the T-DNA into the transforming host organism. DNA containing the fungal selectable marker is thus cloned into this region, along with DNA of interest. Induced *Agrobacterium* is able to transfer DNA into most fungal organisms through an intact cell wall of the host (Abuodeh *et al.*, 2000; Chen *et al.*, 2000[need ref in reference section]; Gouka *et al.*, 1999; de Groot *et al.*, 1998). However, this was not possible with either spores or germlings of *R. oryzae*. Only protoplasts were susceptible to transformation with this method (Michielse *et al.*, 2004a, 2004b).

Protocol 14.3
A. tumefaciens-mediated transformation

Equipment, materials and reagents

Spores from which protoplasts will be generated

Refrigerated centrifuge

Incubator shaker

Light microscope

Haemocytometer

Sonicator (Branson Sonifier 450, Branson, Swampscott, MA, USA)

Sterile water

PBS

PBS containing 0.01% (v/v) Tween 80

PD agar

YPD broth

0.5 M sorbitol

10 mM sodium phosphate buffer, pH 6.4 with 0.5 M sorbitol

Protoplasting enzymes:

Lysing enzyme (Sigma-Aldrich, Poole, Dorset, UK cat. no. L-1412) stock concentration 312.5 mg/ml

Chitnase (Sigma-Aldrich cat. no. C-6137) stock concentration 30.0 mg/ml

Chitosinase (US Biological cat. no. C4163) stock concentration 10.0 mg/ml

Protoplasting solution: (40 µl of lysing enzyme, 250 µl chitnase, 750 µl chitosinase) and 9 ml of 10 mM sodium phosphate buffer, pH 6.4 with 0.5 M sorbitol

Agrobacterium tumefaciens

Electroporater

LB medium

Minimal medium gram/litre [2.05 K_2HPO_4, 1.45 KH_2PO_4, 0.15 NaCl, 0.50 $MgSO_4$-$7H_2O$, 0.1 $CaCl_2$-$7H_2O$, 0.0025 $FeSO_4$-$7H_2O$, 0.5 and $(NH4)_2SO_4$]

Induction medium (minimal medium, 20 mM MES, pH 5.3, 10 mM glucose, 0.5% (w/v) glycerol and 200 μm acetosyringone

Sterile nylon membrane (Hybond-N-Amersham, GE Healthcare, Little Chalfont, UK)

Method

A. Transformation of *A. Tumefaciens* with DNA to be Transformed into *Mucorales*

1. Large loopfuls of *A. tumefaciens* are scraped from an LB plate grown for 2 days at 30 °C.

2. The loopfuls are suspended and washed 3 times with sterile water and then resuspended in a final volume of 80 μl sterile water.

3. Approximately 100–500 ng of the plasmid to be transferred is added to the bacterial suspension in a 0.1 cm cuvette and electroporated using 200 Ω, 25 μF, 2.5 kb.

4. 1 ml of LB medium is added to the electroporated suspension followed by shaking at 30 °C for 3 h.

5. The electroporated suspension is plated on LB plates containing the appropriate antibiotic for colony selection.

B. Transformation of *R. oryzae* with *A. tumefaciens* (adapted from Michielse *et al.* 2004a, 2004b)

6. *A. tumefaciens* carrying the DNA to be transferred is grown in LB medium supplemented with the appropriate antibiotics for overnight at 30 °C in a shaker incubator.

7. 1.5 ml of the overnight culture is aliquoted and centrifuged at high speed using a microfuge and the pellet washed once with 200 μl of induction media and then resuspended in 5 ml of induction medium with 200 mM acetosyringone.

8. The *Agrobacterium* is incubated for 5–6 h at 30 °C in a shaker (200 rpm) after which the optical density of the culture is measured at 600 nm.

9. The culture is diluted to a density of 4–5 × 10^8 cells/ml and an aliquot of 100 μl of these cells are mixed with a 100 μl of *R. oryzae* protoplasts (generated as above).

10. The mixture is spread on nitrocellulose paper and placed on an induction medium agar plate and incubated for 24 h at 28–30 °C.

11. After co-cultivation the paper is transferred to selection plates and incubated further at 30 °C for another 4 days.

C. Other methods of DNA delivery

The above three techniques represent the methods used by the vast majority of researchers working with filamentous fungi. Still, other methods should be mentioned. One interesting technique that has been used successfully with the Mucorales fungus *Phycomyces blakesleeanus* is microinjection, whereby plasmid carrying a selectable marker is injected into young sporangium through the open ends of a freshly cut sporangiophores. This technique has the potential to result in high-efficiency transformation (Ootaki *et al.*, 1991), but can also lead to unexpected phenotypic changes (Obraztsova *et al.*, 2004). Other methods such as electroporation and lithium acetate treatment have been used successfully with other filamentous fungi but still have not been shown to be effective with Mucorales fungi.

14.5 Molecular analysis of transformants

Protoplasts or spores of Mucorales are typically multinucleated, and transformation methods usually result in heterokaryosis, in which genetically different nuclei are associated with the same cell. Growth of the heterokaryon proceeds by hyphal extension on the selective medium. Even though the majority of Mucorales are generally coenocytic, or lacking well-defined septa, the mixed nuclei frequently segregate in the somatic hyphae, such that individual spores within a sporangium will usually be homokaryotic. Therefore, it is often advantageous to allow transformant cultures to proceed through at least one round of sporulation on selective medium before proceeding to single-colony isolation.

As previously mentioned, DNA used for the transformation of Mucorales rarely integrates but is replicated in high molecular weight concatenated structures (i.e. series or chain). Analysis of these structures is often misleading and is easily mistaken for genomic DNA because of the high molecular weight. Autonomous replication of plasmids in Mucorales fungi does not appear to require a defined origin of replication (Skory, 2002; Revuelta and Jayaram, 1986; Wostemeyer *et al.*, 1987; Benito *et al.*, 1995; vanHeeswijck and Roncero, 1984a), as is typical with many other higher eukaryotes. However, these structures typically do not segregate well during nuclear division, and spores obtained from transformant cultures may have less than 1% of total spores containing the recombinant plasmid, even after multiple rounds of sporulation on selective medium. Attempts to isolate sequences that improve the efficiency of stable replication in Mucorales have only been moderately successful (Roncero *et al.*, 1989; Burmester *et al.*, 1992).

Integration is the only effective approach to obtaining a stable homokaryotic transformant. However, this too has proven to be fairly difficult with Mucorales. In *Rhizopus*, circular plasmid will only integrate at homologous loci at frequencies of less than 5%, with the remainder of transformants having autonomous replication of concatenated plasmids. This frequency of integration can increase up to 20% of the total transformants, if plasmids used for transformation are linearized with a double strand break (DSB) within a region having homology with chromosomal DNA. Integration almost always occurs at the homologous locus in a manner frequently described as type I, additive or cross-over integration (Figure 14.2) (Skory, 2002; Hinnen *et al.*, 1978).

One of the reasons that the total number of transformants having integration is not higher is the efficient end-joining of the DSB (Skory, 2002). When plasmids are linearized with a single-restriction endonuclease, the predominant repair event is re-ligation of the plasmid, which is then capable of autonomous replication as described. This type of repair is most common in complex eukaryotes and is often referred to as non-homologous end-joining (NHEJ) (Ray and Langer, 2002; Frank-Vaillant and Marcand, 2002). While end-joining is usually precise, NHEJ also has the ability to fill in and delete nucleotides to allow for the repair of DSB having non-homologous ends (Sandoval and Labhart, 2002). Therefore, attempts to modify the DSB with various methods such as dephosphorylation had little effect in preventing NHEJ in *Rhizopus* (Skory, 2005).

When integration does occur, it is not always by type I (additive integration). A type-III gene replacement will sometimes occur with *Rhizopus* when the selectable marker is similar to a genomic homologue. An example of this might include using a *pyrG* (or *ura3*) gene to complement an auxotrophic mutant (Figure 14.2). Analysis of *Rhizopus* transformants may reveal up to 5–10% of the isolates having restored *pyrG* genomic sequence. While encouraging that a double cross-over gene-replacement event should occur with reasonable frequency, putting this into actual practice has not been very fruitful with Mucorales. Only a single study that utilized a disruption cassette using *pyrG* as a marker flanked by 0.8–1.3 kb sequence corresponding to the upstream and downstream regions of the target gene described successful gene disruption in *Mucor circinelloides* (Navarro *et al.*, 2001). Selection for the gene replacement was facilitated by allowing transformants to sporulate on non-selective medium in an effort to allow plasmid loss with isolates relying on autonomous replication. While this results in sporulation of primarily untransformed isolates, transformants become enriched for those having stable integration, which can then be recovered by growth on selective medium (Arnau *et al.*, 1991).

Agrobacterium-mediated transformation may offer more hope of accomplishing integration at a higher frequency. With *Rhizopus*, this method yielded transformants that showed 100% mitotic stability resulting from an integration of the transformed DNA (Michielse *et al.*, 2004b). In this single study of *A. tumefaciens*-mediated transformation, the authors did not attempt to determine the transformation efficiency,

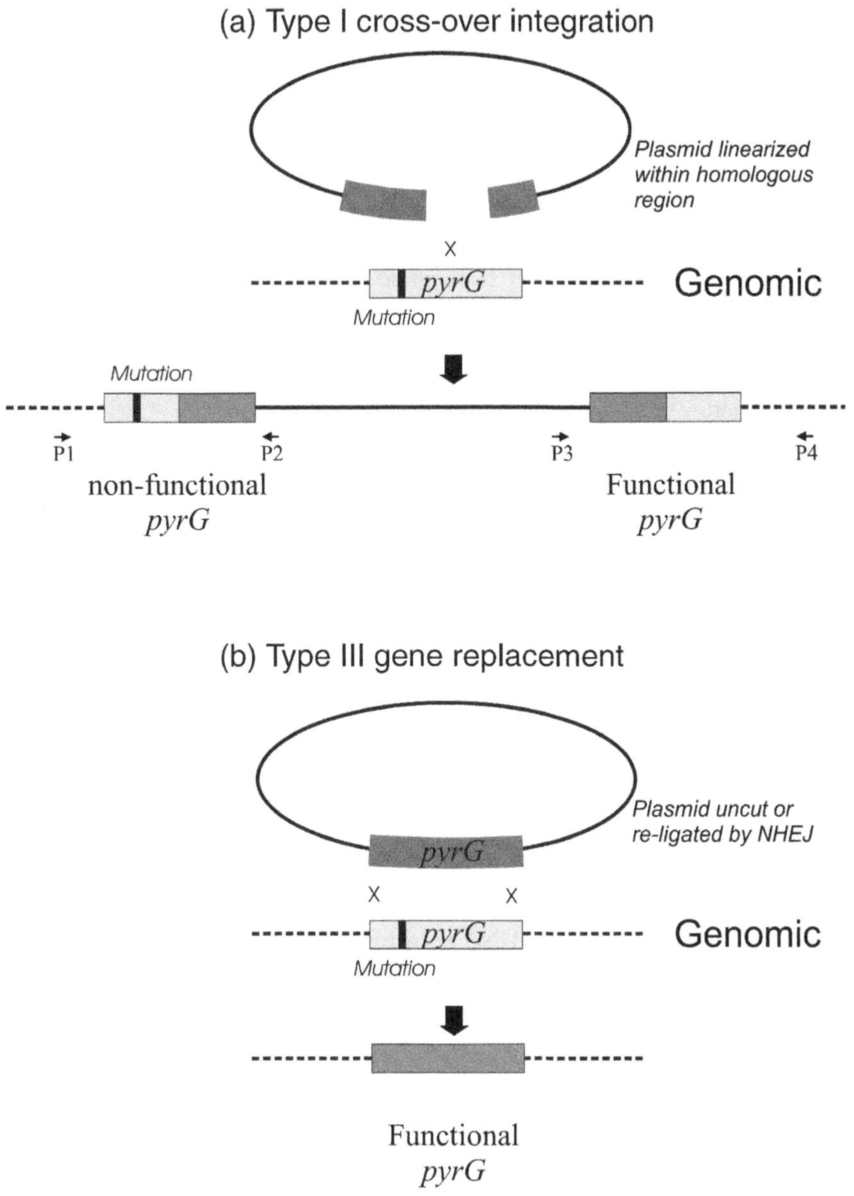

Figure 14.2 Diagram summarizing type of integration of plasmids transformed into *R. oryzae*. Type I (a) or Type III (b) cross over integration occurs when linearized or circular plasmids are transformed into *R. oryzae*, respectively

owing to the vigorous growth of the transformed cells on selection plates. Nevertheless, Southern blot analysis of eight purified transformants revealed that six transformants had an extra copy of *pyrG* at an undetermined ectopic locus, not associated with the *pyrG* gene. This type-II integration occurred at the same locus in the chromosome for all transformants, suggesting the presence of a hot spot for integration that has yet to be identified (Michielse *et al.*, 2004b). In contrast, the other two transformants had undergone *pyrG* gene-conversion replacement through type-III integration since only one copy of the *pyrG* could be detected (Michielse *et al.*, 2004b). It is important to mention that another organism of Mucorales, *Mucor miehei*, has also been transformed via *A. tumefaciens*-mediated transformation using the exogenous *aphI* marker. However, the resultant transformants were mitotically unstable, indicating that the transformed DNA had not stably integrated into the genome (Monfort *et al.*, 2003). It is possible that the Mucorales have a defence mechanism that detects foreign DNA and prevents its integration into the genome. This would explain why the use of an endogenous marker (e.g. *pyrG*) resulted in the high frequency of integration seen with *R. oryzae* transformed with *Agrobacterium*-mediated method.

It becomes evident from these descriptions that it is of the utmost importance to design the DNA analysis experiments for transformant isolates in a manner that is able to detect each of the various replication or recombination events. With Southern blot analysis, this typically involves choosing restriction endonuclease (RE) enzymes that are not able to cleave the plasmid used for transformation. Additionally, enzymes that linearize the DNA used for transformation should be included. Probes used for hybridization frequently include regions of the plasmid associated with selection. If necessary, vector sequences can also be included in additional Southern blot analysis to determine the extent of integration or recombination. When integration is anticipated or detected, PCR amplification of the putative integration junctions is very useful for confirmation. This involves using primers designed to anneal to the plasmid sequence, as well as regions of the chromosome that are not present on the transforming plasmid (Figure 14.2). This type of amplification will be very specific for integration and will not be as easily prone to false positives like any amplification of internal regions of the plasmid. It is only through a careful analysis of transformants that it is possible to conclusively determine the fate of DNA used for transformation.

14.6 Overview

The ability to genetically manipulate the zygomycetes fungus *M. circinelloides* with recombinant DNA techniques was first demonstrated by Van Heeswijck and Roncero (1984b). It quickly became clear with continued work with *Mucor* and other zygomycetes that these fungi were unique in their ability to replicate DNA autonomously in the absence of a defined origin of replication. This realization made it clear why mitotic stability was so low with the generated transformants. While research was somewhat hampered by this unusual trait for a filamentous fungus, significant progress has continued with our understanding of zygomycetes physiology (Navarro *et al.*, 2000, 2001; Roze *et al.*, 1999; Skory, 2004) and genetics, including the heterologous expression of genes (Yamazaki *et al.*, 1999; Houghton-Larsen and Pedersen, 2003). Still, the desire to accomplish double cross-over knockouts in an efficient manner continues because of the necessity of this technique to study certain gene functions and their physiological importance. Alternatively, other new technologies such as gene silencing by RNA interference may prove to be more successful tools for accomplishing this goal (Nicolas *et al.*, 2003). Finally, genome sequencing of filamentous fungi, including zygomycetes, is occurring at an unprecedented rate and new information to overcome many of these obstacles is likely to emerge. With the groundwork for genetic analyses of zygomycetes firmly in place, it will truly be an exciting time for the exploration of knowledge.

14.7 References

Abuodeh, R. O., Orbach, M. J., Mandel, M. A., Das, A. and Galgiani, J. N. (2000) Genetic transformation of *Coccidioides immitis* facilitated by *Agrobacterium tumefaciens*. *J. Infect. Dis.* **181**: 2106–2110.

Anaya, N. and Roncero, M. I. (1991) Transformation of a methionine auxotrophic mutant of *Mucor circinelloides* by direct cloning of the corresponding wild-type gene. *Mol. Gen. Genet.* **230**: 449–455.

Appel, K. F., Wolff, A. M. and Arnau, J. (2004) A multicopy vector system for genetic studies in *Mucor circinelloides* and other zygomycetes. *Mol. Genet. Genomics.* **271**: 595–602.

Arnau, J., Murillo, F. J. and Torres-Martinez, S. (1988) Expression of Tn5-derived kanamycin resistance in the fungus *Phycomyces blakesleeanus*. *Mol. Gen. Genet.* **212**: 375–377.

Arnau, J., Jepsen, L. P. and Stroman, P. (1991) Integrative transformation by homologous recombination in the zygomycete *Mucor circinelloides*. *Mol. Gen. Genet.* **225**: 193–198.

Benito, E. P., Diaz-Minguez, J. M., Iturriaga, E. A., Campuzano, V. and Eslava, A. P. (1992) Cloning and sequence analysis of the *Mucor circinelloides* pyrG gene encoding orotidine-5′-monophosphate decarboxylase: use of pyrG for homologous transformation. *Gene.* **116**: 59–67.

Benito, E. P., Campuzano, V., Lopez-Matas, M. A., De Vicente, J. I. and Eslava, A. P. (1995) Isolation, characterization and transformation, by autonomous replication, of *Mucor circinelloides* OMPdecase-deficient mutants. *Mol. Gen. Genet.* **248**: 126–135.

Burmester, A., Wostemeyer, A., Arnau, J. and Wostemeyer, J. (1992) The SEG1 element: A new DNA region promoting stable mitotic segregation of plasmids in the zygomycete *Absidia glauca*. *Mol. Gen. Genet.* **235**: 166–172.

de Groot, M. J., Bundock, P., Hooykaas, P. J. and Beijersbergen, A. G. (1998) *Agrobacterium tumefaciens*-mediated transformation of filamentous fungi. *Nat. Biotechnol.* **16**: 839–842.

Edwards, J., Jr. (1989) Zygomycosis. In: *Infectious Disease* (P. Hoeprich and M. Jordan, eds.), J.B. Lippincott Co., Philadelphia, pp. 1192–1199.

Frank-Vaillant, M. and Marcand, S. (2002) Transient stability of DNA ends allows nonhomologous end joining to precede homologous recombination. *Mol. Cell.* **10**: 1189–1199.

González-Hernández, G. A., Herrera-Estrella, L., Rocha-Ramirez, V., Roncero, M.I.G. and Gutierrez-Cor (1997) Biolistic transformation of the Zygomycete *Mucor circinelloides*. *Mycological Research* **101**: 953–956.

Gouka, R. J., Gerk, C., Hooykaas, P. J., Bundock, P. *et al.* (1999) Transformation of *Aspergillus awamori* by *Agrobacterium tumefaciens*-mediated homologous recombination. *Nat. Biotechnol.* **17**: 598–601.

Hinnen, A., Hicks, J. B. and Fink, G. R. (1978) Transformation of yeast. *Proc. Natl. Acad. Sci. USA.* **75**: 1929–1933.

Horiuchi, H., Takaya, N., Yanai, K., Nakamura, M. *et al.* (1995) Cloning of the *Rhizopus niveus* pyr4 gene and its use for the transformation of *Rhizopus delemar. Curr. Genet.* **27**: 472–478.

Houghton-Larsen, J. and Pedersen, P. A. (2003) Functional expression of rat adenosine A1 receptor in the dimorphic zygomycete *Mucor circinelloides. Appl. Microbiol. Biotechnol.* **63**: 64–67.

Husain, S., Alexander, B. D., Munoz, P., Avery, R. K. *et al.* (2003) Opportunistic mycelial fungal infections in organ transplant recipients: Emerging importance of non-*Aspergillus* mycelial fungi. *Clin. Infect. Dis.* **37**: 221–229.

Ibrahim, A. S., Edwards, J. E. J. and Filler, S. G. (2003) Zygomycosis. In: *Clinical Mycology* (W. E. Dismukes, P. G. Pappas and J. D. Sobel, eds.), Oxford University Press, New York, pp.241–251.

Kelly, J. M. and Hynes, M. J. (1985) Transformation of *Aspergillus niger* by the amdS gene of *Aspergillus nidulans. EMBO J.* **4**: 475–479.

Klein, B. S. (2000) Molecular basis of pathogenicity in *Blastomyces dermatitidis*: The importance of adhesion. *Curr. Opin. Microbiol.* **3**: 339–343.

Kontoyiannis, D. P., Wessel, V. C., Bodey, G. P. and Rolston, K. V. (2000) Zygomycosis in the 1990s in a tertiary-care cancer center. *Clin. Infect. Dis.* **30**: 851–856.

Kontoyiannis, D. P., Lionakis, M. S., Lewis, R. E., Chamilos, G. *et al.* (2005) Zygomycosis in a tertiary-care cancer center in the era of *Aspergillus*-active antifungal therapy: A case-control observational study of 27 recent cases. *J. Infect. Dis.* **191**: 1350–1360.

Kwon-Chung, K. J. and Bennett, J. E. (eds) (1992) Mucormycosis. In: *Medical Mycology*, Lea & Febiger, Philadelphia, pp. 524–559.

Kwon-Chung, K. J., Goldman, W. E., Klein, B. and Szaniszlo, P. J. (1998) Fate of transforming DNA in pathogenic fungi. *Med. Mycol.* **36**(Suppl 1), 38–44.

Liou, C. M., Yanai, K., Horiuchi, H. and Takagi, M. (1992) Transformation of a Leu-mutant of *Rhizopus niveus* with the leuA gene of *Mucor circinelloides. Biosci. Biotechnol. Biochem.* **56**: 1503–1504.

Marr, K. A., Carter, R. A., Crippa, F., Wald, A. and Corey, L. (2002) Epidemiology and outcome of mould infections in hematopoietic stem cell transplant recipients. *Clin. Infect. Dis.* **34**: 909–917.

McNulty, J. S. (1982) Rhinocerebral mucormycosis: Predisposing factors. *Laryngoscope.* **92**: 1140–1143.

Michielse, C. B., Ram, A. F., Hooykaas, P. J. and Hondel, C. A. (2004a) Role of bacterial virulence proteins in *Agrobacterium*-mediated transformation of *Aspergillus awamori. Fungal Genet. Biol.* **41**: 571–578.

Michielse, C. B., Salim, K., Ragas, P., Ram, A. F. *et al.* (2004b) Development of a system for integrative and stable transformation of the zygomycete *Rhizopus oryzae* by *Agrobacterium*-mediated DNA transfer. *Mol. Genet. Genomics.* **271**: 499–510.

Monfort, A., Cordero, L., Maicas, S. and Polaina, J. (2003) Transformation of *Mucor miehei* results in plasmid deletion and phenotypic instability. *FEMS Microbiol. Lett.* **224**: 101–106.

Navarro, E., Ruiz-Perez, V. L. and Torres-Martinez, S. (2000) Over-expression of the crgA gene abolishes light requirement for carotenoid biosynthesis in *Mucor circinelloides. Eur. J. Biochem.* **267**: 800–807.

Navarro, E., Lorca-Pascual, J. M., Quiles-Rosillo, M. D., Nicolas, F. E. *et al.* (2001) A negative regulator of light-inducible carotenogenesis in *Mucor circinelloides*. *Mol. Genet. Genomics.* **266**: 463–470.

Nicolas, F. E., Torres-Martinez, S. and Ruiz-Vazquez, R. M. (2003) Two classes of small antisense RNAs in fungal RNA silencing triggered by non-integrative transgenes. *EMBO J.* **22**: 3983–3991.

Obraztsova, I. N., Prados, N., Holzmann, K., Avalos, J. and Cerda-Olmedo, E. (2004) Genetic damage following introduction of DNA in *Phycomyces*. *Fungal Genet. Biol.* **41**: 168–180.

Ootaki, T., Miyazaki, A., Fukui, J., Kimura, Y. *et al.* (1991) A high efficient method for introduction of exogenous genes into *Phycomyces blakesleeanus*. *Japan. J. Genet.* **66**: 189–195.

Panepinto, J. C., Oliver, B. G., Fortwendel, J. R., Smith, D. L. *et al.* (2003) Deletion of the *Aspergillus fumigatus* gene encoding the Ras-related protein RhbA reduces virulence in a model of invasive pulmonary aspergillosis. *Infect. Immun.* **71**: 2819–2826.

Pochanavanich, P. and Suntornsuk, W. (2002) Fungal chitosan production and its characterization. *Lett. Appl. Microbiol.* **35**: 17–21.

Pillsbury, H. C. and Fischer, N. D. (1977) Rhinocerebral mucormycosis. *Arch. Otolaryngol.* **103**: 600–604.

Ray, A. and Langer, M. (2002) Homologous recombination: Ends as the means. *Trends Plant. Sci.* **7**: 435–440.

Revuelta, J. L. and Jayaram, M. (1986) Transformation of *Phycomyces blakesleeanus* to G-418 resistance by an autonomously replicating plasmid. *Proc. Natl. Acad. Sci. USA.* **83**: 7344–7347.

Ribes, J. A., Vanover-Sams, C. L. and Baker, D. J. (2000) Zygomycetes in human disease. *Clin. Microbiol. Rev.* **13**: 236–301.

Ruiz-Hidalgo, M. J., Eslava, A. P., Alvarez, M. I. and Benito, E. P. (1999) Heterologous expression of the *Phycomyces blakesleeanus* phytoene dehydrogenase gene (carB) in *Mucor circinelloides*. *Curr. Microbiol.* **39**: 259–264.

Roncero, M. I., Jepsen, L. P., Stroman, P. and van Heeswijck, R. (1989) Characterization of a leuA gene and an ARS element from *Mucor circinelloides*. *Gene.* **84**: 335–343.

Roze, L. V., Mahanti, N., Mehigh, R., McConnell, D. G. and Linz, J. E. (1999) Evidence that MRas1 and MRas3 proteins are associated with distinct cellular functions during growth and morphogenesis in the fungus *Mucor racemosus*. *Fungal Genet. Biol.* **28**: 171–189.

Sandoval, A. and Labhart, P. (2002) Joining of DNA ends bearing non-matching 3′-overhangs. *DNA Repair.* (Amst) **1**: 397–410.

Skory, C. D. (2002) Homologous recombination and double-strand break repair in the transformation of *Rhizopus oryzae*. *Mol. Genet. Genomics.* **268**: 397–406.

Skory, C. D. (2004) Lactic acid production by *Rhizopus oryzae* transformants with modified lactate dehydrogenase activity. *Appl. Microbiol. Biotechnol.* **64**: 237–242.

Skory, C. D. (2005) Inhibition of non-homologous end joining and integration of DNA upon transformation of *Rhizopus oryzae*. *Mol. Genet. Genomics.* **274**: 373–383.

Staab, J. F. and Sundstrom, P. (2003) URA3 as a selectable marker for disruption and virulence assessment of *Candida albicans* genes. *Trends Microbiol.* **11**: 69–73.

Suarez, T., Orejas, M. and Eslava, A. P. (1985) Isolation, regeneration, and fusion of *Phycomyces blakesleeanus* spheroplasts. *Exp. Mycol.* **9**: 203–211.

Sugar, A. M. (1995) Agent of mucormycosis and related species. In: *Principles and Practices of Infectious Diseases* (G. Mandell, J. Bennett and R. Dolin, eds), Churchill Livingstone, New York, pp. 2311–2321.

Takaya, N., Yanai, K., Horiuchi, H., Ohta, A. and Takagi, M. (1996) Cloning and characterization of the *Rhizopus niveus* leu1 gene and its use for homologous transformation. *Biosci. Biotechnol. Biochem.* **60**: 448–452.

van Heeswijck, R., Roncero, M. I. and Jepsen, L. P. (1998) Genetic analysis and manipulation of Mucor species by DNA-mediated transformation. In: *Modern Methods of Plant Analysis* (H. F. Linskens, and J. F. Jackson, eds), Vol. 7, Springer Verlag, New York, pp. 207–220.

van Heeswijck, R. (1984a) The formation of protoplasts from *Mucor* species. *Carlsberg Res. Commun.* **49**: 597–609.

vanHeeswijck, R. and Roncero, M. I. G. (1984b) High frequency transformation of *Mucor* with recombinant plasmid DNA. *Carlsberg Res. Commun.*, **49**: 597–609.

Wada, M., Beppu, T. and Horinouchi, S. (1996) Integrative transformation of the zygomycete *Rhizomucor pusillus* by homologous recombination. *Appl. Microbiol. Biotechnol.* **45**: 652–657.

Waldorf, A. R., Levitz, S. M. and Diamond, R. D. (1984) *In vivo* bronchoalveolar macrophage defense against *Rhizopus oryzae* and *Aspergillus fumigatus. J. Infect. Dis.* **150**: 752–760.

Waldorf, A. R., Ruderman, N. and Diamond, R. D. (1984) Specific susceptibility to mucormycosis in murine diabetes and bronchoalveolar macrophage defense against *Rhizopus. J. Clin. Invest.* **74**: 150–160.

Wostemeyer, J., Burmester, A. and Weigel, C. (1987) Neomycin resistance as a dominantly selectable marker for transformation of the zygomycete *Absidia glauca. Curr. Genet.* **12**: 625–627.

Yamazaki, H., Ohnishi, Y., Takeuchi, K., Mori, N. *et al.* (1999) Genetic transformation of a *Rhizomucor pusillus* mutant defective in asparagine-linked glycosylation: Production of a milk-clotting enzyme in a less-glycosylated form. *Appl. Microbiol. Biotechnol.* **52**: 401–409.

Index

ABC transporters 102, 105
Absidia 120, 305
Acid phosphatase 21
Actin curve 178
Adherence 13, 51
Adhesion 70–73
Agrobacterium-mediated transformation 217, 316
Amphotericin B 44, 94, 101, 120, 201, 306
Amplicons 182
Animal models of infection 115–135
Antibodies 28, 30, 32, 34, 35
  Monoclonal 3, 4, 5, 66
  Polyclonal 9
Antifungal(s) 43, 44, 45, 67, 93, 96, 110, 115, 120, 130 200
Antigen detection 33, 37
Aspergillosis 2, 115, 125, 211
Aspergillus fumigatus 43, 44, 47, 67, 95, 115, 211–229, 231
  Conidia 23, 44, 47, 211, 218, 225
  Virulence 258
Aspergillus nidulans 217, 246
Aspergillus niger 232, 253
Azoles 94

Beta-galactosidase activity 237, 239
Beta-lymphocytes 51
Blastospores (*Candida albicans*) 5, 21,
Blood (whole) 48
Buccal epithelial cells (BEC) 70, 72
Bud scar 7
Blastospore 5

Candida albicans 67, 72
  Commensal 8
  Differentiation 77
  Microarrays 181
  Morphological forms 6, 7, 21
  Oral cavity 2, 10, 11
  Virulence factors 69
Candida *dubliniensis* 94
Candida *glabrata* 4, 11, 16, 95
Candida *krusei* 4, 11, 101
Candida *lusitaniae* 4, 11
Candida *parapsilosis* 4, 11
Candida *tropicalis* 4, 11, 76
Candidosis 8, 43
Capillary transfer 166, 167
Capsule 276, 297
Caspofungin 96, 120
Cathether surfaces 76
cDNA synthesis 62, 188, 190
CDR 102
Cell wall (fungal) 13, 14, 16, 23, 28, 258
Chemically assisted fragmentation 150
Chemiluscent detection 65
Chitin 15, 28
Coenzyme C 53
Coenzyme Q 52
Colony forming units (CFU) 129
Cryofixation 36
Cryptococcosis 277, 298
Cryptococcus neoformans 275
Cryptococcus 95
Cyanine labeled DNA 190, 195, 263
Cytochemistry 30, 38

Medical Mycology: Cellular and Molecular Techniques. Edited by Kevin Kavanagh

Cytokines 44, 66
Cytoskeletal changes in *Aspergillus* 227

Data validation 199
Databases 141
Deletion strains (screening) 292
Denture acrylic 75
Diagnosis (periodontal disease) 7
Differentiation of strains 77
 Phenotypic switching 81
 Resistogram 77
 Serotyping 79
DNA 159
 Extraction 161, 164, 287, 291
 Hybridization 194
 Microarrays 182
 Purification 162, 289
 Quality/Quantity 163
Double jointed PCR gene deletion 215
Drug administration 123
Drug resistance 93, 100

Ectopic integration 296
Electronmicroscopy 8, 13, 20, 23
Electrophoresis 61
Endoplasmic reticulum 16, 19
Enzymes 15, 83, 258
 Cell wall degrading 13, 159, 244, 289, 308
Epithelial cells 72, 74
Erg genes 106
Erythrocytes 50

Ferritin 23–25
Ficoll 49, 50
Filamentous fungi 115, 117, 122
Filter membrane 73
Flow cytometer 55, 80
Fluconazole 94, 97, 99
Fluorocytosine 94, 201
Fungal infections 43, 69
Fusarium 43, 116

Galactose 28
Gene disruption cassettes 278
Gene disruption in *Aspergillus* 213
Gene expression 58, 63, 105, 174–179, 203, 251
Germ tube (*C. albicans*) 31, 88
Gingival tissue 3
Glucans 15, 29
Glucuronoxylomannan (GXM) 298, 299
Glycoconjugates 26, 27, 31
Gold complexes 15, 30, 33, 36
Gram (+)/Gram(−) bacteria 7
Green fluorescent protein (GFP) 231

Heparin 48, 49
Histopaque (Ficoll) 49
Host tissue colonization 13, 38
Hybridization 65, 266, 288
Hydrophobicity 71, 225, 258
Hygromycin B 232, 246
Hyphae 6, 47, 53, 54

Immuno-detection 36, 38
Immuno-histochemical staining 5, 8, 13, 15, 38
Immuno-modulators 57
In gel digestion 142
In solution digestion 143
Inanimate surfaces 75
Infection 116
Inflammatory mediators 8
Innate immunity 67
Invasive fungal infections 43
Isoelectric focusing 138
Itraconazole 95, 98, 120, 124

Ketoconazole 201
Knock-out vectors 212, 231

Lac-Z fusion 231, 236–241
Lanosterol demethylase 106
Lanosterol 202
Lectins 15
Lethal dose ($LD_{90\%}$, $LD_{10\%}$) 122
Leukocyte 128
LR white 33, 35, 36
Lymphocytes 51, 56
Lysis buffer 66
Lytic enzymes 13, 159, 244, 289, 308

Macrophages 44
MALDI-TOF 140
  Interpretating data 152
  Linear 145
  Reflectron 146
Mannoproteins 3, 7, 15
Mass spectrometery 138, 145
Matrix 145, 147
Membrane sterol 108
Microarrays 63,
  A. fumigatus 257
  Antifungals drugs 200
  Biofilms 204
  C. albicans 181
  Image analysis 196, 270–272
  Morphogenesis 202
  Pathogenicity 196
Minimum fungicidal concentration (MFC) 100
Minimum inhibitory concentration (MIC) 96
Mitochondria 15, 37
Monocytes 48
Mucor 116, 305
Mucorales 305
Mucosal surfaces 43, 69, 71

NADPH-oxidase 44
Nested primers 216
Neurospora 156
Neutropenia 126
Neutrophils 44
Northern blotting 160
Nuclear morphology 226

Oligonucleotide microarrays 183
Organ transplantation 181
Osmium tetroxide 16

Parafin wax 4, 5
Pathogenicity and microarrays 196
PCR 62, 212, 234, 235, 278
Peridontal disease 3, 7, 10
Peridontal macrophages 44
Peroxidase activity 5
Phagocytes 43, 47, 54
Phagocytosis assay 55, 56
Phenotypic switching 81
Phospholipase production 87
Plasma 50
Plasmid pUChph 247
Polysacchardides 14, 26, 28, 29
Posconazole 120
Probe synthesis 64, 107, 165, 190–194
Proteinase production 85, 86
Proteomics 137–156
  Peptide mass fingerprinting 139, 149
  Post source decay 149
Protoplasts 309
Psedohyphae 6, 9, 10
pyr G as a selection marker 249

Reporter genes 231
Resistogram typing 77
Rhizopus 45, 115, 305
Rhodamine 6G 102
Rhodamine assay 57
RNA 159
  Housekeeping rules 169
  Interference 212
  Isolation 59, 64, 169–172, 185–187, 259
RT-PCR 57, 62, 176

Saccharomyces cerevisiae 75, 107, 156, 181
Scedosporium 43
SDS-PAGE (2-dimensional) 137–139
Secretory pathway 18–20
Selectable markers 307
Serotyping 79
Serum 6
Site directed mutagenesis 233
Southern blotting 160, 165, 220
Spurr's resin 17
Sterol content 108
Subgingival plaques 8
Superoxide anion 53

Taq polymerase 59, 62
Tetrazolium 51
Time of flight (TOF) 145
Tissue sections 2
T-lymphocytes 51
TOLL-like receptors (TLR)
Toxins(*Aspergillus*) 258
Transformant analysis 221–229, 319

Transformation 217
  Aspergillus 241–246, 249
  Biolistic 283, 311–313
  Chemically induced 218
  Cryptococcus neoformans 283, 293
  Electroporation 286
  Zygomycetes 308, 318–321
Transmission electron-microscopy 13–39
Transposon-based gene disruption 214
Trichosporon 43, 45

Ultrastructural studies 36, 38
Ultrastructure of cell 13, 14, 16, 33
Ultrathin sections 18, 33
URA-blaster 233, 251
Uranyl acetate 30, 35

Virulence
  Aspergillus 257
  C. neoformans 275, 276
  Candida 69–89
Voriconazole 98, 117

XTT metabolic conversion 51, 52

Yeast 2, 3, 18, 51, 72

Zygomycetes 43, 53, 118, 305, 322
Zygomycosis 118, 125
Zymolyase 4, 13

Printed and bound by CPI Group (UK) Ltd, Croydon, CR0 4YY
09/06/2025
14685995-0001